Receptors:
Structure and function

The Practical Approach Series

Related **Practical Approach** Series Titles

Functional Genomics

Bioinformatics: sequence, structure and databanks

Protein Purification Techniques

Please see the **Practical Approach** series website at
http://www.oup.com/pas
for full contents lists of all Practical Approach titles.

No. 253

Receptors: Structure and function

A Practical Approach

Edited by

Clare Stanford

Department of Pharmacology, University Collge
London, Gower Street, London WC1E 6BT, UK

and

Roger Horton

Department of Pharmacology, St. George's
Hospital Medical School, Cranmer Terrace,
London SW17 0RE, UK

OXFORD
UNIVERSITY PRESS

*This book has been printed digitally and produced in a standard specification
in order to ensure its continuing availability*

OXFORD
UNIVERSITY PRESS

Great Clarendon Street, Oxford OX2 6DP

Oxford University Press is a department of the University of Oxford.
It furthers the University's objective of excellence in research, scholarship,
and education by publishing worldwide in

Oxford New York

Auckland   Cape Town Dar es Salaam Hong Kong   Karachi
Kuala Lumpur Madrid Melbourne Mexico City Nairobi
New Delhi Shanghai Taipei Toronto
With offices in
Argentina Austria Brazil Chile Czech Republic France Greece
Guatemala Hungary Italy Japan South Korea Poland Portugal
Singapore Switzerland Thailand Turkey Ukraine Vietnam

Oxford is a registered trade mark of Oxford University Press
in the UK and in certain other countries

Published in the United States
by Oxford University Press Inc., New York

© Oxford University Press, 2001

ISBN 0-19-963882-9

Printed and bound by CPI Antony Rowe, Eastbourne

Preface

In choosing the topics to cover in this book, we decided that there were so many exciting developments across the whole spectrum of receptor research that it would be folly to restrict ourselves to one particular approach, or class of receptors. This explains why we have included studies ranging from G protein-coupled surface receptors, to the delivery of antisense DNA inside living cell systems. We have been extremely fortunate to recruit a distinguished team of experts, who collectively could cover these diverse areas and all of whom have proved willing to reveal both the difficulties we might expect to encounter when embarking on our own experiments and the 'tricks' that will help us to overcome them.

Wherever appropriate, the theoretical basis of each topic is explained first so that the results emanating from the practical procedures can be fully understood. Also, to ensure that each of the chapters provides a comprehensive guide to successful experiments, some inevitably incorporate protocols for well established techniques (e.g. radioligand binding). Yet, even here, all but the most seasoned researcher will benefit from insight into recent developments in the ligands that can be used and detectors for measuring their binding.

The series of chapters starts on the cell surface. Using G protein-coupled peptide receptors as a model, Gerardo Turcatti covers recent developments in fluorescence spectroscopy: from the design and labelling of the ligands, through analysis of their spectral properties, to measurements of the kinetics of their interactions with receptors and their use in elucidating the receptor structure. Moving on from this topic, Olivier Civelli and colleagues explain the 'tried and tested' series of procedures that should be adopted in order to identify and characterize a native ligand, once a new G protein-coupled receptor emerges from homology screening of the genome. Next, we focus on the receptors themselves. Dianne Perez and her colleagues explain how the introduction of point mutations and the construction of receptor chimeras can help to identify the elements of α_1-adrenoceptors that are fundamental to ligand binding and their coupling with second messenger systems. On a similar theme, Anne Stephenson and Lynda Hawkins explain how to elucidate the structure of NMDA receptors, which have a complex subunit composition, including how to make selective antibodies and the use of epitope tagging in the determination of receptor sub-unit stoichiometry. Moving on from this, Chapter 5 (Israel Pecht and colleagues) describes how receptor stoichiometry can be evaluated by using fluorescence resonance energy transfer.

Thomas Hughes and his colleagues cover the use of fluorescent fusion proteins for studying the intracellular trafficking of receptors: from the design of fusion constructs to the techniques required to deduce their movement through cells. Staying with the general theme of the life cycle of receptors, Michael Edwardson explains the theoretical basis of the analysis of receptor kinetics and goes on to describe how to estimate the rate constants for the movement of G protein-coupled receptors to and from the surface membrane.

We then concentrate on studying cell signalling systems. Microphysiometry, which monitors changes in cellular metabolic activity is covered by Martyn Wood and Darren Smart. This is a technique that will be available in only a few, well-endowed laboratories but which is nevertheless proving an invaluable resource in the study of receptors and their cell signalling systems. Chapter 9 (Richard Huganir and Bernard McDonald) explains how to study the regulation of cell signalling through identification of the phosphorylation sites that determine post-translational modification of ionotropic receptors.

We then move to studies of intracellular function. Ghassan Bkaily and colleagues write on confocal microscopy: from recent developments in fluorescent ion and organelle probes to their use in studies of cell structure and function as well as the coupling between cells. Nuclear receptors are covered next: Scott Nunez and Wayne Vedeckis warn us of the special difficulties encountered in studies of these receptor systems and explain how to overcome them. We end with a chapter on *in vivo* transfection of receptor antisense because this could ultimately become a viable treatment for some medical disorders. Factors relevant to the design of the antisense are described before explaining the process of its delivery to the desired DNA target.

We are gratefully indebted to all these authors for their enthusiasm for this project, their regard for deadlines, and for their patient and willing compliance with our, often inane, requests for more details. Such teams are hard to find and we are most fortunate to have been granted their support. Finally, we wish to thank Rupert Cousens at OUP: we could not have hoped for a more helpful, supportive, and patient colleague.

London   R. H.
February 2001   S. C. S.

Contents

List of protocols *page xv*

Abbreviations *xix*

1 Probing G protein-coupled receptors by fluorescence spectroscopy *1*

Gerardo Turcatti

1 Introduction *1*

2 Fluorescent reporter groups *2*

3 Ligands *3*

    Labelling of ligands *3*

    Strategies for the generation of labelled peptide ligands *6*

    Characterization of fluorescent ligands *9*

4 Receptors *14*

5 Fluorometric assays *17*

    Ligand binding assays *17*

    Kinetics of ligand binding and dissociation *17*

    Fluorescence quenching experiments *19*

    Steady-state fluorescence anisotropy measurements *24*

6 Applications *27*

    Probing ligand–receptor interactions using fluorescent ligands *27*

7 Mapping receptor architecture using fluorescent receptors and
fluorescent ligands *29*

    Labelling receptors *29*

    Characterization of fluorescent receptors *31*

    Fluorescence resonance energy transfer (FRET) between a specific bound
fluorescent ligand and a fluorescently labelled receptor *36*

Acknowledgements *38*

References *38*

2 From receptor to endogenous ligand *41*

Hans-Peter Nothacker, Rainer K. Reinscheid, and Olivier Civelli

1 Introduction *41*

2 The orphan receptor strategy *41*

    What is the orphan receptor strategy? *41*

    The dilemma of the orphan receptor strategy *42*

3 Identification and characterization of orphan GPCRs 43
      Cloning of novel orphan GPCRs based on homology 43
      Computer-based orphan GPCR discovery 44
      Classification of the novel orphan GPCRs 44

4 Expression of orphan GPCRs 45
      Vectors for use of expression in mammalian cells 45
      Transient versus stable expression of GPCRs 46
      Confirmation of receptor expression 48
      Co-expression of promiscuous and chimeric G protein α-subunits 51

5 Functional assays 52
      cAMP assays 52
      Ca^{2+} assay 55
      Reporter gene assay 57

6 Ligand screening 60
      Preparation of tissue extracts as sources for ligand screening 60
      Purification and structure elucidation of endogenous ligands for orphan
         receptors 62

   Acknowledgements 63

   References 63

3 Point mutations and chimeric receptors in studies of ligand binding domains and second messenger activation of α_1-adrenoceptors 65

Dianne M. Perez, Michael J. Zuscik, Sean A. Ross, and David J. J. Waugh

1 Introduction 65

2 Applications 65
      Random *versus* chimeric 65
      Directed *versus* chimeric 66

3 Methods for creating chimeric receptors 67
      Cassette replacement 67

4 Methods for creating point mutant receptors 68
      PCR mutagenesis 68

5 Manipulation of DNA for expression 72
      Subcloning the PCR or chimeric fragments 72
      Identification and propagation of mutants in bacteria 75

6 Expression of mutant receptors 77
      Culturing of cells 77
      Transient transfection 80

7 Ligand binding 82
      Membrane preparation 82
      Saturation binding 83
      Competitive binding 85

8 Second messenger assays 87
      Total inositol phosphates 87

9 Interpretation 89

   References 90

4 Determination of the subunit composition of native and cloned NMDA receptors 91

F. Anne Stephenson and Lynda M. Hawkins

1 Introduction 91

2 Production and characterization of amino acid sequence-directed anti-NMDA receptor subunit antibodies 92

3 Determination of the subunit composition of native NMDA receptors by immunoaffinity purification using anti-NMDA receptor subunit-specific antibodies 96

4 Determination of NMDA receptor subunit stoichiometry using an epitope tagging strategy 101
   Epitope tagging of NMDA receptor subunits 102
   Epitope tagging mammalian expression vectors 106
   Critical assessment of the epitope tagging strategy for the determination of the number of NR1 and NR2 subunits per NMDA receptor 107

5 Conclusions 110

Acknowledgements 111

References 111

5 Evaluating receptor stoichiometry by fluorescence resonance energy transfer 113

Dmitry M. Gakamsky, Richard G. Posner, and Israel Pecht

1 Introduction 113

2 Theoretical basis of fluorescence energy transfer 114
   Theory of FRET 114
   Factors affecting the efficiency of fluorescence energy transfer 115

3 Labelling of biomolecules with fluorescent probes 116
   Random labelling 116
   Site-specific labelling 118
   Indirect labelling using fluorescent antibodies 120
   Labelling with green fluorescent protein constructs 121
   Specific covalent labelling via 4-cysteine's tag 122

4 Protocols for measuring fluorescence energy transfer 122
   Monitoring donor quenching 123
   Monitoring the acceptor-sensitized fluorescence 125
   Studying stoichiometry of a complex composed of several subunits 126
   Photobleaching FRET techniques 131
   Homogeneous time-resolved FRET assay 132

5 Determination of a complex stoichiometry 132

6 Conclusions 133

Acknowledgements 133

References 134

6 Monitoring intracellular trafficking of receptors 137

Raymond Molloy, Antoine Robert, James R. Howe, and Thomas E. Hughes

1 Introduction 137

2 Planning *139*
    Getting started *139*
    Sequence analysis *139*

3 Building the construct *143*
    Initial considerations *143*
    Creating fusion proteins by restriction and ligation *144*
    Creating a fusion protein by PCR *145*

4 Expression *146*
    Introduction *146*
    Protocol for HEK 293 transfection *148*
    Neuronal transfection *150*

5 Time-lapse microscopy *152*

References *153*

7 Monitoring the turnover kinetics of G protein-coupled receptors *155*

J. Michael Edwardson

1 Introduction *155*

2 Mathematical modelling of receptor turnover *156*

3 Measurement of rate constants for endocytosis and recycling *159*
    Use of insurmountable receptor antagonists *160*
    Comparing rate constants for transfected and native receptors *162*

4 Relationship between agonist intrinsic activity and receptor endocytosis and internalization *163*

5 Manipulating receptor turnover using inhibitors of endocyctosis or recycling *165*
    Inhibitors of endocytosis *165*
    Inhibitors of recycling *167*

6 Role of receptor trafficking in desensitization and resensitization *167*

7 Pharmacological manipulation of the receptor reserve *171*

8 Future prospects *172*

Acknowledgements *173*

References *173*

8 Real time receptor function *in vitro*: microphysiometry and the fluorometric imaging plate reader (FLIPR) *175*

Martyn Wood and Darren Smart

1 Introduction *175*

2 Microphysiometry *175*
    Principles of microphysiometry *175*
    Practical aspects *177*
    Experimental data *183*

3 Fluorometric imaging plate reader (FLIPR) *184*
    Principles of the FLIPR *184*
    Practical aspects *185*
    Experimental data *190*

4 Conclusions *191*

Acknowledgements *191*

References *191*

9 Regulation of receptors and receptor-coupled ion channels by protein phosphorylation *193*

Bernard J. McDonald and Richard L. Huganir

1 Introduction *193*

2 Biochemical determination of receptor phosphorylation *194*
   Examination of endogenous full-length receptor proteins *194*
   Examination of recombinant full-length receptor proteins *197*
   Phosphorylation of recombinant receptor fragments *in vitro* *198*

3 Biochemical characterization of receptor phosphorylation *199*
   Proteolytic digestion of purified phosphoproteins *199*
   Phosphopeptide fingerprint mapping *201*
   Phosphoamino acid identification *202*

4 Phosphorylation site identification *205*
   Site-directed mutagenesis *205*
   Phosphorylation site-specific antibodies *206*
   Candidate protein kinase identification *207*

5 Conclusions *207*

Acknowledgements *208*

References *208*

10 Using confocal imaging to measure changes in intracellular ions *209*

Ghassan Bkaily, Danielle Jacques, Pedro D'Orléans-Juste, Ghada Hassan, and Sanaa Choufani

1 Introduction *209*

2 Preparation of tissue *209*
   Choice of cell types for confocal microscopy *209*
   Co-culture of cells *210*

3 Choice and properties of intracellular markers *211*
   Ca^{2+} probes *211*
   Na^+ fluorescent probes *216*
   pH fluorescent probes *216*
   Membrane potential dyes: Di-8-ANEPPS and JC-1 probes *218*
   Quantification of free intracellular ions *219*

4 Volume rendering and nuclear Ca^{2+}, Na^+, and pH measurements *220*

5 Organelle, receptor, and channel fluorescent probes *222*
   Endoplasmic reticulum staining *223*
   Mitochondria staining *224*
   Channel and receptor dyes *225*

6 Introducing non-permeable fluorescent macromolecules into cultured mammalian cells by osmotic lysis *225*

7 Fluorescent labelling for identification of cell types and communication in co-culture *227*

8 Settings of the confocal microscope *228*

9 Conclusion *229*

Acknowledgements *230*

References *230*

11 Monitoring nuclear receptor function *233*

B. Scott Nunez and Wayne V. Vedeckis

1 Introduction *233*

2 Cell culture assays of transfected gene expression constructs *234*
Creation and isolation of eukaryotic gene expression constructs *234*
Use of Tet-'Off' and Tet-'On' systems *236*
Use of reporter constructs in promoter characterization *237*

3 *In vitro* analysis of protein–DNA interaction *237*
Isolation of nuclear extract *237*
In vitro DNase I footprinting *238*
Electrophoretic mobility shift assay *240*

4 Assays of endogenous gene expression *243*
Northern blot analysis *243*
RNase protection assay *244*
Reverse transcription-polymerase chain reaction *244*
DNA array technology *246*
Western blot analysis *248*

5 Analysis of protein–protein interaction *251*
Immunoprecipitation *251*
Pull-down assays *253*
Far Western analysis *253*
Yeast two-hybrid assay *253*

6 Conclusions *254*

References *255*

12 *In vivo* gene transfer of antisense to receptors: methods for antisense delivery with adeno-associated virus *257*

M. Ian Phillips, Keping Qian, and Dagmara Mohuczy

1 Introduction *257*

2 Antisense oligodeoxynucleotides *258*
Designing antisense molecules *258*
Stability of oligonucleotides *261*
Cellular uptake of oligonucleotides *261*
Pharmacology of AS-ODNs *261*
Toxicity *262*
Delivery of antisense *262*

3 Viral vectors for antisense DNA delivery *263*
Retroviruses *264*
Adenoviruses *264*
Adeno-associated virus (AAV) *265*

4 Methods for antisense delivery by AAV *267*
    Construction of plasmids *267*
    Preparation of pAAV-AT$_1$R-AS *270*
    Method to prepare recombinant adeno-associated virus (rAAV) *270*

  Acknowledgements *275*

  References *275*

A1 *List of suppliers* *279*

  Index *287*

Protocol list

Ligands

Labelling and purification of fluorescent peptide derivatives: NK_2 receptor antagonists *7*

Concentration determination of peptide conjugates (ex: $(S\text{-}Dns)HCy^{11}\text{-}SP$): extinction coefficient calculations *10*

Receptors

Cells or cell membranes: incubation with fluorescent ligands (CHO/NK_2 cells) *15*

Fluorometric assays

Steady-state fluorescence monitoring of binding and dissociation of NBD antagonists of NK_2 receptors *18*

Collision quenching experiments with the anionic quencher iodide: $^aNK_2$ receptor probed with NBD antagonists *23*

Steady-state anisotropy measurements: fluorescent ligands bound to NK_1 receptors *25*

Mapping receptor architecture using fluorescent receptors and fluorescent ligands

Preparation of oocyte membranes enriched with NK_2 receptors for fluorescence microscopy assays *31*

FRET measurements by fluorescence microscopy using NK_2 receptor NBD fluorescent mutants expressed in oocytes and fluorescent ligand TMR antagonist *33*

Expression of orphan GPCRs

Transient transfection of orphan receptors and chimeric G protein into CHO cells *47*

Analysis of receptor expression in HEK 293T cells by immunofluorescence microscopy *49*

Functional assays

Extraction and quantification of cAMP from mammalian cells using immunoassays *52*

Detection of cellular cAMP increase in CHO and HEK 293 cells *53*

Detection of cellular cAMP decrease in stably transfected CHO cells *54*

Intracellular Ca^{2+} measurement using the FLIPR™ (fluorometric imaging plate reader) and GPCR transfected CHO dhfr⁻ *56*

Measurement of luciferase activity *59*

Ligand screening

Peptide extracts from total rat brains *61*

Methods for creating chimeric receptors
    Oligonucleotide phosphorylation and annealing *68*

Methods for creating point mutant receptors
    PCR amplification of mutant DNA *70*

Manipulation of DNA for expression
    Restriction digestion of DNA *72*
    Preparation of expression vector for subcloning *73*
    Ligation of PCR fragments into an expression plasmid *74*
    Transformation into bacteria *75*
    Plasmid preparation for double-stranded DNA sequencing *76*

Expression of mutant receptors
    Culturing of COS-1 cells *78*
    DEAE–dextran transfection *80*

Ligand binding
    Preparation of COS cell membranes *82*
    Radioligand binding isotherms at α_1-adrenergic receptors *83*
    Competition binding isotherms at α_1-adrenergic receptors *86*

Second messenger assays
    Inositol phosphate assays in COS-1 cells *88*

Production and characterization of amino acid sequence-directed anti-NMDA receptor subunit antibodies
    The *m*-maleimidobenzoic acid *N*-hydroxysuccinimide ester (MBS) method for the coupling of peptides and carrier protein via the cysteine of the peptide to the primary amine of the carrier protein *94*
    Purification of anti-peptide antibodies by peptide affinity chromatography *95*

Determination of the subunit composition of native NMDA receptors by immunoaffinity purification using anti-NMDA receptor subunit-specific antibodies
    Immunoaffinity purification of native NMDA receptors *99*

Determination of NMDA receptor subunit stoichiometry using an epitope tagging strategy
    Preparation of NMDA receptor subunit epitope tagged constructs *108*

Labelling of biomolecules with fluorescent probes
    Conjugation of amino reactive probes to proteins *117*
    Conjugation of thiol reactive probes with proteins *119*
    Indirect labelling proteins with fluorescent antibodies *121*

Protocols for measuring fluorescence energy transfer
    Determination of equilibrium binding constant and reaction stoichiometry by monitoring donor quenching *124*
    Determination of protein proximity by photobleaching FRET *131*

Building the construct
    PCR of GFP to add a restriction site in-frame *145*
    Creating a fusion by SOEing *146*

Expression
    Preparation of HEK 293 cells for transfection *148*

Transfection of HEK 293 cells with pEGFP *150*
Transfecting neurons *151*

Time-lapse microscopy
Time-lapse microscopy of transfected neurons *153*

Mathematical modelling of receptor turnover
Measurement of muscarinic acetylcholine receptor density at the plasma
  membrane of cultured cells (e.g. NG108-15) *159*

Measurement of rate constants for endocytosis and recycling
Alkylation of muscarinic acetylcholine receptors at the surface of cultured cells
  using PrBCM *160*

**Relationship between agonist intrinsic activity and receptor endocytosis and
internalization**
Measurement of Ins(1,4,5)P_3 generation by receptors that couple to $G_{q/11}$ *165*

Role of receptor trafficking in desensitization and resensitization
Population measurements of changes in intracellular Ca^{2+} concentration receptors
  coupled to $G_{q/11}$ *168*
Measurement of the inhibition of forskolin-stimulated adenylyl cyclase activity by
  receptors coupled to G_i *170*

Microphysiometry
Cell preparation for microphysiometry using CHO cells *178*
Typical experimental details for Cytosensor microphysiometer using human
  muscarinic receptors expressed in CHO cells *179*
Microphysiometer practical advice *181*

Fluorometric imaging plate reader
Typical experimental details for FLIPR studies using Chinese hamster ovary
  (CHO) cells *186*

Biochemical determination of receptor phosphorylation
Examination of protein phosphorylation in cells pre-labelled with [32]P *195*
In vitro phosphorylation of receptor fragments *198*

Biochemical characterization of receptor phosphorylation
Trypsin cleavage of purified phosphoproteins *200*
Peptide map analysis of trypsin digested phosphoproteins *201*
Phosphoamino acid analysis of trypsin digested phosphoproteins *203*

Preparation of tissue
Co-culture of vascular endothelial cells (VECs) and vascular smooth muscle cells
  (VSMCs) *211*

Choice and properties of intracellular markers
Loading with Fluo-3 AM *214*
Loading with the ratiometric pH indicator, carboxy-SNARF-1 AM *217*
Loading with the voltage-dependent potentiometric dye, Di-8-ANEPPS *218*
Ca^{2+} calibration curve for a non-ratiometric dye such as Fluo-3 *220*

Organelle, receptor, and channel fluorescent probes
Dual-labelling of SR (ER) or mitochondria and the nucleus of cells loaded or not with
  Fluo-3 Ca^{2+} probe *223*

Introducing non-permeable fluorescent macromolecules into cultured mammalian cells by osmotic lysis

Cell loading procedures for adherent cells on glass coverslips using Influx pinocytic cell loading reagent kit 226

Fluorescent labelling for identification of cell types and communication in co-culture

DiI fluorescent labelling for identification and study of communication between same and different cell types 227

Cell culture assays of transfected gene expression constructs

Freeze-shock transformation of bacteria 235

In vitro analysis of protein–DNA interaction

Electrophoretic mobility shift assay 242

Assays of endogenous gene expression

cDNA array hybridization using membrane filter arrays and radiolabelled probes 247

SDS–PAGE analysis 249

Western blot analysis using enhanced chemiluminescence 250

Analysis of protein–protein interaction

Identification of protein–protein interaction by immunoprecipitation 252

Methods for antisense delivery by AAV

Large scale plasmid preparation 267

Production of rAAV by calcium phosphate transfection 271

Production of rAAV by modified AAV helper plasmid 271

Iodixanol density gradient followed by heparin affinity chromatography 273

Virus titre assay 274

Abbreviations

AAV	adeno-associated virus
AMV	avian myeloblastosis virus
APS	ammonium persulfate
AS-ODN	antisense oligodeoxynucleotide
ATCC	American Type Culture Collection
ATP	adenosine 5′-triphosphate
AVP	arginine vasopressin
BFP	blue fluorescence protein
β-gal	β-galactosidase
BME	β-mercaptoethanol
BSA	bovine serum albumin
CamKII	Ca^{2+}/calmodulin-dependent protein kinase type II
cAMP	3′5′-cyclic adenosine monophosphate
CAT	chloramphenicol acetyltransferase
CCR	CC chemokine receptor
cGMP	guanosine 3′:5′-cyclic monophosphate
CHO	Chinese hamster ovary (cells)
CFP	cyan fluorescence protein
CIP	calf intestine phosphatase
CMV	cytomegalovirus
Con A	concanavalin A
COSHH	control of substances hazardous to health
CRE	cAMP response element
CREB	cAMP response element-binding protein
Dab	L-2,4-diaminobutyric acid
DABCYL	4-dimethylaminoazobenzene
DAG	diacylglycerol
Dap	L-2,3-diaminopropionic acid
DEAE	diethylaminoethane
Dhfr⁻	dihydrofolate reductase (minus)
DMEM	Dulbucco's modified Eagle's medium
DMF	*N,N*-dimethylformamide

ABBREVIATIONS

DMS	dimethyl sulfate
DMSO	dimethylsulfoxide
Dns	5-(dimethylamino)-1-naphtalenesulfonyl
DsFP583	(Coral-derived) Discosoma species fluorescent protein with 583 emission peak
DsRed	Discosoma-like red fluorescent protein with humanized codon usage
DTT	1,4-dithio-L-threitol
ε	molar extinction coefficient
ECFP	enhanced cyan fluorescent protein
ECL	enhanced chemiluminescence (reagents)
EDT	1,2-ethanedithiol
EDTA	ethylenediamine tetraacetic acid
EEO	electric-endo-osmosis
EGF	epidermal growth factor
EGFP	enhanced green fluorescent protein
EGTA	ethyleneglycol-bis(β-aminoethyl ether)-N,N,N',N'-tetraacetic acid
EMSA	electrophoretic mobility shift assay
ES-MS	electrospray mass spectrometry
EYFP	enhanced yellow fluorescent protein
FACS	fluorescence-assisted cell sorting
FBS	fetal bovine serum
FCS	fetal calf serum
FITC	fluorescein isothiocyanate
FLIPR	fluorometric imaging plate reader
Flu	5-fluoresceinthiocarbamyl
FRET	fluorescence resonance energy transfer
GABA	γ-amino butyric acid
GABA$_A$	hetero-oligomeric-γ-aminobutyric acid (receptor)
GC	glucocorticoid
GFAP	glial fibrillary acid protein
GFP	green fluorescent protein
GFPuv	green fluorescent protein with intact UV absorption peak and bacterial codon usage
GPCR	G protein-coupled receptors
GR	glucocorticoid receptor
GRE	glucocorticoid responsive element
GRK	G protein-coupled receptor kinase
GST	glutathione-S-transferase
HBSS	Hank's balanced salt solution
HEAT	2-[β-(4-hydroxyl-3-[^{125}I]-iodophenyl)ethylaminomethyl]tetralone
HEK	human embryonic kidney (cells)
Hepes	N-[2-hydroxyethyl]piperazine-N'-[2-ethanesulfonic acid]
HPLC	high pressure (performance) liquid chromatography
HSP 90	heat shock protein 90

Hydroxy-TEMPO	4-hydroxy-2.2.6.6-tetramethylpiperidine-1-oxyl
IBMX	3-isobutyl-1-methylxanthine
IgG	immunoglobulin G
IMAC	immobilized metal affinity chromatography
Ins(1,4,5)P$_3$	inositol (1,4,5) triphosphate
IPB	immunoprecipitation buffer
ITR	inverted terminal repeat
K_D	dissociation constant
K_i	inhibition constant
LAPS	light addressable potentiometric sensor
LTR	long terminal repeat
Luc	firefly luciferase
mAbs	monoclonal antibodies
MAP	mitogen-activated protein (kinase)
MBS	m-maleimideobenzoic acid N-hydroxysuccinimide (ester)
MCH	melanin-concentrating hormone
MCS	multiple cloning site
MEM	minimal essential medium
MHC-I	class I major histocompatibility complex
MIP-1α	macrophage inflammatory protein 1α
MMLV	Moloney murine leukaemia virus
MOI	multiplicity of infection
MS	mass spectrometry
Nb	neurobasal medium
Nb/B27	neurobasal medium supplemented with B27
NBD	7-nitrobenz-2-oxa-1,3-diazol-4-yl
NGRE	negative glucocorticoid responsive element
NK	neurokinin
NKA	neurokinin A
Nle	L-norleucine
NMDA	N-methyl-D-aspartate
NMS	N-methylscopolamine
NSE	neuron-specific enolase
NVOC	6-nitroveratryloxycarbonyl
ODN	oligodeoxynucleotide
Orn	L-ornithine
OVA	ovalbumin
PAGE	polyacrylamide gel electrophoresis
PAO	phenylarsine oxide
PBS	phosphate-buffered saline
PCR	polymerase chain reaction
pdCpA	5′-phospho-2′-deoxycytidyl (3′-5′) adenosine
PEG	polyethylene glycol
Ph	phenyl
PKA	protein kinase A

PKC	Ca^{2+}/phospholipid-dependent protein kinase
PKG	GMP-dependent protein kinase
PLB	phospholamban
PLC	phospholipase C
PMSF	phenylmethylsulfonyl fluoride
PMT	photomultiplier tube
PrBCM	propylbenzilylcholine mustard
rAAV	recombinant adeno-associated virus
RAMP	receptor accessory membrane protein
RANTES	regulated on activation normal T cell expressed and secreted
RAS	renin angiotensin system
RPA	RNase protection assay
RP-HPLC	reverse-phase high performance liquid chromatography
RSV	Rous sarcoma virus
RT	room temperature
RTK	receptor tyrosine kinase
RT-PCR	reverse transcriptase polymerase chain reaction
SBP	systolic blood pressure
SDS	sodium dodecyl sulfate
SDS–PAGE	sodium dodecyl sulfate–polyacrylamide gel electrophoresis
SEAP	secreted alkaline phosphatase
SHR	spontaneously hypertensive rats
SNAP	S-nitroso-N-acetylpenicillamine
SP	substance P
TAE (buffer)	Tris/acetate/EDTA
TEMED	tetramethylethylenediamine
TetR	Tet repressor
TFA	trifluoroacetic acid
TLC	thin-layer chromatography
TM	transmembrane (domain)
TMR	tetramethylrhodaminethiocarbamyl
TPCK	N-tosyl-L-phenylalanine chloromethyl ketone
TRE	TetR responsive element
Tris	Tris[hydroxymethyl]aminoethane
TRITC	tetramethylrhodamine isothiocyanate
UTP	uracil triphosphate
VEC	vascular endothelial cell
VSMC	vascular smooth muscle cell
v-src	transforming tyrosine kinase of the Rous sarcoma virus
WT	wild-type
YFP	yellow fluorescence protein

Chapter 1

Probing G protein-coupled receptors by fluorescence spectroscopy

Gerardo Turcatti

GenInEx S.A. Zone Industrielle, 1267 Coinsins, Switzerland.

1 Introduction

The broad development of approaches based on fluorescence detection in receptor research in recent years can be seen as a consequence of two main scientific advances. First, is the development and optimization of systems for the overexpression of receptors (1) that started early in the 1990s. Classic fluorescent techniques, that were not sensitive enough to detect receptors from natural sources or poorly expressed recombinant receptors, became available for this purpose with the important advantage of several years of experimentation behind them. Secondly, are the methodological and instrumental improvements in the sensitivity of fluorescence measurements that can now detect single molecules (for a review see ref. 2).

In addition, there has been much progress in the generation of fluorescent molecules with better spectral properties. Examples of these advances are:

(a) Brighter cyanine dye probes with emission spectra from green to the infrared (3).

(b) Alexa dyes, obtained by sulfonation of the aminocoumarin or rhodamine molecules, that resulted in a new generation of dyes with improved photostability (4).

(c) Fluorescent proteins used to label receptors using molecular biology-based techniques (for reviews see refs. 5 and 6; for a recent application to G protein-coupled receptors (GPCRs), see ref. 7).

Interest in the use of fluorescent techniques in receptor research arises not only from the sensitivity of detection that can be achieved but also from the different aspects of fluorescence outputs; anisotropy, lifetime, brightness, and energy transfer.

The procedures developed in our laboratory are of general applicability in receptor studies using fluorescent probes and standard instrumentation. These studies are aimed at probing ligand–receptor interactions, characterization of binding sites, measurement of intermolecular distances between labels at different positions on a receptor and labelled ligands, and measurement of receptor conformational changes upon ligand activation.

In this chapter, I describe some of these steady-state fluorescence-based approaches and methods applied to the following GPCRs: the tachykinins NK_1 and NK_2, the V_{1A} vasopressin and chemokine receptors.

2 Fluorescent reporter groups

Large varieties of fluorescent molecules bearing different reactive groups for labelling peptides are available from different suppliers. The functional groups used most frequently to attach reactive fluorophore to a peptide are those that react with primary amines and thiols. The peptide amino terminal group, lysine epsilon amino group, and thiol group from the side-chain of cysteine are the sites of choice for labelling natural amino acids. To react with amines, active esters like succinimidyl esters or isothiocyanates are the most popular reactive species in the fluorophore moiety. Alternatively, *in situ* synthesis of active esters (i.e. hydroxybenzotriazole) can be performed using carboxylic acids of fluorescent molecules (8). This represents an advantage over preformed hydroxysuccinimide active esters of fluorophores which hydrolyse during storage. Maleimide esters and iodoacetamide are widely used to label thiol groups.

Several fluorescent reporter molecules have been successfully applied to receptor research and their choice is dictated by the spectral and chemical properties of the chromophore. These probes are then evaluated for the type of information needed and the type of measurement to be performed. Some of the criteria affecting the choice of the fluorescent reporter group are summarized as follows:

(a) Fluorescence sensitivity: defined by the extinction coefficient and quantum yield of the fluorophore under the experimental conditions of the assay.

(b) Molecular weight and hydrophobicity of the fluorophore.

(c) Desired range of emission wavelengths.

(d) Fluorescence properties upon labelling. Some probes do not retain the same fluorescence properties on conjugation and spectral characteristics have to be evaluated with the conjugated molecules.

(e) Environmental sensitivity of probes to solvent polarity, pH, and presence of quencher molecules.

(f) Spectral overlap requirements for covering the range of distances separating a donor–acceptor pair for its use in fluorescence energy transfer measurements in a given system.

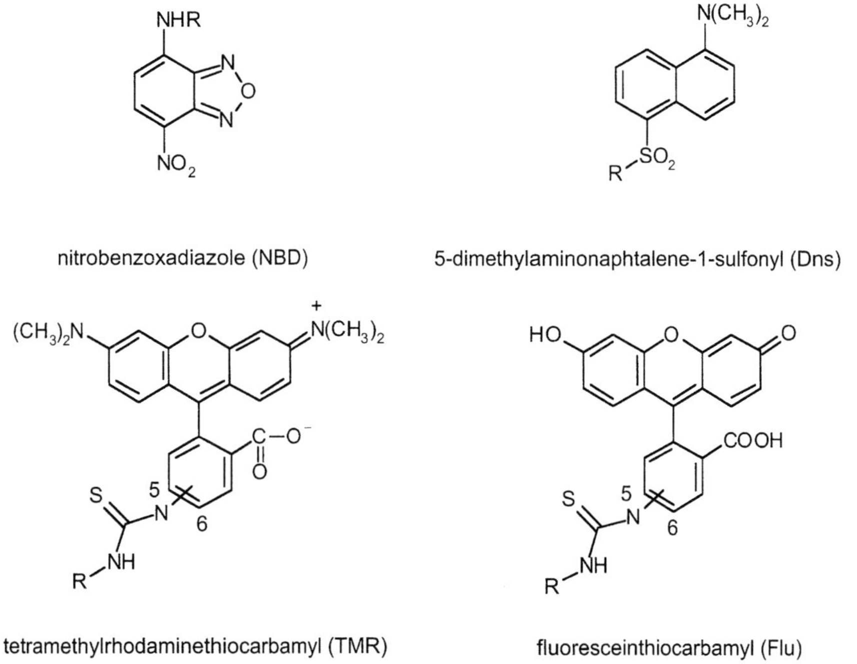

nitrobenzoxadiazole (NBD)

5-dimethylaminonaphtalene-1-sulfonyl (Dns)

tetramethylrhodaminethiocarbamyl (TMR)

fluoresceinthiocarbamyl (Flu)

Figure 1 Structure of fluorescent probes.

For most of the applications described in this chapter the fluorescent molecules used are: fluorescein (Flu), the polarity sensitive probes 5-(dimethylamino)-1-naphtalenesulfonyl (Dns) and 7-nitrobenz-2-oxa-1,3-diazol-4-yl (NBD), and tetramethylrhodamine (TMR) (*Figure 1*). Cyanine dyes (Cy5), linked to substance P (SP), neurokinin A (NKA), and galanin have been used to measure ligand–receptor binding constants in a homogeneous assay (9). Recently the tachykinin NK_2 receptor has been investigated using a bodipy-labelled NKA ligand (7).

As shown in *Figure 1*, for some fluorophores, isomers are possible. In fluorescein and rhodamine moieties the reactive functionality can be at position 5 or 6 and spectral differences can be observed for each isomer (8). We prefer to work with single isomers to facilitate the purification of derivatives. Unless described elsewhere in this chapter, the isomer 5 has been used for these fluorophores.

3 Ligands

3.1 Labelling of ligands

Natural peptide agonists are often taken as a starting point for derivatization at positions where reactive amino acids are present and also for introducing mutation sites with amino acids containing reactive side chains, either from natural or non-natural amino acids. The same strategy is followed when a peptide antagonist is available (*Figure 2*). In the less frequent case of non-peptidic, small molecular weight molecules, their fluorescent labelling represents an important

(A) Antagonists

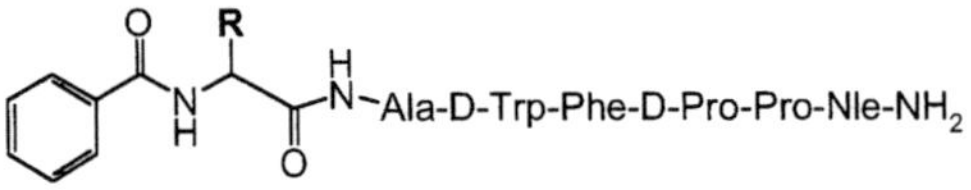

NK2 receptor antagonist [a]	amino acid at position 1	R	d (Å) [b]	pKi [c]
GR94800	Ala	—	-	9.81
1	Dab	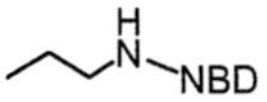 (structure: —N(H)—NBD)	4	8.87
2	Orn	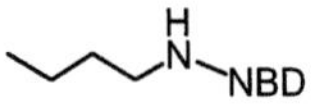 (structure: —N(H)—NBD)	5	8.84
3	Lys	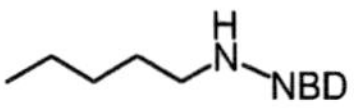 (structure: —N(H)—NBD)	6.5	8.83
4	Lys	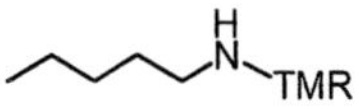 (structure: —N(H)—TMR)	6.5	8.80 [27]
5	Lys	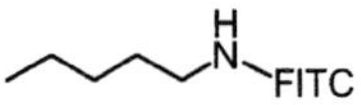 (structure: —N(H)—FITC)	6.5	7.61 [12]
6	Lys-Gly	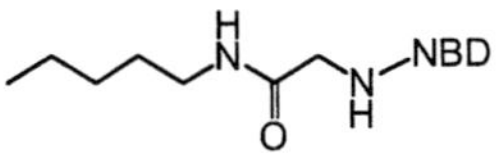 (structure: —N(H)—C(O)—CH2—N(H)—NBD)	10.5	8.80
7	Lys(ε-ah)	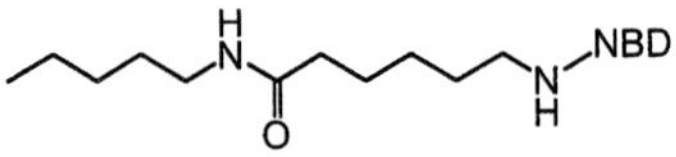 (structure: —N(H)—C(O)—(CH2)—N(H)—NBD)	15	8.62
8	Lys(ε-bah)	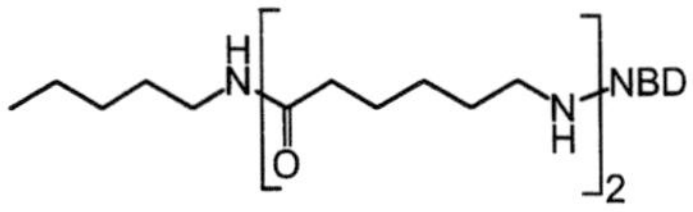 (structure: [—N(H)—C(O)—...—N(H)]2—NBD)	24	8.32

(B) Agonists

$$F^* - \text{His-Lys-Thr-Asp-Ser-Phe-Val-Gly-Leu-Met-NH}_2$$
positions: 1 4 10

$$F^* - \text{Asp-Ser-Phe-Val-Gly-Leu-Nle-NH}_2$$

Compound	F*	pKi
NKA	NBD	8.23
NKA	FITC	8.80
NKA	TMR	8.65
NKA[4-10]	NBD	8.08
NKA[4-10]	FITC	6.51

Figure 2 Structure of fluorescent NK$_2$ receptor ligands and pharmacological constants. (A) ah, 6-aminohexanoyl; bah, bis(6-aminohexanoyl). (B) d represents the extended distance between the fluorophore and the Cα of the amino acid at position 1. [c] pK$_i$ values from competition assays using [^{3}H]GR 100679 in CHO/NK$_2$ cells as described (13). Reported data are averaged from at least three separate assays performed in triplicate.

Substance P agonists derivatives:

$$1 \quad\quad 3 \quad\quad\quad\quad\quad\quad 8 \quad\quad\quad 11$$

Substance P: Arg-Pro-Lys-Pro-Gln-Gln-Phe-Phe-Gly-Leu-Met-NH$_2$

Non-peptidic dansylated antagonist:

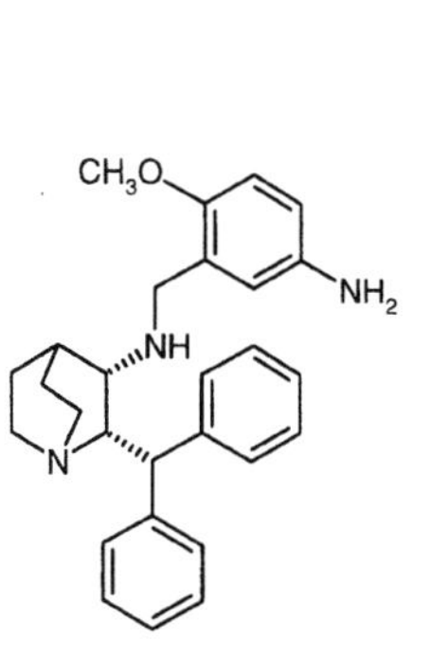

CP96345 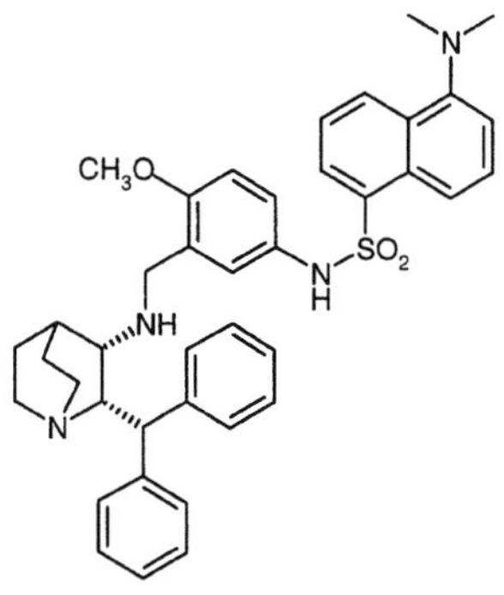   Dansyl-CP94345

Ligand	Pharmacological constants for SP derivatives on NK$_1$ receptors		Spectral characteristics of NK$_1$ receptor ligands in PBS pH 7.2	
	Competition binding assay [a] IC$_{50}$ (nM)	Ca^{2+} assay in CHO cells [b] EC$_{50}$ (nM)	ε at λ_{max} Abs (M^{-1} cm^{-1})	λ_{max} Abs/ λ_{max} Em (nm)
SP	0.29 ± 0.06	0.60 ± 0.16		
(ε–Dns)-Lys^3-SP	1.1 ± 0.2	n.d.	4700±450	330 / 549
(β–Dns)-Dap^8-SP	5150 ± 1440	n.d.	3550±150	330 / 549
(S-Dns)-Hcy^{11}-SP	29.2 ± 6.4	1.51 ± 0.24	3900±370	330 / 554
N-α-Flu-SP	21.1 ± 3.7	13.6 ± 3.6	79000±2000	494 /521
CP96345	0.42 ± 0.03	Antagonist [c]		
Dns-CP96345	3.3 ± 0.2	Antagonist [c]	3600±340	330 / 475, 542

Figure 3 Structure of fluorescent NK$_1$ receptor ligands. Pharmacological constants and spectral characteristics. (a) Fluorescent ligands were assayed for NK$_1$ receptor binding by competitive binding analysis with monoiodinated [^{125}I]Bolton-Hunter labelled substance P ([^{125}I]BH-SP) using COS-7 cells. Binding experiments were performed for 3 h at 4 °C with 50 pM [^{125}I]BH-SP plus variable amounts of fluorescent ligands in 0.5 ml of 50 mM Tris–HCl pH 7.4, containing 150 mM NaCl, 5 mM MnCl$_2$, and 0.1% (w/v) BSA supplemented with 0.1 mg/ml bacitracin. All determinations were performed in triplicate and non-specific binding was determined as the binding in the presence of 1 μM SP. (b) EC$_{50}$ calculated from Ca^{2+} mobilization assays using CHO cells as reported (10). (c) As expected for antagonists, no Ca^{2+} mobilization was detected. n.d. not determined.

challenge because the comparable size of both molecules can lead to derivatives with altered receptor biospecific recognition. However, a dansylated non-peptide antagonist of NK_1 receptors resulted in a potent and selective antagonist shown in *Figure 3* (10).

A prerequisite for the use of labelled molecules to study intermolecular interactions is that they conserve the biospecificity of the parent molecule. For this reason it is imperative to perform an extensive pharmacological character- ization of the derivatized peptides. Only those molecules that conserve pharma- cological properties similar to its parent compound are useful probes.

The polarity and size of the fluorophore and the site of labelling in the peptide sequence might affect ligand–receptor recognition by inducing non- favourable conformations for its binding and result in altered affinities, altered function of the receptor, and loss of selectivity. The characterization of the agonist or antagonist properties of the modified ligand must be rigorously deter- mined. An example of this change in peptide–receptor triggered activity is detected with amino terminally-modified Rantes proteins where the full agonist properties of some derivatives is lost (11).

3.2 Strategies for the generation of labelled peptide ligands

3.2.1 Peptide antagonists and agonists at NK_2 receptors

The ligand binding domain of the NK_2 receptor was studied using fluorescent antagonists and agonists. The binding sites for agonists and antagonists were compared using fluorescence techniques described in Section 5 of this chapter.

Fluorescent antagonists were derived from the potent and selective hepta- peptide ligand GR 94800 (*Figure 2*). Two modifications were investigated in order to introduce a fluorophore into the structure while the affinity and selectivity for the NK_2 receptor were preserved. The first was the effect of replacement of the bulky hydrophobic benzoyl group by the fluorescent groups fluorescein and NBD. In the second, Ala^1 was replaced by diamino acids and the NBD fluorophore reacted to the amino group in the side-chain (*Figure 2*). This resulted in ligands with increasing length spacer between the $C\alpha$ at position 1 and the NBD group. The linearized distances of the tethers ranged from approximately 4 Å to 24 Å.

A general method used to synthesize the NBD and fluorescein derivatives described above is reported in *Protocol 1*.

The NBD group was selected because its fluorescence is sensitive to the polarity of the environment. This feature is then exploited to investigate the inter- actions with the NK_2 receptor and to assess the hydrophobicity of the binding site. In addition NBD is more compact than fluorescein and therefore less likely to perturb receptor binding affinity.

The importance of the lipophilic N-terminal benzoyl substituent in the GR 94800 compound (pK_i = 9.81) was confirmed by the reduced affinity deriv- atives produced from the replacement of this group by NBD (pK_i = 7.06) or fluores- cein (pK_i = 5.99) (12). In contrast, NBD labelling of the side-chain of the diamino acids at position one resulted in a moderate decrease in affinity (*Figure 2*).

Protocol 1

Labelling and purification of fluorescent peptide derivatives: NK$_2$ receptor antagonists (*Figure 2*)

Equipment and reagents

- HPLC instrumentation (Waters-Millipore): two pumps (Model 510), UV-VIS detector (Model 481), gradient controller
- Analytical HPLC: column, Nucleosil 300-7 C8 (Macherey Nagel ET 250/8/4); flow rate, 1 ml/min; gradient, 0–75% B over 60 min
- HPLC-purified and characterized peptides (synthesized according to refs 12 and 13)
- RP-HPLC buffer A: 0.1% TFA (w/v) in HPLC grade water

- Semi-preparative RP-HPLC: Nucleosil 300-7 C18 (Macherey Nagel SS 250/0.5″/10); flow rate, 6 ml/min
- RP-HPLC buffer B: 0.09% TFA (w/v) in 80% (v/v) acetonitrile/water (Baker)
- 10% (w/v) TFA (Fluka) in water, DMF (Fluka), FITC (Molecular Probes), NBD fluoride (Sigma or Molecular Probes), 50 mM sodium borate pH 9.5

Method

1 Dissolve 5 μmol of peptide in 2.0 ml of 50 mM sodium borate pH 9.5/DMF (1:1).[a]

2 Add to the peptide solution, 0.5 ml of 0.1 M FITC in DMF (50 μmol) or 0.03 M NBD fluoride in DMF (15 μmol).

3 Incubate the mixture in the dark at 4°C under continuous stirring.

4 Take aliquots (5 μl, 10 nmol) over time for analytical HPLC monitoring of the reaction.

5 Stop the reaction by acidifying with 10% (w/v) TFA to a final concentration of 0.1% TFA.

6 Inject the crude mixture directly into a small scale preparative RP-HPLC and run a gradient, 0–35% B over 35 min, 35–75% B over 80 min, 75–85% B over 5 min at 6 ml/min.

7 Monitor the peak elution at 214 nm, collect the peak fractions, and lyophilize them.

8 Pool fractions containing pure material and take aliquots for extensive analytical characterization (*Protocol 2*). Calculate the yield of the reaction.[b]

[a] The peptide NK$_2$ receptor antagonists are not soluble in aqueous buffer; for this reason the fluorescence labelling reaction is performed in the presence of DMF. For aqueous soluble peptides like substance P (*Figure 3*) the fluorescent derivatives were synthesized in borate buffer (10).

[b] Typical yields of derivatization reactions after HPLC purification for NBD-labelled compounds reported in *Figure 2* ranged from 55–75%. Antagonist 1, 2, and 3 yields were 57%, 66%, and 73%, respectively (12). For FITC derivatives, yields ranged from 45–65%. Some yields of fluorescein peptide derivatives are reported here: Flu-MIP-1α, 62% (31); Nα-Flu-SP, 47%; NK$_2$ antagonist 5, 58% (12).

The fluorescent agonists were prepared from modification of the $N\alpha$ position of the natural ligand NKA with fluorescein and NBD according to *Protocol 1* (13). The potent and selective heptapeptide agonist derived from NKA, Nle^{10}-NKA[4–10] (14) was also labelled at the $N\alpha$ position with NBD and fluorescein. In addition, another NBD peptide was synthesized from this truncated NKA where Val^7 was replaced by Dap and the NBD attached to its amino side-chain. The affinity of this derivative for the NK_2 receptor was strongly reduced. In another study, a 2000-fold decrease in affinity was found for a NKA derivative when Val^7 was replaced by *p*-nitro-Phe (Turcatti, unpublished results). This strongly suggests an unfavourable steric/electronic interaction of these derivatives and the receptor in the binding pocket.

Both N-terminal NKA derivatives and the NBD of the truncated version conserved a relatively high affinity for the NK_2 receptor compared to NKA whereas the FITC derivative of NKA[4–10] was much less potent (*Figure 2*). All of them remained agonists as evidenced by their activation of Ca^{2+} mobilization that can be blocked by the specific NK_2 receptor antagonist GR 94800 (12).

3.2.2 Peptide agonists and non-peptide antagonists at NK_1 receptors

Undecapeptide NK_1 receptor agonists were prepared by site-specific dansylation of SP at positions 3, 8, 11 in the peptide backbone. Derivatives at positions 8 and 11 were obtained by attaching the dansyl group to the diamino acids, Dap and HCy, that replaced the residues present in the natural sequence of SP, Phe^8 and Met^{11}, respectively. Another fluorescent peptide analogue was obtained by derivatization of $N\alpha$-SP with fluorescein (10) (*Figure 3*). A fluorescein derivative at the ε-amino of Lys^3 has also been reported (15).

A dansylated derivative of a non-peptide NK_1 receptor antagonist (16) was prepared (10) (*Figure 3*). In this study we compared the interaction of fluorescent peptide agonists and a non-peptide dansylated antagonist with NK_1 receptors using spectrofluorometric techniques described in Section 5.

Labelling of SP with dansyl at position 3 maintained a high affinity for NK_1 receptors comparable with that of the parent peptide. Modification of SP with fluorescein at the N-terminal position or with Dns at position 11 resulted in a 70–100-fold decrease in affinity compared to SP. More dramatically, the derivative Dns-Dap^8-SP showed a 18 000-fold decrease in affinity when compared to SP that prevented its further use as a fluorescent probe for NK_1 receptors (see *Figure 3*).

3.2.3 Linear vasopressin receptor antagonists

Fluorescent analogues of linear vasopressin receptor antagonists have been prepared by derivatization of Lys ε-amino group (8). To add the fluorophore at different positions of the ligand backbone, the two Arg residues in the sequence were replaced successively with Lys at positions 6 and 8, respectively. Derivatives were prepared using two fluorophores, fluoresceinyl and rhodamyl, using an amide link instead of the thiocarbamyl bond. The antagonist properties of the ligand were not affected by the addition of fluorophores.

3.2.4 Peptide CCR$_1$ and CCR$_5$ receptor agonist

Fluoresceinthiocarbamoyl-MIP-1α (Flu-MIP-1α) was synthesized from MIP-1α by reaction of 1.2 equivalents of FITC at pH 8.0 using *Protocol 1*. Under these conditions, isothiocyanate electrophiles react preferentially with the low pK$_a$ N-terminal group of peptides rather than with the high pK$_a$, ϵ-amino groups of lysil side-chains. The presence of a monoderivative containing the Flu group was indicated by mass spectrometry. Chemical characterization of the derivative was carried out by comparative Edman sequencing of the chemokine and its fluorescent derivative by analysis of phenylthiohydantoin released at each cycle. This analysis confirmed that more than 95% of the Flu group was in the N-terminal position of the peptide (31).

An N-terminal dansylated MIP-1α has been prepared in our laboratory according to the general procedure described in *Protocol 1*. This derivative had an affinity for the CCR$_5$ receptor expressed in CHO cells which was similar to that of its parent peptide. The Nα-Dns-MIP-1α retained full agonist efficacy at the same potency of the MIP-1α peptide as assessed by measuring release of Ca^{2+} from intracellular stores in CHO cells loaded with the dye, fura-2, as described previously (17).

3.3 Characterization of fluorescent ligands

3.3.1 Analytical characterization

Fluorescently labelled peptide ligands have to be chemically characterized to determine:

(a) Peptide composition.

(b) Presence of fluorophore at the desired position.

(c) Molar ratio of fluorophore/peptide.

Molecular mass determination by electrospray mass spectrometry (ES-MS) will give the mass of the fluorescent peptide enabling points (a) and (c) to be answered, but gives no information about the labelling position. N-terminal sequencing by Edman degradation will detect the amino acid at which the fluorophore is attached in general giving a non-assigned amino acid for the cycle corresponding to the derivatized amino acid.

We also run compositional amino acid analysis to obtain another criterion of purity and to determine accurately the concentration of the peptide derivative that allows further determination of extinction coefficients in different solutions (see *Protocol 2*). Internal standard calibration is performed with a mixture of all natural amino acids and Nle as the internal standard, except when this amino acid is present in the synthetic peptide sequence, as for NK$_2$ receptor ligands described in *Figure 2*, where Leu is employed.

It is of paramount importance to determine accurately the concentration of fluorescent ligand in order to calculate the pharmacological constants with precision.

Protocol 2

Concentration determination of peptide conjugates (ex: (S-Dns)HCy[11]-SP): extinction coefficient calculations (*Figures 3, 4* and *Table 1*)[a]

Equipment and reagents

- Spectrophotometer UV-VIS (Jasco Model 550)
- Amino acid analyser: Waters AccQ.Tag chemistry
- Hewlett Packard 1090 liquid chromatograph equipped with a fluorescence detector HP1046A (Agilent Technologies)
- Protein hydrolysis workstation (Pico-Tag-Waters)

- Fluorescent peptides
- Amino acid standard mixture, unnatural amino acids that enter in the composition of peptides to be analysed
- Internal standard amino acid[b] (Nle if this amino acid if absent in the peptide composition) (Pierce)
- Methanol (Fluka)
- Constant boiling HCl 6 M (Pierce) containing 1 mg/ml phenol

Method

1. Estimate the concentration of the peptide solution by the absorption of the fluorescent reporter group using the extinction coefficient reported in the literature. For this dansylated compound, the extinction coefficient of 3000 was taken at 330 nm and the stock solution in water estimated to be 2 mM.

2. Make accurate diluted solutions of dansylated peptide in different solutions: Dilute the 2 mM stock solution of dansylated peptide to 100 μM with water (solution 1) and in methanol to give a final methanol/water ratio of 95:5 (solution 2).

3. Record absorption spectra for each solution and tabulate its maximal absorption values and wavelength.

4. Add 15 μl of stock solution (estimated 1.5 nmol) to three tubes for gas phase acid hydrolysis. Add 2 nmol of the internal standard Nle in water to each tube and lyophilize.

5. Put the tubes on a hydrolysis reactor containing 6 M HCl and 1 mg/ml phenol. Purge with nitrogen and evacuate to remove oxygen from the reactor (repeat twice). Perform the reaction for 16 h at 112°C under vacuum.

6. Remove the tubes from the reactor and evaporate the HCl. Dissolve the sample in amino acid analysis loading buffer and inject into the analyser.

7. Using internal standard calculation mode, quantify the amount of each amino acid. Average the amount per amino acid and calculate the concentration of the initial stock solution.

Protocol 2 continued

8 Use this concentration to calculate an accurate extinction coefficient using the absorbance values measured in step 3.

[a] Extinction coefficients of fluorescent peptides calculated using this protocol are reported in *Figure 3* and *Tables 1, 2,* and *3.*

[b] For peptides containing Nle in their peptide backbone, an amino acid absent in the peptide composition has to be taken as an internal standard. Peptides described in *Figure 2* contain Nle, therefore compositional amino acid analyses for these peptides were performed using Leu as internal standard, an amino acid which is absent in the peptide sequence.

3.3.2 Characterization of spectral properties

The determination of accurate concentrations by compositional amino acid analysis allows accurate calculation of extinction coefficients by acquiring absorption spectra under different experimental conditions (*Protocol 2*). In *Figure 3* are reported the extinction coefficients for SP fluorescent derivatives in phosphate-buffered saline (PBS) solution at pH 7.2 with the wavelengths of maximum absorption and emission for these derivatives. *Table 1* illustrates the increase in quantum yield of a dansylated ligand by decreasing the polarity of the solvent while the extinction coefficient remains unchanged.

For dansyl and NBD-labelled peptides, both the quantum yield of their emission increases at a low-polarity environment and their maximum of emission is blue-shifted. This property has been utilized to study the environment of the ribosome–nascent chain–membrane complexes using NBD incorporated into the signal sequence (18) and to study the environment of NBD NK_2 receptor antagonists as described in Section 6.1.1.

To evaluate the utility of NBD or dansyl as probes of polarity of the surrounding medium, the emission of ligands labelled with these probes is examined in solvents of different polarities: generally water/dioxane mixtures at different proportions. This dependence of solvent polarity is shown in *Figure 4* for the dansyl group attached to the peptide substance P. Similar traces were generated for NBD-labelled NK_2 receptor ligands (13).

The absorption and emission spectra of TMR ligands of NK_2 receptor antagonist 4 and the agonist NKA derivative shown in *Figure 2*, were also solvent dependent. It has been reported that the absorption spectra of TMR

Table 1 Spectral characteristics of (S-Dns)-HCy[11]-SP as a function of the polarity of the solvent

Solvent	λ_{max} Abs (nm)	Absorbance at λ_{max} Abs	λ_{max} Em (nm)	Emission at λ_{max} Em	ε[a] (M^{-1} cm^{-1})
H_2O	330	0.3706	554	477	3956 ± 30
MeOH/H_2O 95:5 (v/v)	338	0.3714	531	3929	3964 ± 30

[a] The concentration determined by compositional amino acid analysis according to *Protocol 2* was 93 μM.

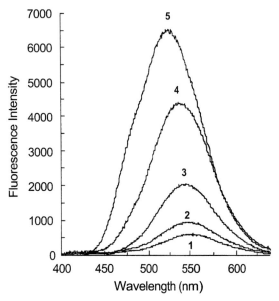

Figure 4 Dansyl a as probes of the polarity of the environment. Fluorescence spectra of a representative dansylated NK_1 receptor ligand in solvents of different polarity. Emission spectra of (S-Dns)-HCy[11]-SP in PBS pH 7.2, containing varying amounts of dioxane (%, v/v) **1**: 0%, **2**: 10%, **3**: 25%, **4**: 50%, **5**: 90%. All the dansyl peptide derivatives described in *Figure 3* have a similar spectral behaviour.

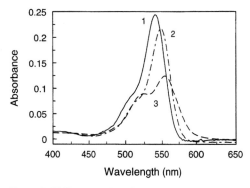

Figure 5 TMR antagonist of NK_2 receptors. Absorption spectrum variations as a function of the solvent. 1.7 μM solution of (ε-TMR)-Lys antagonist of NK_2 receptors (compound 4 in *Figure 2*) in **1**: methanol; **2**: 50 mM Tris–HCl pH 7.4; **3**: 5:95 dioxane/50 mM Tris–HCl pH 7.4.

conjugates are frequently complex, usually split into two absorption peaks at about 520 and 550 nm. From *Figure 5*, the absorption maximum of the TMR antagonist was shifted from 540 nm to 554 nm when the solvent was changed from methanol to aqueous buffer at pH 7.4. The extinction coefficient at the absorption maximum was also affected by this solvent change, being reduced to 73 000 at 550 nm. The quantum yield of the TMR derivative also changed with the solvent composition, but the λ_{max} of emission remained unchanged at 562 ± 1 nm (*Table 2*).

Table 2 Spectroscopic data of ε-Lys-(TMR) antagonist of NK_2 receptors (*Figure 2*) in different solvents

Solvent	ε (M^{-1} cm^{-1}) at λ_{max} Abs (nm)[a]	Relative fluorescence (λ_{ex} = 530 nm)
Methanol	145 000, 540	9240
50 mM Tris–HCl pH 7.4	73 200, 554	1524
5:95 dioxane/50 mM Tris–HCl pH 7.4 (v/v)	129 000, 548	7715

[a] Reported values are the average of three independent determinations. The mean standard deviation was 3000.

3.3.3 Pharmacological characterization

The affinities of the fluorescent ligands for their receptor are determined by competition binding experiments using receptor transfected cell or membrane preparations. Displacement binding curves of a radiolabelled ligand by the different fluorescent ligands are generated as previously reported (12). Reported K_i values for NK_2 and NK_1 receptor ligands are given in *Figures 2* and *3* respectively.

Competition binding curves of [^{125}I]NKA with NKA or its derivative Nα-Flu-NKA on human NK_2/CHO cell receptors are shown in *Figure 6A*. The K_i values for the NK_2 receptor obtained for these ligands were very similar. This type of assay reveals affinity changes for the labelled ligands but is not indicative of functional activity. In contrast, changes in intracellular Ca^{2+} concentration evoked by agonist stimulation of GPCR can be used to obtain information about the ability of modified ligand analogues to activate the receptor. The agonist efficacy and potency of ligands is determined by calculation of EC_{50} values in Ca^{2+} mobilization experiments in transfected cells using the fluorescent Ca^{2+}-chelating agent fura-2 (19). Calcium assay experiments in CHO/NK_2 cells of NKA and Flu-NKA are reported in *Figure 6B* where the calculated values of EC_{50} are 0.8 nM and 3 nM, respectively.

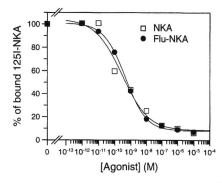 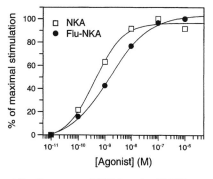

Figure 6 Competitive binding analysis and Ca^{2+} mobilization assay of NKA ligands. (A) NKA ligands were assayed for NK_2 receptor binding by competitive binding analysis with monoiodinated [^{125}I]NKA CHO cells (13). (B) Functional assay. Ca^{2+} mobilization in CHO/NK_2 cells as a function of increasing concentrations of either NKA or Flu-NKA using the fluorescent chelating agent fura-2 (19). The fluorescence was recorded at 480 nm to avoid overlap with the fluorescein fluorescence.

4 Receptors

Membrane receptor proteins have been expressed in a variety of expression systems with some examples of high overexpression (1). In our different studies we used stably transfected CHO cells with expression yields ranging from 30 000 to 700 000 receptors/cell.

A CHO cell line stably expressing a high number of NK_2 receptors per cell has been established by fluorescence-activated cell sorting (FACS) and subcloning (20). Saturation binding assays (Scatchard plot) for this NK_2/CHO receptor were performed with the potent radiolabelled NK_2 receptor antagonist [³H]GR 94800. Analysis of four separate experiments using triplicate points showed a single class of saturable binding sites with a K_D of 2.0 ± 0.6 nM and a maximum binding capacity B_{max} of 27 ± 4 pmol/mg of total membrane protein. This is equivalent to an expression level of approximately 700 000 receptors/cell. This high express-ing cell line allowed extensive fluorometric characterization of fluorescent NK_2 ligand interactions with the receptor.

The ability of a fluorescent ligand to recognize and label its receptor can be determined by flow cytometry. Saturation binding assays can be performed by incubating the cells at different concentrations of the fluorescent ligand in the

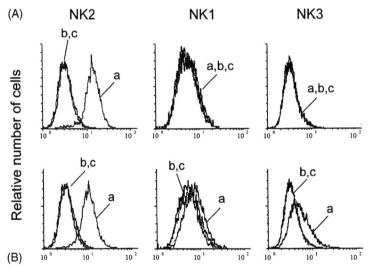

Figure 7 Flow cytometry analysis of tachykinin receptors using fluorescent peptide ligands. Cells (about 10^6) transfected with the tachykinin receptors (NK_1, NK_2, and NK_3) were incubated in PBS pH 7.2 containing $MnCl_2$ (3 mM) and BSA (0.2 mg/ml) in the presence of 20 nM fluorescent peptide ligand (see below) at 4 °C for 3 h or 20 °C for 1.5 h. The cells were washed twice with ice-cold PBS pH 7.2, and kept on ice. Non-specific binding was measured in the presence of the non-fluorescent competitor antagonist GR 100679 (20 µM). The cell fluorescence histograms were acquired at room temperature with a Becton Dickinson FACSCAN instrument using either linear or logarithmic modes with excitation/emission at 488/518 nm. (A) NBD antagonist 1. (B) Flu-NKA. (Both ligands are described in *Figure 2*.) a, 20 nM fluorescent ligand; b, 20 nM fluorescent ligand and 20 µM GR 100679; c, no ligand (autofluorescence of cells). Reproduced with permission from ref. 12. Copyright 1994, Am. Chem. Soc.

presence or absence of non-fluorescent ligands and measurement of the mean fluorescence intensity (MFI) for each cell population.

Fluorescent ligand binding can be compared for the same receptor and selectivity determination carried out using receptors of the same family as shown in *Figure 7*. The NK_2 receptor agonist Flu-NKA was more potent than the NBD antagonist 1. In contrast, the antagonist is more selective than Flu-NKA for NK_2 over the two other tachykinin (NK_1 and NK_3) receptors using stable CHO cell clones selectively expressing each receptor at more than 10^5 receptors/cell. At various concentrations of the NBD antagonist ligand, only CHO cells expressing the NK_2 receptor were labelled, whereas the same concentrations of Flu-NKA gave positive results for all three tachykinins.

We used entire cells or membranes in suspension for fluorescence measurements. *Protocol 3* describes the preparation of membranes and cell culture conditions for such measurements. We found that when cells were grown up to about 80% confluence we obtained better signal-to-noise ratios than total coverage of the culture flask.

Protocol 3

Cells or cell membranes: incubation with fluorescent ligands (CHO/NK₂ cells)

Equipment and reagents

- Spectrofluorometer equipped with a thermostatically controlled cuvette holder unit and magnetic stirrer
- CHO cells expressing NK_2 receptors at 500 000 to 800 000 receptors per cell, untransfected cells (cell controls)
- Or cell membrane preparations for the receptor of interest and control membranes from untransfected cells

- Sonic bath.
- Binding buffer: PBS pH 7.2 containing 3 mM $MgCl_2$, 0.2 mg/ml BSA, and 0.5 mM sodium azide
- 2 mM stock solutions in DMSO of fluorescent ligands (described in *Figure 2*) and non-fluorescent competitor ligand GR 100679

Method[a]

1   Culture cells as monolayers in a humidified 5% CO_2 atmosphere at 37°C in a Dulbecco's MEM/NUT MIX F12 medium (Gibco BRL, Life Technologies) supplemented with 10% fetal calf serum, 2% (w/v) Pen-Strep, and 0.5% (w/v) Nystatin.

2   Harvest the cells with PBS containing 1 mM EDTA when they are at about 80% confluence. Wash the cells with ice-cold PBS.

3   Suspend the cells in PBS and aliquot 10^6 cells in each Eppendorf tube (9 tubes for triplicate analyses). Centrifuge and suspend the cells in 200 µl binding buffer.

4   Incubate the cells (triplicates) for 1 h at 25°C in incubation buffer containing:
    (a) 20 nM fluorescent ligand (total binding).

    (b) 20 nM fluorescent ligand and 5 μM of non-fluorescent antagonist GR 100679 (non-specific binding).

    (c) Without any addition (control for autofluorescence of cells).

5 Remove the excess ligand by washing the cells twice with ice-cold PBS pH 7.2 immediately before use and resuspend in 400–500 μl for fluorescence measurements.

6 Record emission spectra from 500–600 nm for each sample (*Figure 8*).

ᵃ Variations of the protocol for fluorometric binding assay using cell membranes are described in the legend of *Figure 8*.

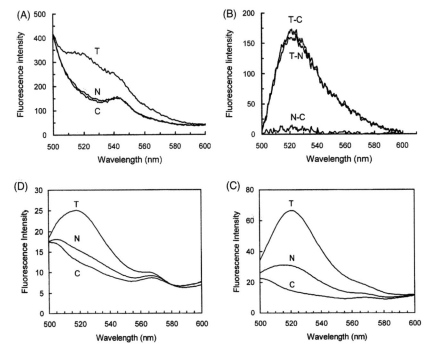

Figure 8 Spectrofluorometric analysis of binding of the fluorescent ligand. (A, B) Nα-Flu-SP to NK₁ receptors in CHO cell membrane suspensions. Membranes from NK₁/CHO cells were suspended in PBS pH 7.2, containing MgCl₂ (3 mM), BSA (0.2 mg/ml), at a concentration of about 5 nM [³H]SP binding sites and incubated with 50 nM ligand.Nα-Flu-SP in the presence (non-specific binding; trace N) or absence (total binding; trace T) of 50 μM non-fluorescent ligand GR 147696 at 20 °C for 1 h. Membrane controls (trace C) were incubated in the absence of fluorescent ligand. After binding, the membranes were washed twice with ice-cold PBS, and homogenized by sonication in an ice-cold bath for 3 min immediately prior to fluorometric measurements. Samples were resuspended in 200 μl PBS at a receptor concentration of about 1.2 nM and fluorescence was recorded under continuous stirring by irradiating at 475 nm. (A) Spectra obtained for the different membrane suspensions. (B) Traces were corrected from control samples as described in the figure, T-N accounts for specific binding fluorescence. (C, D) Nα-Flu-MIP-1α binding to CCR₁ and CCR₅ receptors in CHO cell membrane suspensions. Raw emission spectra obtained for Nα-Flu-MIP-1α binding to CHO cells transfected with CCR₁ (C) and CCR₅ (D) receptors, respectively (31). Total binding (trace T), non-specific binding (trace N), and membrane autofluorescence control (C).

5 Fluorometric assays

5.1 Ligand binding assays

In general, the assays using fluorescent ligands are performed with cells or membranes in suspension using a spectrofluorometer or a cell sorter (FACS). Control samples, incubated under the same conditions are included in every experiment. Fluorescence from untransfected cells or membranes accounts for autofluorescence and Rayleigh scattering under the experimental irradiation conditions.

Protocol 3 describes a standardized protocol for measuring fluorescence associated with binding events. In general, for spectrofluorometric assays, transfected cell membranes at a receptor concentration 2–5 nM are suspended in binding buffer and incubated with a concentration $10 \times K_D$ of fluorescent ligand in the presence (non-specific binding, NSB) or in the absence (total binding, T) of 500- to 1000-fold excess of non-fluorescent competitor ligand. As an additional control, cells or membranes are incubated in the absence of fluorescent ligand (control, C). Cells or membranes are then washed and centrifuged prior to final homogenization for fluorometry experiments. Typical traces of emission spectra associated with a fluorescent ligand bound to its receptor are then generated as illustrated in *Figure 8*. The raw traces (*Figure 8A*) are subtracted from the contribution of cells to give the net total bound fluorescence (specific + non-specific) and net non-specific fluorescence. Subtraction of traces T-N enables determination of the fluorescence associated with the specific binding of the ligand to the receptor (traces in *Figure 8B*). The non-specific fluorescence contribution represents less than 5% of the total bound fluorescence giving a signal-to noise ratio of 21.

Representative raw traces from the binding experiments of Nα-Flu-MIP-1α with two receptors of the same family, CCR_1 and CCR_5, are illustrated in panels C and D of *Figure 8*. The specific fluorescein-bound fluorescence has its maximal emission at 522 nm and represents 72% of the total fluorescence for both receptors.

The binding of fluorescent ligands to the receptor in cell membranes at conditions of receptor saturation described below is required for further fluorescence studies to probe the fluorophore in the ligand–receptor bound state. These spectrofluorometric methods, polarity of the binding site, accessibility to quenchers, anisotropy, and FRET measurements are described in the next sections.

5.2 Kinetics of ligand binding and dissociation

Fluorescence labelled ligands can serve as probes for monitoring receptor binding events in real time (13, 21, 22). The quantum yield of the NBD group of the NBD peptide antagonists of NK_1 and NK_2 receptors (*Figure 2*) is increased more than fivefold and accompanied by a blue-shift of the emission when these antagonists are bound to NK_2/CHO receptor. This indicates that the fluorophore in the receptor-bound state is in a less polar environment than in the unbound state. This can be interpreted as an increase of the hydrophobicity of the fluorescent group microenvironment. A similar increase in fluorescence intensity is obtained by dissolving the NBD ligand in a 5% (v/v) dioxane/water mixture.

The discrimination between bound and free NBD antagonist can be used to monitor the time-course of the association of the ligand to the receptor and its dissociation after addition of excess competitor antagonist.

Protocol 4

Steady-state fluorescence monitoring of binding and dissociation of NBD antagonists of NK$_2$ receptors (*Figure 9*)

Equipment and reagents

- Spectrofluorometer equipped with a thermostatically controlled cuvette holder unit and magnetic stirrer
- CHO cells expressing 500 000 to 800 000 NK$_2$ receptors per cell, untransfected cells (cell controls)
- Or cell membrane preparations for the receptor of interest and control membranes from untransfected cells

- Binding buffer: PBS pH 7.2 containing 3 mM MgCl$_2$, 0.2 mg/ml BSA, and 0.5 mM sodium azide
- 2 mM stock solutions of fluorescent ligands (described in *Fig. 2*) and non-fluorescent competitor ligand GR 100679 or GR 94800

Method

1 Resuspend 1-2 × 10^6 CHO/NK$_2$ cells in 500 μl binding buffer at 20°C. The final concentration of receptors is about 1-2 nM [^{3}H]GR 94800 binding sites.

2 Transfer the suspension to a 0.5 × 1.0 cm quartz cuvette (Hellma) with a small magnet and the cuvette holder maintained at 20°C.

3 Measure the emission of the sample at 520 nm using an excitation wavelength of 476 nm with continuous stirring.[a] Select excitation and emission bandwidths.

4 After a stable recording of the cells autofluorescence, add 10 nM of fluorescent antagonist to initiate the binding measurement and continue the recording at 540 nm until a constant value of fluorescence emission is obtained indicating the completion of binding.

5 To initiate the dissociation, add to the suspension a 1000-fold molar excess of non-fluorescent competitor antagonist GR 94800 or GR 100679 (2.5 μl of 2 mM solution).

6 Stop the experiment when only a small fluorescence decrease is detected over time.

7 Repeat the same experiment, starting at step 1, and pre-incubate the samples with 5 μM GR 100679 in binding buffer for 1 h. Continue as described in steps 2-4 to register the fluorescence that accounts for unbound ligand.

[a] Fluctuations of the fluorescence recordings are possible due to cell aggregation, in this case gentle pipetting helps to re-homogenize the suspension.

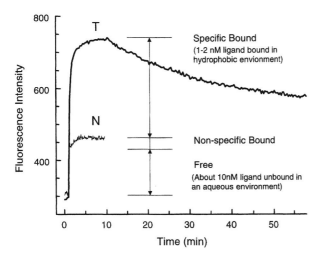

Figure 9 Real time monitoring of ligand–receptor association and dissociation. Trace T: About 10^6 CHO/NK$_2$ cells are suspended in 500 μl of binding buffer in a fluorescence cuvette. At time about 1 min 2.5 μl of a 2 μM solution of fluorescent NK$_1$ receptor antagonist (Dab-NBD, in *Figure 2*) were added to the cell suspension. At 10 min 2.5 μl of 2 mM of non-fluorescent competitor antagonist GR 94800 were added to initiate the dissociation. Trace N: 10 nM of fluorescent ligand was added to cells pre-incubated with 5 μM of the non-fluorescent competitor antagonist GR 94800.

Protocol 4 gives experimental details for performing this type of time-course measurement. *Figure 9* shows the fluorescence at 540 nm as a function of time for suspended CHO/NK$_2$ cells. When a stable baseline is obtained, 10 nM of antagonist 1 is added in the presence (N) or in the absence of 500-fold molar excess of GR 94800. There is a rapid increase in fluorescence for both traces due to the fluorescence of the ligand free in solution. In the absence of competitor, a further increase in fluorescence (T) accounts for the specific binding to NK$_2$ receptors. This is demonstrated by the small increase seen in samples saturated with non-fluorescent antagonist (N). Equilibrium is reached at about 7–8 min. The contribution of free ligand is calculated from control experiments. The binding is reversible; addition of an excess of competitor antagonist GR 94800 or GR 100679 causes a slow decrease in fluorescence which is explained by dissociation of the fluorescent ligand from the NK$_2$ receptor. The kinetic parameters calculated for the association and a biphasic dissociation are similar to those obtained using a radiolabelled ligand. The calculated constants were $k_{ass} = 0.53$ min^{-1}, $k_{diss}^1 = 0.022$ min^{-1}, and $k_{diss}^2 = 0.006$ min^{-1} (13).

5.3 Fluorescence quenching experiments

Collision quenching of fluorescence is a powerful technique that can be used to probe solvent accessibility of receptor-bound ligand (23). This approach has been used to characterize the binding determinants for NK$_2$, NK$_1$, CCR$_1$, and CCR$_5$

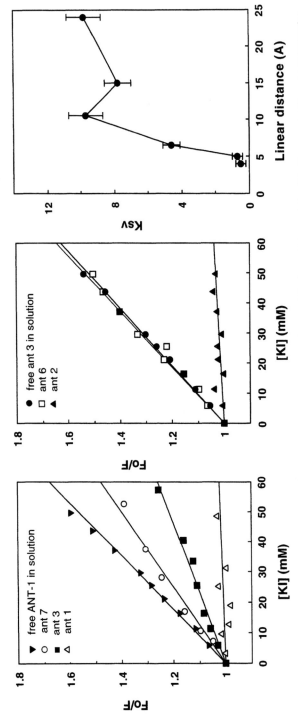

Figure 10 Probing accessibility of bound ligands by quenching experiments (see *Protocol 6*). Stern–Volmer plots of Fo/F *versus* iodide concentration for quenching of the fluorescence of NK₂-bound NBD antagonist ligands (described in *Figure 2*) by iodide (A and B). (C) Stern–Volmer quenching constant Ksv as a function of spacer length (assuming an extended conformation) between NBD and the peptide backbone of the antagonists. (C) Reproduced with permission from ref. 13. Copyright 1995, Am. Chem. Soc.

receptor ligands (10, 13). For basic principles and description of quenchers see ref. 24.

The anion iodide and the cation Co^{2+} have been shown to quench NBD fluorescence (25, 26) and this has been applied to NBD antagonists of NK_2 as described in *Protocol 5*. Quenching curves obtained for the peptide antagonists listed in *Figure 2* are plotted in *Figure 10*. The method was adapted and applied also to NK_1 fluorescent ligands (*Figure 3*).

Some points concerning the choice of quenchers and precautions are listed below:

(a) The concentration range of the quencher where appreciable quenching is observed.

(b) It has to be determined whether a quencher reacts with or perturbs the conformation of a protein leading to changes in the biomolecular recognition of the system. To confirm that the affinity of the ligand for the receptor is not altered by the presence of the quencher, binding and functional assays should be performed in the presence of the quencher at concentrations used in quenching experiments.

(c) When using ions as quenchers it is important to maintain a constant ionic strength of the medium throughout the whole experiment so as to exclude ionic interactions that may alter the fluorescence related to the ligand–receptor bound state.

(d) To exclude any charge interaction, a cation and an anion can be used as quenchers successively and quenching constants (*Ksv*, defined in *Protocol 5*, point 6) compared. For instance, iodide and Co^{2+} have been used as NBD quenchers (13).

(e) Solvents and chemicals have to be free from fluorescent contaminant that could act as a quencher or alter the background signal during the experiment.

(f) If convenient, use solutions degassed prior to their use to avoid quenching by oxygen.

(g) Some of the quenchers of fluorescence absorb light in the range of excitation or emission fluorescence giving rise to inner filter effects that could be confused with quenching because the emission of the fluorophore decreases as quencher concentration increases. In this case the experimental points have to be corrected using reported methods (27). In general, the absorbance of the sample at the excitation wavelength should be kept below 0.2 to ensure a linear response of the signal.

(h) The accessibility of fluorophores by quenchers can be assessed by the use of water soluble quenchers (iodide, Co^{2+}, hydroxy-TEMPO) or apolar lipophilic quenchers that partition into the lipid bilayer (TEMPO) (see *Table 3*).

The measurement of accessibility constants for the ligand free in solution has to be accompanied by the control experiment where the ligand is free but in the

Table 3 Stern–Volmer constants for different fluorescent ligand–receptors pairs and various quenchers (M^{-1})

(a)

NK$_2$ receptor antagonists (from *Figure 2*)	Iodide		Co^{2+}	
	*K*sv (free)	*K*sv (bound)	*K*sv (free)	*K*sv (bound)
(γ-NBD)-Dab (1)	10.9 ± 0.4	0.5 ± 0.3	16 ± 2	50 ± 2
(δ-NBD)-Orn (2)	10.8 ± 0.4	0.7 ± 0.3	14 ± 3	51 ± 2
(ε-NBD)-Lys (3)	10.7 ± 0.3	4.6 ± 0.5	n.d.[a]	n.d.[a]
(NBD)-Lys-Gly (6)	10.7 ± 0.5	9.8 ± 1.0	51 ± 6	52 ± 1
(ε-TMR)-Lys (4)	16.5 ± 1.5	5.0 ± 1.0	n.d.[a]	n.d.[a]

(b)

NK$_1$ receptor ligands (from *Figure 3*)	Tempo		Hydoxy-tempo	
	*K*sv (free)	*K*sv (bound)	*K*sv (free)	*K*sv (bound)
Dns-CP 96345	28 ± 2	0.6 ± 2.0	28 ± 32	−1 ± 2
(S-Dns)-HCy^{11}-SP	28 ± 2	0.7 ± 1.0	27 ± 2	25 ± 3
Nα-Flu-SP	n.d.[a]	n.d.[a]	21 ± 1	24 ± 3

(c)

Chemokine	Iodide		
	*K*sv (free)	*K*sv (bound) CCR$_1$	*K*sv (bound) CCR$_5$
Nα-Flu-MIP-1α	11.0 ± 0.9	11.2 ± 1.5	11.9 ± 1.9

[a] n.d. = not determined.

presence of a membrane environment (untransfected cells or transfected cells blocked with a non-fluorescent ligand; see *Protocol 5*). The quenching experiment is performed after binding of the ligand to the receptor and subsequent removal of excess ligand. It is important to run control experiments to confirm that the quenching curves obtained account for accessibility measurements of fluorescent ligand bound to the receptor rather than dissociated ligand during the time allocated to the measurements.

Single point measurements using a fixed concentration of quenchers, instead of successive additions to the cuvette, served as controls for the undesired quenching of labelled ligands dissociated from the receptor. As another control for ligand dissociation, samples were incubated during the times typically allocated to quenching experiments, centrifuged, and the measurement repeated. A lack of any changes in the fluorescence values is indicative of negligible dissociation of ligand (10, 13). Due to the low value of $K_D = k_{off} / k_{on}$ for the receptor–ligand studied, the dissociation was only significant in the presence of competitors in the time-scale discussed.

Protocol 5

Collision quenching experiments with the anionic quencher iodide:[a] NK$_2$ receptor probed with NBD antagonists (listed in *Figure 2*)

Equipment and reagents

- Spectrofluorometer equipped with a thermostatically controlled cuvette holder unit and a magnetic stirrer
- Quartz fluorescence cuvettes 0.5 × 1 cm (Hellma)
- NBD ligands as described in *Figure 2*
- Cells or membranes containing the NK$_2$ receptor prepared and incubated as described in *Protocol 4*
- Fresh solution of 0.15 M KI in water containing 1 mM Na$_2$S$_2$O$_3$ to prevent I$_3^-$ formation

Method

1 Incubate cells or cell membranes with fluorescent ligand as described in *Protocol 4*, steps 3–5 (total binding, non-specific binding, and control).

2 Start the quenching assay immediately after removing the excess ligand to limit dissociation. Resuspend the cells in 500 µl PBS at a density of 1–2 × 10^6 (about 1–2 nM [^{3}H]GR 94800 binding sites) in a fluorescence cuvette equipped with a small magnet (0.3 × 0.3 mm) and monitor the emission of the sample at 540 nm[b] by irradiating at 475 nm as a function of time at 20 °C.

3 When stable recording is obtained, add increasing amounts of KI solution to the initial volume of 500 µl and continue the recording. Stop the assay at a final concentration of iodide of 50–60 mM.

4 Repeat the experiment with samples incubated in the presence of an excess non-fluorescent competitor (N), control cells not incubated with fluorescent ligand (C), and ligand in aqueous solution (F).

5 Plot the fluorescence values as a function of the final quencher concentration in the cuvette for each time point. Correct the fluorescence values to take into account the dilution after each addition of quencher. For receptor containing samples subtract the values obtained for the control sample for each given concentration to account for possible cell autofluorescence quenching.

6 Calculate the slope of the fluorescence curve versus quencher concentration using the Stern–Volmer equation (28):

$$F_o / F = 1 + Ksv \text{ [Quencher]}$$

where F_o / F is the ratio of fluorescence intensities in the absence and presence of quencher (here, iodide). The Stern–Volmer quenching constant Ksv is determined from the slope of F_o / F as a function of the quencher concentration.

7 Average at least three Ksv values obtained from separate experiments.

Protocol 5 continued

8 In *Figure 10*, the quenching curves obtained for NBD antagonists bound to the NK_2 receptor are represented. *Table 3* reports quenching constants for various fluorescent ligands / receptor pairs using different quenchers.

[a] Quenching experiments with Co^{2+}, were performed as for iodide but 50 mM $CoCl_2$ was added to cells suspended in 20 mM Hepes/Na^+ pH 7.4, 130 mM NaCl. Fluorescence was corrected for the inner filter effect, as previously described (27). Results are reported in *Table 3a*.

[b] The fluorescence is recorded at the wavelength corresponding to the maximum of emission intensity for a given ligand. The environment of the NBD group changed as a function of the spacer arm being blue-shifted for longer distances. The value of 540 nm corresponds to the emission maximum for the NK_2 antagonist ligand 1 bearing the shortest linker (extended distance 4 Å, *Figure 2*).

A solvent influence has been observed for some quenchers and *Ksv* constants could differ. It has been shown that absolute values of Stern–Volmer constants for Co^{2+} were solvent dependent (29). Therefore, comparisons of quenching efficiencies have to be done using data obtained under the same experimental conditions.

5.4 Steady-state fluorescence anisotropy measurements

The fluorescence anisotropy of a chromophore, attached to macromolecules in general and to peptides in particular, reflects directly the molecular mobility of the fluorescent reporter group. In our present case, the mobility of the peptide ligands will, upon binding to membrane receptors, be considerably restricted compared with that of the free state in solution. In principle, the mobility of the ligand-attached fluorophore will reflect the mobility of the receptor or the ligand binding site of the receptor, the internal mobility of the ligand, and finally, the motional freedom of the chromophore relative to that of the ligand molecule. These different kinds of motions are expected to occur in different time regimes and to a certain extent might be distinguished by time-resolved fluorescence anisotropy measurements, depending on the fluorescence lifetime of the chromophore (22). It is the sum of all these effects which finally influences the steady-state fluorescence anisotropy. In particular, the relative mobility of the chromophore *versus* the peptide ligand molecule will determine the extent of the steady-state fluorescence anisotropy change upon receptor binding.

Protocol 6 describes a general method for measuring the anisotropy of fluorescent ligands in cells or cell membranes a containing the receptor and the construction of Perrin plots. Results obtained using this method are graphically represented in *Figure 11*. Fluorescence anisotropy for the ligand Nα-Flu-SP bound to NK_1 receptors and free in solution was measured in the 2–37 °C temperature range. Comparable plots were obtained at 20 °C by varying the viscosity using increasing amounts of glycerol (0–80%, w/v) at a constant concentration of ligand. Limiting anisotropy A_0, defined as the anisotropy in the absence of all rotational freedom, can be calculated from the intercepts of Perrin plots to the y-axis. The A_0 values obtained for the ligand free in solution or bound to the NK_1 receptor were similar.

Protocol 6

Steady-state anisotropy measurements: fluorescent ligands bound to NK_1 receptors

Equipment and reagents

- Spectrofluorometer (Jasco FP-777) equipped with a fluorescence polarization accessory (Jasco ADP-301)
- CHO cell membranes containing the NK_1 receptors prepared and incubated as described in the legend of *Figure 7*

- Fluorescent labelled NK_1 receptor ligands as reported in *Figure 3*: unlabelled SP as a non-fluorescent competitor NK_1 receptor ligand

A. Anisotropy measurements at 20 °C

1   Incubate CHO/NK_1 cell membranes as a suspension in PBS pH 7.2, containing 3 mM $MgCl_2$, 0.2 mg/ml BSA, with 50 nM fluorescent ligand in the presence (non-specific binding) or absence (total binding) of 50 μM non-fluorescent ligand at 20 °C for 1 h. For controls, also incubate membranes in the absence of fluorescent ligand. Wash membranes twice with ice-cold PBS and homogenize by sonication in an ice-cold bath for 3 min prior to fluorescence measurements.

2   Measure the emission intensity by setting the excitation-side polarizer in the vertical position (V) and the emission-side polarizer in either the horizontal (H) or vertical position.

3   Correct the emission intensities in respective V and H positions by subtracting the corresponding background signals from the cells or membrane suspensions.

4   Convert the data to anisotropy (A) with the following equation:

$$A = [(I_V/I_H)/G - 1] / [(I_V/I_H)/G + 2]$$

where I_V/I_H is the ratio of the vertical and the horizontal emission intensities when the excitation polarizer is in the vertical position, and G is the same ratio when the excitation light is horizontally polarized (G is an instrumental correction factor).

5   Anisotropy for a series of fluorescent ligands both free and bound to their receptors were measured at 20 °C and results reported in *Table 4*.

B. Construction of Perrin plots

1   For construction of Perrin plots, measure the anisotropy of free ligand; 0.1–10 μM in PBS solution (F), bound to the receptor in cell membranes at a concentration of 5 nM [^{3}H]SP binding sites (T). Also unbound ligand in the presence of membranes and 1000-fold excess non-fluorescent ligand (free ligand in membrane environment) as a function of temperature (over a range 2–37 °C) for 25–30 min.

Protocol 6 continued

2   Alternatively, vary the viscosity of the solution at constant temperature by addition of increasing amounts of glycerol (0–80%). Determine the viscosity at each temperature from standard tables (30).

3   Run control experiments where separate total bound samples are measured at a given temperature immediately after removal of excess ligand. Results that are similar to those of complete Perrin plots (at this temperature) indicate that the measured anisotropy accounts for receptor bound fluorescent ligand rather than dissociated ligand.

4   Plot 1/A as a function of $T/1000\eta$ and calculate A_0, the limiting anisotropy in the absence of rotation and the rotation correlation times (ϕ) using the Perrin equation for the case of a spherical molecule (28):

$$A^{-1} = A_0^{-1}\,(1 + \tau/\phi) = A_0^{-1}\,(1 + \tau kT/V_{h\eta})$$

τ is the fluorescence lifetime of the NBD chomophore attached to the peptide, k is the Boltzmann's constant, T is the temperature, V_h is the partial volume of a hydrated sphere, and η is the viscosity.

5   Representative Perrin plots for NK_1 receptor ligands are illustrated in *Figure 11*.

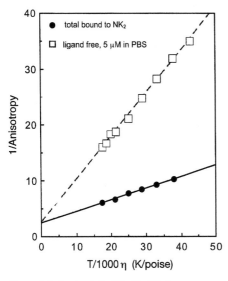

Figure 11 Perrin plot of Nα-Flu-SP. Open squares: Nα-Flu-SP, 5 μM in PBS solution (free). Filled circles: Nα-Flu-SP, total bound to NK_1/CHO cell suspensions. The non-specific fluorescence was too low for anisotropy measurements. Adapted from ref. 10.

Table 4 summarizes data obtained with different fluorescent groups for different ligand-GPCRs. In general, all these fluorescent groups have a reduction in mobility in the presence of cells or membranes (no receptor or blocked receptor). This increase in anisotropy is more important for the peptide NK_2

Table 4 Anisotropy at 20 °C for free and bound fluorescent ligands for different ligand–receptor systems

Ligand	Free in aqueous solution	In membrane environment (blocked receptor)	Specifically bound to the receptor	Limiting anisotropy A_0[a]
NK$_2$ receptor antagonists (from *Figure 2*)				
(γ-NBD)-Dab (1)	0.13 ± 0.05	0.23 ± 0.02	0.38 ± 0.02	0.36 ± 0.03
(ε-TMR)-Lys (4)	0.166 ± 0.005	0.228 ± 0.005	0.364 ± 0.005	0.38 ± 0.01
NK$_1$ receptor antagonist				
Dns-CP 96345	0.049 ± 0.001	0.15 ± 0.01	0.22 ± 0.02	0.32 ± 0.02
NK$_1$ receptor agonists				
(S-Dns)-HCy^{11}-SP	0.035 ± 0.005	0.090 ± 0.003	0.15 ± 0.01	0.24 ± 0.02
Nα-Flu-SP	0.039 ± 0.005	0.049 ± 0.005	0.12 ± 0.01	0.35 ± 0.05

[a] Calculated from Perrin plots (total bound).

receptor antagonists than the non-peptide NK$_1$ receptor antagonists, suggesting that these ligands partition non-specifically into the membrane due to their hydrophobic nature (10, 13). This reduction in mobility is not as strong as in the case of binding to the receptor, where the limiting anisotropy values were reached for some of them, demonstrating that ligand motions are restricted at the receptor binding site(s) in the nanosecond time range (*Table 4*).

6 Applications

6.1 Probing ligand–receptor interactions using fluorescent ligands

In this section, the application of fluorescent-based methodologies is described for three receptors using fluorescentyl labelled ligands already described above. These examples will illustrate the broad applicability of the methodology for investigating ligand–receptor recognition.

6.1.1 Peptide agonists and antagonists at NK$_2$ receptors

The investigation of changes in properties of environmentally-sensitive fluorescent groups on ligands, upon interaction with their receptor, is a valuable approach to characterization of binding sites. Spectrofluorometric methods have been used to study the interaction of fluorescent antagonist or agonist peptide ligands with the NK$_2$ receptor in transfected CHO cell lines. From analysis of the spectra of NBD derivatives bound to NK$_2$ receptors it was concluded that the environment around the amino terminal end of ligands bound to NK$_2$ receptors is more hydrophobic for antagonists than agonists.

Collision quenching experiments showed that labelled agonists were fully accessible to the soluble quenchers. From antagonist quenching curves in *Figure 10*, it can be concluded that the binding pocket for heptapeptide antagonists is buried in the protein or the membrane approximately 5–10 Å away from the membrane–water interface.

Using the methods described in Section 5, it was demonstrated that the ligand binding site on the NK_2 receptor for heptapeptide antagonists is in a hydrophobic pocket, shielded from the solvent, whereas the agonist binding site is accessible to the solvent (13). These results confirmed the idea that agonist-binding determinants are located in the extra-membranous regions of the receptor.

6.1.2 Agonist peptides and non-peptide antagonists at NK_1 receptors

Fluorescent peptide derivatives of SP and a dansylated non-peptidic antagonist (*Figure 3*) were used to investigate the polarity and solvent accessibility of NK_1 receptor binding pockets as well as to measure the motional freedom of these receptor-bound ligands (10). Moreover this report demonstrated the complementarity of site-directed mutagenesis and spectrofluorometric techniques in investigating ligand–receptor recognition in the absence of high resolution structural data.

The main differences between SP receptor agonist derivatives and the fluorescent antagonist CP 96345 using the fluorescence techniques described in Section 5 are summarized:

(a) Dns-CP 96345 occupies a hydrophobic pocket in the NK_1 receptor while, in sharp contrast, all labelled sites on SP were located in a more hydrophilic environment.

(b) The N-terminal moiety of SP (positions 1 and 3) was in a more hydrophilic environment than the C-terminal part.

(c) By fluorescence anisotropy measurements it was demonstrated that all the ligands have their mobility restricted at the receptor binding site. The immobilization of the dansylated antagonist was more important than for the SP analogues (*Table 4*).

(d) From collision quenching experiments it was concluded that the antagonist binds in a buried pocket that is shielded from the solvent and from the lipids in the bilayer surrounding the receptor (see *Table 3* for quenching constants with different quenchers). In contrast all the fluorescent groups in the SP derivatives were fully accessible to aqueous soluble quenchers. This is consistent with binding of substance P in the extracellular part of NK, confirmed also by photocrosslinking experiments where contact sites were identified.

6.1.3 Peptide agonists at CCR_1 and CCR_5 receptors

Fluorescence techniques have been used to study recognition modes of the chemokine MIP-1α by CCR_1 and CCR_5 in transfected CHO cells. In a recent report, fluorescence spectroscopic methods were combined with photoaffinity labelling to provide valuable insights into ligand–receptor recognition patterns (31).

The analysis of the binding of a fluorescein derivative of the agonist ligand

MIP-1α to CCR_1 and CCR_5 receptors was performed as described in the standardized method described in *Protocol 4*. The emission spectra obtained in *Figure 8C* and *8D* were obtained with membrane suspensions from CHO cells transfected with CCR_1 and CCR_5 receptors, respectively. The *Ksv* constants obtained from collision quenching experiments (as described in *Protocol 6*) with iodide are about 11 M^{-1} (see *Table 3c*) for the fluorescent ligand free in solution or bound to both receptors. Therefore, the N-terminal part of the labelled ligand was fully accessible to the soluble quencher iodide, suggesting that the N-terminus of bound MIP-1α interacts with the extracellular domain and loops or water–membrane interface parts of the transmembrane helices. From these results, it is unlikely that the N-terminal of MIP-1α penetrates the hydrophobic pocket formed by the seven transmembrane helices of the receptor.

7 Mapping receptor architecture using fluorescent receptors and fluorescent ligands

The utilization of fluorescently labelled ligands (as shown in previous sections) can provide valuable information about ligand–receptor interactions. The range of applications for structural and functional studies is increased by the introduction of a second label in the receptor molecule. The absence of a reliable and general method for the introduction of a unique fluorescent reporter into the polypeptide chain of GPCRs (or intrinsic membrane receptors in general) has limited the scope of structural studies of these proteins using spectrofluorimetry techniques.

Conformational changes of the receptor could be analysed by the technique of FRET (32, 33), that requires the presence of two fluorophores acting as a couple donor–acceptor in the same molecule or in two different molecules of the ternary ligand–receptor–G protein complex.

The fluorometric characterization of the NBD-labelled NK_2 receptor mutant in oocytes, the binding of tetramethylrhodamine labelled peptides to this fluorescent receptor, and the measurement of the fluorescence energy transfer between these two fluorophores has been published (34–36). Practical aspects of the methodology concerning fluorescence measurements are reported in the following sections.

7.1 Labelling receptors

Three main methods have been applied for obtaining a fluorescent receptor:

(a) Chemical modification of Cys of mutated receptors having a limited number of Cys with a thiol reactive fluorophore (IANBD). β_2-Adrenoreceptors have been labelled using this strategy and conformational changes have been measured upon receptor activation (37). This method needs previous knowledge of optimal conditions for the difficult task of membrane receptor purification and reconstitution, and so the approach suffers from limited applicability.

(b) Fusion constructs of receptor and fluorescent protein (i.e. green fluorescent protein) represents a molecular biology-based alternative. Bodipy-TMR labelled NKA ligand and NK_1-GFP expressed in CHO cells has been used to monitor FRET changes between the donor moiety attached to the receptor and the acceptor linked to the N-terminal part of the agonist ligand (7).

(c) Site-specific stop codon suppression of mutant receptors in *Xenopus* oocytes using suppressor tRNA misacylated with a fluorescent amino acid. The overall strategy is described in *Figure 12* (34, 35).

The methodology for the site-specific incorporation of fluorescent amino acids into the NK_2 receptor has been described in detail (38). For functionally active, fluorescently labelled NK_2 receptor mutants expressed in *Xenopus* oocytes, distance measurements between labelled receptors and a labelled ligand partner can thus be performed by FRET (Section 7.4).

The generation of misacylated tRNA (*Figure 12*) is the most critical part of the methodology of stop codon suppression mutagenesis and this has limited its use.

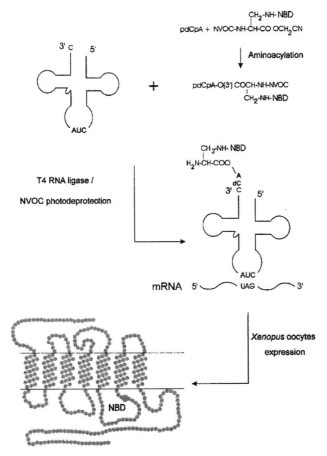

Figure 12 Outline of the approach for biosynthetic incorporation of fluorescent amino acids by stop codon suppression mutagenesis.

In our laboratory we optimized and adapted some of the critical steps with the objective of introducing fluorescent amino acids into site-selected positions of the GPCR NK_2 receptor (38).

7.2 Characterization of fluorescent receptors

Fluorescent NK_2 receptors were produced by site-specific suppression mutagenesis in *Xenopus laevis* oocytes. Oocytes were microinjected with RNA and misacylated suppressor tRNA as previously described (39). Typically, functional assays were performed after 24 h and biological activity was recorded electrophysiologically for the large number of GPCRs that are capable of mobilizing Ca^{2+} from intracellular stores after agonist stimulation. Elevation of the intracellular Ca^{2+} concentration activates Ca^{2+}-dependent Cl^- channels. The inward Cl^- currents are detectable by standard voltage clamp techniques (34). For saturation binding assays and epifluorescence microspectrofluorometry, membranes from transfected fluorescent NK_2 receptor mutants were prepared from oocytes where functional activity was detected (see *Protocol 7*).

Protocol 7

Preparation of oocyte membranes enriched with NK_2 receptors for fluorescence microscopy assays (performed using a method reported in ref. 39)

Equipment and reagents

- Beckman TL-100 ultracentrifuge
- *Xenopus* oocytes prepared as described in ref. 39 injected with 25 ng of mRNA and 100 ng of suppressor tRNA were incubated for 24 h at 18°C (34)
- Buffer A: 75 mM Tris–HCl pH 7.4, 12.5 mM $MgCl_2$, 1 mM EDTA, 30% sucrose, 0.15 mg/ml benzamidin, 0.10 mg/ml bacitracin

Method

1 Homogenize about 100 oocytes 24 h after mRNA + tRNA microinjection in 800 μl ice-cold buffer A by passing the mixture ten times through a Pipetteman equipped with a 200 μl tip.

2 Centrifuge the homogenate at 3000 g for 10 min to remove pigment granules. Take the supernatant fraction and centrifuge it at 400000 g for 30 min at 4°C in an ultracentrifuge.

3 Resuspend the resulting pellet (referred as the membrane fraction) in ice-cold buffer A and sonicate it in an ice-cold water bath sonicator for 5 min.

4 Take an aliquot of the membrane suspension for total protein determination using the method of Lowry with BSA as a standard.[a]

Protocol 7 continued

5 Prepare aliquots of the membrane fraction of about 50 μg total protein content (yielded from five oocytes) and store them at −80°C for radioligand and fluorescence binding assays.[b]

[a] Typically, the yield was about 10 μg of total protein per oocyte in the membrane enriched fraction.

[b] Radioligand binding assays as described in ref. 34.

From saturation binding assays of different NBD-Dap mutants, the K_D values for binding of a radiolabelled antagonist to membrane fractions from oocytes were similar for wild-type NK_2 receptors and fluorescent mutants. In addition these values were comparable to the K_D for binding to NK_2 receptors expressed in CHO cells (1.32 nM). The efficiency of stop codon suppression was calculated from B_{max} values obtained for suppression mutants compared to wild-type NK_2 receptors. The expression of the fluorescent mutants relative to wild-type NK_2 receptors ranged from 13–30% (35). Oocyte batch-to-batch variability was important due to different expression/suppression levels of the fluorescent NK_2 receptor mutants; this levelled off in binding assays that used membrane fractions averaged from 100–200 oocytes.

The NBD fluorescence incorporated into the NK_2 receptor was first measured in a standard spectrofluorometer using aqueous suspensions of oocyte membranes in microcuvettes of 20 μl. A low specific signal with the spectral characteristics of the NBD group was detected. Unfortunately, such a small volume cannot be stirred, but this proved to be necessary to avoid fluorescence signal fluctuations due to inhomogeneities in the membrane suspension. To overcome this problem, fluorescence microscopy was used to detect and characterize the fluorescence incorporated into the NK_2 receptor. The fluorescent NK_2 receptor mutants expressed in oocytes were measured using oocyte membrane fractions immobilized on silanized quartz slides.

For the detection of the NBD fluorescence of the specifically labelled receptor the following procedure was applied:

(a) The membranes were measured and the background fluorescence of oocyte membranes which did not contain expressed receptors, subtracted.

(b) Quartz slides were preferred over ordinary glass slides, commonly used in fluorescence microscopy, to avoid interference in the red region of the spectrum.

(c) Membrane immobilization was performed by hydrophobic interaction of the sample to a $SiCl_2(CH_3)_2$ derivatized quartz slide surface. This procedure is described in *Protocol 8A*.

Oocyte membrane samples showed a high fluorescence signal, characteristic of NBD emission, confirming the presence of this fluorophore in the polypeptide chain of the receptor. This is illustrated in the traces (D: donor alone) reported in *Figure 15* (see *Protocol 8B*).

Protocol 8

FRET measurements by fluorescence microscopy using NK$_2$ receptor NBD fluorescent mutants expressed in oocytes and fluorescent ligand TMR antagonist

Equipment and reagents

- Inverted microscope Axiovert 100 TV (Carl Zeiss S.A.)
- Filter sets: No. 1, BP450–490:FT510:LP520; No. 2, BP450–490:FT510:LP515–565; No. 3, BP546/12:FT580:LP590 (Omega)
- Short arc mercury lamp HBO 100 W/AC (Carl Zeiss S.A.)
- CCD spectrometric detector TE/CCD-576EUV (Princeton Instruments, Inc., USA)
- Tunable argon ion laser (Omnichrome series 532-AP, MAP) (Newport Instruments AG-Sclieren, Switzerland)
- Quartz slides for microscopy 76 × 25 × 1 mm (VQT, Neuchatel-Switzerland)
- Silanization solution: 20% (v/v) dimethyldichlorosilane (Merck) in dry dichloromethane (Aldrich)

- Desiccator and vacuum pump
- 50% HNO$_3$ in deionized water (Fluka)
- NK$_2$ and NK$_2$ fluorescent receptor mutants in oocyte membranes: aliquots, prepared as described in *Protocol 7* (each aliquot is the yield from five oocytes; 50 μg total protein content)
- Fluorescent NK$_2$ receptor ligand (TMR antagonist 4 in *Figure 2*): 2 mM stock solution in DMSO
- Non-fluorescent NK$_2$ receptor antagonist GR 94800: 2 mM stock solution in DMSO
- Ice-cold 50 mM Tris–HCl pH 7.4 solution
- Binding buffer: PBS pH 7.2 containing 3 mM MgCl$_2$, 0.2 mg/ml BSA, and 0.5 mM sodium azide

A. Silanization

1 Prior to the silanization treatment, to render the slide surface hydrophobic, the slides are subjected to the following cleaning procedure: place ten quartz slides in a flask containing 100 ml of 50% HNO$_3$ and incubate them at room temperature for 6 h. Wash three times with deionized water and three times with acetone.

2 Dry the slides in air, first at room temperature to remove most of the acetone, then at 130°C in an oven for 6 h. Store the slides in an evacuated desiccator before use.

3 Prepare the silanization solution immediately before its use by mixing 20% (v/v) dimethyldichlorosilane in dry dichloromethane and incubate the clean slides in this solution for 3 h.

4 Wash the slides three times with dichloromethane, twice with acetone, and store them in a desiccator.

B. Fluorescence detection of NK$_2$(NBD)/oocyte membrane immobilization on silanized slides

1 Apply 5 μl of the oocyte membrane suspension at 0.5–2 oocyte/μl on the hydrophobic quartz surface and dry the slide in a desiccator using a water pump for 5 min. Rehydrate the membranes with 5 μl of ice-cold 50 mM Tris–HCl pH 7.

Protocol 8 continued

2 Irradiate the membranes corresponding to oocytes injected with stop mutant-NK_2 mRNA fluorescent suppressor tRNA for 10×1 sec using either the mercury lamp or the laser set at 476 nm and the filter set No. 1.

3 Run control samples corresponding to membranes from the same number of oocytes (from the same batch) injected with wt-NK_2 receptor mRNA and fluorescent suppressor tRNA to account for intrinsic autofluorescence and contribution of non-incorporated NBD. Include also membranes obtained from non-injected oocytes.[a]

C. Tetramethylrhodamine-antagonist (TMR-ANT) binding to NBD-NK₂ receptor mutants in oocyte membranes:[a] monitoring TMR-ligand bound emission

1 Aliquot 10–20 µl of oocyte membranes suspension at 0.5 oocytes/µl for oocytes transfected with wt-NK_2 receptor, NBD-labelled NK_2 receptors, and untransfected oocytes.

2 For each receptor, incubate the membranes (triplicates) for 1 h at 25°C in incubation buffer containing:

   (a) 10 nM TMR-labelled NK_2 compound 4 in *Figure 2* (total binding).

   (b) 10 nM fluorescent ligand and 1 µM of non-fluorescent antagonist GR 100679 (non-specific binding).

   (c) Without any addition (control for autofluorescence of oocyte membranes).

3 Remove the excess ligand by centrifugation for 15 min at 4°C and washing the membrane pellet twice with ice-cold 50 mM Tris–HCl pH 7.4 and repeating the 15 min centrifugation.

4 Immediately before the fluorescence microscopy experiments resuspend the membrane pellet in 5–10 µl of the same buffer and apply 5 µl of the homogeneous suspension to the surface of a silanized quartz slide. Allow it to dry and rehydrate with 5 µl water as described in *Protocol 8B*, step 1.

5 Irradiate the samples 0.5–1 sec using the arc mercury lamp and the filter set No. 3.

6 Record emission spectra for each sample using the CCD detection system (see *Figure 13*).

D. Measurement of fluorescence energy transfer efficiency

1 Use *Protocol 8C*, steps 1–4, described for measuring the TMR antagonist bound to the receptor, and the efficiency of donor quenching due to non-radiative energy transfer between the NBD (donor) and the TMR group (acceptor).

2 Irradiate the sample with the 476 nm line[b] of the argon laser (laser output power = 10 mWatts) combined with the filter set No. 1, for 10 sec. Correct for CCD background during the same irradiation time.

3 Record the emission spectrum for each sample, correct the spectra for membrane autofluorescence, and compare traces of NBD-receptor (donor alone), NBD-receptor

Protocol 8 continued

with bound TMR-peptide (donor in the presence of acceptor), and free TMR ligand in the presence of untransfected membranes (acceptor alone). See *Figure 14*.

4 Calculate the extent of donor quenching by fitting the donor–acceptor spectrum with fractional contributions of donor alone and acceptor alone. Values of NBD quenching efficiency and the intermolecular distances calculated for different NBD-labelled NK_2 receptor mutants are reported in *refs. 34, 35, 36*.

[a] The binding analysis conditions are similar to those reported for CHO/NK_2 cells in *Protocol 4*.

[b] The 476 nm line of the argon laser has been chosen because the direct excitation of the TMR acceptor is less than that of the more widely used 488 nm line.

7.2.1 Binding of TMR antagonist to NBD-labelled NK_2 receptor mutants

The method used for the detection of the fluorescence associated with the binding of a TMR-labelled antagonist 4 (*Figure 2*) to NK_2 receptors expressed by the suppression mutagenesis technique in oocytes is reported in *Protocol 8C*.

Figure 13 illustrates a typical spectrofluorometric binding analysis of TMR-ligand to the NK_2 receptor. By comparing the specific receptor-bound fluorescence for the different fluorescent mutants to the wild-type NK_2 receptor (from the same oocyte batch preparation) the relative expression levels can be estimated. These values were in agreement with those obtained by radioactive saturation binding traces for binding of TMR-ligand to NK_2 receptors and validate the use of fluorescence microscopy to analyse the ligand binding of fluorescent labelled receptors in oocyte membranes (35).

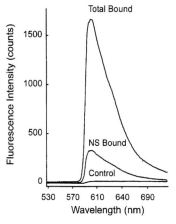

Figure 13 Spectrofluorometric analysis of binding of TMR-labelled antagonist of NK_2 receptor in wild-type NK_2 oocyte membranes by fluorescence microscopy. NK_2 receptors in *Xenopus* oocyte membranes were incubated with TMR antagonist 4 (*Figure 2*) as described in *Protocol 8C*. Fluorescence emission spectra were collected from samples immobilized on silanized quartz slides irradiated 0.5 sec with a mercury arc lamp and the filter set BP546/12:FT580:LP590 (Omega).

7.3 Fluorescence resonance energy transfer (FRET) between a specific bound fluorescent ligand and a fluorescently labelled receptor

FRET can reveal absolute intermolecular distances between two fluorophores and conformational changes in the receptor that occur after interaction with a

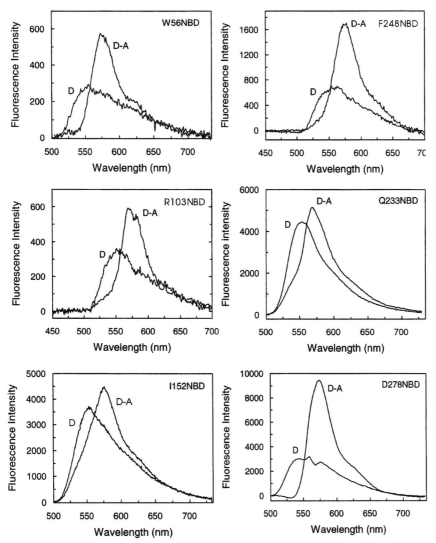

Figure 14 FRET between NBD-labelled NK_2 receptor mutants and a TMR-labelled antagonist in oocyte membranes by fluorescence microscopy. Representative FRET between receptor-bound TMR-labelled antagonist (compound 4 in *Figure 2*) and mutants as indicated in each panel. The 476 nm argon laser line was used for NBD excitation in order to reduce the direct excitation of the TMR fluorophore. In each panel, traces **D** (donor) represent specific fluorescence emission spectra of NBD-labelled NK_2 receptor mutants. Traces **DA** (donor in the presence of acceptor) represent fluorescence emission spectra in the presence of receptor-bound TMR antagonist peptide. The distance R between the NBD and TMR groups was calculated as described (34–36).

specific ligand. This constitutes a valuable method for obtaining structural and dynamic information on GPCRs. FRET has been applied to the determination of a structural model for interaction of TMR-labelled heptapeptide ligand with the NK$_2$ receptor in the native oocyte membrane environment. By placing the NBD fluorescent probe at different sites in the receptor and measuring inter-molecular distances between these sites and the TMR group on bound ligands, a structural model of the receptor–ligand complex was built (34–36) (*Figure 15*). Full details of the theory and practice of the technique of FRET are described in Chapter 5.

The efficiency of energy transfer was quantified as the fractional decrease of the NBD donor fluorescence due to the binding of TMR-labelled acceptor and was expressed by $E = 1 - F_{DA}/F_D$, where F_{DA} and F_D are the relative yield of fluorescence of the donor in the presence and absence of the acceptor, respectively. An increase in the emission spectrum at the λ_{max} of emission of the TMR antagonist (575 nm) is expected as a result of energy transfer. In addition, the emission of the donor (NBD) will be quenched in the presence of the acceptor (TMR), and the measured intensity of the donor emission will decrease (540 nm). Both these effects are seen in *Figure 14*. The procedure to measure FRET efficiencies is described in *Protocol 8D*.

These distance measurements confirmed the hepta-helical structure of the GPCR NK$_2$ receptor and suggested the structural model for NK$_2$ ligand–receptor interactions (36). There is only one possible orientation for the peptide ligand that is inserted between the fifth and sixth transmembrane domain, suggesting that antagonist binding may prevent proper helix packing required for receptor function (*Figure 15*).

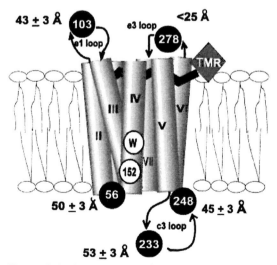

Figure 15 A schematic model of NK$_2$ receptor–ligand interactions. FRET experimental distances between sites on NK$_2$ receptors labelled with NBD (circles with residue number) and bound heptapeptide antagonist labelled with TMR. The receptor is viewed from the side; loops are symbolized by arrows. Reprinted from ref. 35 with permission from Gordon and Breach Publishers. Copyright 1997, Overseas Publishers Association.

Future improvements of unnatural suppression mutagenesis of GPCRs in particular, and proteins in general, are to extend its applicability to mammalian cell lines. For this purpose, the main challenge is to develop an efficient system for the introduction of suppressor tRNA and cRNA into these cells.

Acknowledgements

I thank Dr Alexander Scheer for critical reading of the manuscript and the co-authors of our published articles cited here.

References

1. Grisshammer, R. and Tate, C. G. (1995). *Q. Rev. Biophys.*, **28**, 315.
2. Nie, S. and Zare, R. N. (1997). *Annu. Rev. Biophys. Biomol. Struct.*, **26**, 567.
3. Mujumdar, R. B., Ernst, L. A., Mujumdar, S. R., Lewis, C. J., and Waggoner, A. S. (1993). *Bioconjug. Chem.*, **4**, 105.
4. Panchuk-Voloshina, N., Haugland, R. P., Bishop-Stewart, J., Bhalgat, M. K., Millard, P. J., Mao, F., *et al.* (1999). *J. Histochem. Cytochem.*, **47**, 1179.
5. Tsien, R. Y. (1998). *Annu. Rev. Biochem.*, **67**, 509.
6. Pollok, B. A. and Heim, R. (1999). *Trends Cell Biol.*, **9**, 57.
7. Vollmer, J. Y., Alix, P., Chollet, A., Takeda, K., and Galzi, J. L. (1999). *J. Biol. Chem.*, **274**, 37915.
8. Durroux, T., Peter, M., Turcatti, G., Chollet, A., Balestre, M. N., Barberis, C., *et al.* (1999). *J. Med. Chem.*, **42**, 1312.
9. Mellentin-Michelotti, J., Evangelista, L. T., Swartzman, E. E., Miraglia, S. J., Werner, W. E., and Yuan, P. M. (1999). *Anal. Biochem.*, **272**, 182.
10. Turcatti, G., Zoffmann, S., Lowe, J. A. III, Drozda, S. E., Chassaing, G., Schwartz, T. W., *et al.* (1997). *J. Biol. Chem.*, **272**, 21167.
11. Offord, R. E., Gaertner, H. F., Wells, T. N., and Proudfoot, A. E. (1997). In *Methods in enzymology, Chemokines*. Vol. 287, (Ed. R. Horuk), p. 348. Academic Press, London, UK.
12. Bradshaw, C. G., Ceszkowski, K., Turcatti, G., Beresford, I. J., and Chollet, A. (1994). *J. Med. Chem.*, **37**, 1991.
13. Turcatti, G., Vogel, H., and Chollet, A. (1995). *Biochemistry*, **34**, 3972.
14. Rovero, P., Pestellini, V., Rhaleb, N. E., Dion, S., Rouissi, N., Tousignant, C., *et al.* (1989). *Neuropeptides*, **13**, 263.
15. Tota, M. R., Daniel, S., Sirotina, A., Mazina, K. E., Fong, T. M., Longmore, J., *et al.* (1994). *Biochemistry*, **33**, 13079.
16. Snider, R. M., Constantine, J. W., Lowe, J. A. III, Longo, K. P., Lebel, W. S., Woody, H. A., *et al.* (1991). *Science*, **251**, 435.
17. Gurney, A. M. (1990). In *Receptor effector coupling: a practical approach* (ed. E. C. Hulme), p. 117. IRL Press, Oxford.
18. Crowley, K. S., Reinhart, G. D., and Johnson, A. E. (1993). *Cell*, **73**, 1101.
19. Capponi, A. M., Lew, P. D., Jornot, L., and Vallotton, M. B. (1984). *J. Biol. Chem.*, **259**, 8863.
20. Turcatti, G., Ceszkowski, K., and Chollet, A. (1993). *J. Recept. Res.*, **13**, 639.
21. Carraway, K. L. III and Cerione, R. A. (1993). *Biochemistry*, **32**, 12039.
22. Tairi, A. P., Hovius, R., Pick, H., Blasey, H., Bernard, A., Surprenant, A., *et al.* (1998). *Biochemistry*, **37**, 15850.
23. Tota, M. R. and Strader, C. D. (1990). *J. Biol. Chem.*, **265**, 16891.

24. Eftink, M. R. (1991). In *Biophysical and biochemical aspects of fluorescence spectroscopy* (ed. E. T. G. Dewey), p. 1. Plenum, New York.

25. Allegrini, P. R., Sigrist, H., Schaller, J., and Zahler, P. (1983). *Eur. J. Biochem.*, **132**, 603.

26. Homan, R. and Eisenberg, M. (1985). *Biochim. Biophys. Acta*, **812**, 485.

27. London, E. (1986). *Anal. Biochem.*, **154**, 57.

28. Lakowicz, J. R. (1983). In *Principles of fluorescence spectroscopy*, p. 131. (ed. J. R. Lakowicz). Plenum Press, New York.

29. Morris, S. J., Bradley, D., and Blumenthal, R. (1985). *Biochim. Biophys. Acta*, **818**, 365.

30. Weast, R. C. (ed.) (1980). *Handbook of chemistry and physics*, 61st edn, p. F-51. CRC, Boca Raton, FL.

31. Zoffmann, S., Turcatti, G., Dahl, M., Galzi, J. L., and Chollet, A. (2000). *J. Med. Chem*, **44**, 215.

32. Stryer, L. (1978). *Annu. Rev. Biochem.*, **47**, 819.

33. Steer, B. A. and Merrill, A. R. (1994). *Biochemistry*, **33**, 1108.

34. Turcatti, G., Nemeth, K., Edgerton, M. D., Meseth, U., Talabot, F., Peitsch, M., *et al.* (1996). *J. Biol. Chem.*, **271**, 19991.

35. Turcatti, G., Nemeth, K., Edgerton, M. D., Knowles, J., Vogel, H., and Chollet, A. (1997). *Recept. Channels*, **5**, 201.

36. Chollet, A. and Turcatti, G. (1999). *J. Comput. Aided Mol. Des*, **13**, 209.

37. Kobilka, B., Gether, U., Seifert, R., Lin, S., and Ghanouni, P. (1998). *Life Sci.*, **62**, 1509.

38. Chollet, A. and Turcatti, G. (1999). In *Structure-function analysis of G protein-coupled receptors* (ed. J. Weiss), p. 335. Wiley-Liss, Inc.

39. Nemeth, K. and Chollet, A. (1995). *J. Biol. Chem.*, **270**, 27601.

Chapter 2

From receptor to endogenous ligand

Hans-Peter Nothacker, Rainer K. Reinscheid, and Olivier Civelli

University of California, Irvine, Department of Pharmacology, Irvine, CA 92697-4625, USA.

1 Introduction

Homology screening and advances in the human genome project have spawned an explosion in the field of G protein-coupled receptors (GPCRs) resulting in the discovery of hundreds of new ones. A large number of these bind unknown natural ligands and are called 'orphan' GPCRs. The major challenge lies in the discovery of their functions, defining their physiological roles, and possibly their therapeutic potentials. The key to this daunting task lies in the identification of the natural endogenous ligand(s) that bind to and activate the orphan GPCRs. These ligands can then be used in behavioural assays to attempt to define the functional role of the orphan receptor systems. They can also be used to develop specific agonists and antagonists ultimately to provide therapeutic drugs.

This chapter outlines the multistep process that we have developed to identify the natural ligands of orphan GPCRs, from the discovery and cloning of the GPCRs to the isolation of their endogenous ligands from tissue extracts. We illustrate the entire strategy by choosing GPCRs belonging to the subfamily of peptide receptors, while recognizing that it can, in principle, be applied to any type of GPCR.

The following sections are intended as a guide to the multiple step procedure. Special attention is given to functional assays that have proved to be crucial to the identification of novel ligands.

2 The orphan receptor strategy

2.1 What is the orphan receptor strategy?

In contrast to other receptor classes, most GPCRs are functional and able to bind to their activating endogenous ligands only when embedded in the appropriate membrane environment. Therefore, the use of affinity probes made of particular

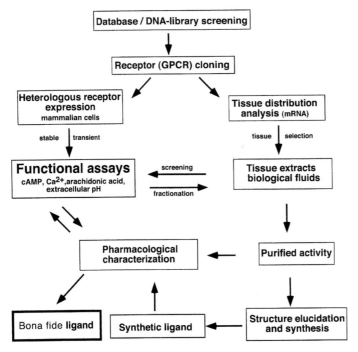

Figure 1 Outline of the orphan receptor strategy. Arrows indicate the direction of flow of information.

domains, instead of intact receptors, has thus far failed to lead to the identification of the natural ligand of an orphan GPCR. The only approach that has been successful in this respect is the 'orphan receptor strategy' (1, 2).

The orphan receptor strategy relies on the expression of an orphan GPCR in a cell in culture that provides the necessary environment (i.e. heterotrimeric GTP-binding proteins and effector molecules) and thus enables receptor signalling upon binding of the cognate ligand. The cells are subsequently exposed to tissue extracts expected to contain the orphan receptor's natural ligand. The presence of the natural ligand activates the receptor resulting in changes in second messenger levels. By monitoring these changes (e.g. cAMP, Ca^{2+}) the receptor activation can be detected and used as a bioassay to purify the natural ligand (3). The entire strategy is outlined in *Figure 1*.

2.2 The dilemma of the orphan receptor strategy

The orphan receptor strategy faces two major obstacles. First, the GPCR multi-gene family comprises a heterogeneous mixture of receptors that can bind to a vast array of diverse molecules ranging from large protein hormones to small molecule ligands. Secondly, receptor stimulation activates a number of different effector molecules. So far, no structural analysis method is able to predict which second messenger will be modulated by a given orphan receptor. The investigator faces a dilemma of having to identify simultaneously both the effector and the ligand of the orphan GPCR. Neither the signalling activity of the receptor nor the concentration of the ligand in tissue extracts can be assessed. Yet, the orphan receptor strategy has led to the discoveries of novel ligands with an increasing success rate (for review see ref. 4). This success can probably be traced to three reasons:

(a) First, the strategy was shown to be viable through its first success.

(b) Comparative structural analyses of orphan GPCR sequences permit rational guesses about the chemical nature of an orphan GPCR ligand.

(c) Technologies have been developed that make the monitoring of several signalling pathways an easy task and thus ensure that the one induced by an orphan GPCR will be detected.

Moreover, a useful alternative to solving the last problem is to force an orphan receptor to induce a given second messenger system via specifically engineered chimeric G protein (see Section 4.4).

3 Identification and characterization of orphan GPCRs

The modular nature of protein structures created by evolution over millions of years has enabled us to identify orphan GPCRs on the basis of their structural similarities with known members of the gene superfamily. Molecular biological techniques based on low stringency hybridization were the first techniques used in discovering orphan GPCRs. Presently, in view of the explosive number of sequences identified by the sequencing of the human genome, computer-based analyses have become the method of choice in the discovery of orphan GPCRs. The following sections provide a short overview of common techniques applied to the discovery of novel GPCRs. It is beyond the scope of this chapter to provide detailed cloning protocols, instead relevant cross-references are provided.

3.1 Cloning of novel orphan GPCRs based on homology

This technique relies on the hybridization of a probe to a related nucleotide sequence. It is used to identify GPCRs from cDNA or genomic libraries. The selection of the probe is crucial for the output of the screen. Usually, probes are selected primarily from conserved transmembrane (TM) regions where the highest sequence conservation between GPCRs exists. This method was applied successfully to many GPCRs after the cloning of the β_2-adrenoceptor in 1986. It became clear that this receptor shared significant sequence homologies with the rhodopsin receptor, another G protein-coupled receptor, suggesting that homology screening could be exploited to clone new members of the GPCR family. Two approaches were devised; one based on low stringency hybridization, the other was based on PCR when it became evident that many members of the GPCR gene family share a particularly high sequence homology in their putative

TM domains. Until recently, these methods led to the discovery of most new GPCRs, all orphans at the time of their discovery. The low stringency approach is now used only to find GPCRs that are part of a subfamily and that share an important degree of sequence homology or to find ortholog GPCRs. The PCR-based approach is still in use to identify novel orphan GPCRs. A comprehensive overview of these approaches is described in Marchese *et al.* (5). However, the availability of the entire human genome sequence makes computer-based homology screening a much easier and faster method to discover novel orphan GPCRs than the traditional ones and has now supplanted all the other approaches.

3.2 Computer-based orphan GPCR discovery

The joint effort to sequence the entire human genome and the availability of the entire genome of several invertebrates (6, 7) led to an enormous amount of sequence information becoming publicly available through the NCBI server (GenBank database, www.ncbi.nlm.nih.gov), the EMBL server in Europe, and the DDBJ in Japan (www.ddbj.nig.ac.jp). These data can be searched easily by sending a query to the server. The query can be a string as 'GPR', a protein, or a DNA sequence. For similarity searches, protein sequences are more effective than nucleotides. The GenBank database is divided into several subdivisions that are either collections of protein (swissprot) or nucleotide (nr, dbest, htgs, and all other available subdivisions) sequences. All subdivisions can be queried with both DNA and protein sequences using the BLAST algorithm. Five different options for BLAST searches are available. BLASTP is used when a protein database is queried with an amino acid sequence, BLASTN for querying DNA subdivisions. BLASTN is usually used for finding identical sequences, but lacks the sensitivity to find similar sequences. The most important program for similarity searches is TBLASTN which compares amino acid sequences against a nucleotide database. The database is thereby translated 'on the fly' into all six reading frames and compared to the query protein sequence. BLASTX is used to perform the opposite task: comparing all reading frames of a DNA sequence to a protein database. It is noteworthy that the only protein database available at NCBI, swissprot, is less frequently updated than the non-redundant nucleotide database (nr). TBLASTN can be used to query all other databases including the monthly update (month) which includes the most recent sequence submissions. Finally, TBLASTX compares all reading frames of a nucleotide sequence to all reading frames of a DNA database and cannot be recommended as regards the extended computing times necessary for this kind of task. A highly recommended introduction to the GPCR informatics has been written by Lynch (8). For more sophisticated similarity searches the gene panning strategy described by Retief *et al.* (9) will be the method of choice for database mining with the goal to identify novel members of gene families.

3.3 Classification of the novel orphan GPCRs

Once the sequence of a novel, putative GPCR has been discovered, classification of this new receptor with regard to its potential ligand is the first step in identifying its functional role. GPCRs are known to bind a vast variety of ligands ranging from photons, volatile odorants, purines, peptides, biogenic amines, lipids, and proteins. Based on their primary structures, mammalian GPCRs fall into three main gene families (for a complete overview visit the G protein-coupled receptor database at http://www.gpcr.org/7tm/). The class A or rhodopsin-like receptor family comprises the majority of GPCRs and represents its most heterogeneous group with regard to the physicochemical properties of their ligands. The structurally distinct class B or secretin-like family includes mostly receptors for structurally related peptides, 30–40 amino acids in length. Recently a subfamily of class B receptor has emerged containing large amino terminal sequences with a modular structure reminiscent of cell adhesion molecules. The class C receptor family consists of the metabotropic glutamate and GABA$_B$ receptors, the Ca^{2+}-sensing receptors and a group of putative pheromone receptors.

A useful strategy in classifying an orphan receptor is to compare the amino acid sequence to a database as described in Section 3.2. This generates a list of most similar GPCR sequences, which can be retrieved and subjected to an in-depth analysis by a multiple sequence alignment program. These programs can be found either as part of sequence analysis software packages (e.g. LaserGene, DNASTAR, Inc.) or on the World Wide Web as site-specific computing services (e.g. www.ddbj.nig.ac.jp).

A convenient way to visualize the relationship between groups of different GPCRs is the generation of phylogenetic trees as shown in *Figure 2*. The former orphan receptors GPR10, GPR14, GPR24 showed a clear homology with particular subfamilies of peptide receptors all belonging to the class A family. GPR10, later identified as prolactin-releasing peptide receptor (10) branched the subfamily of receptors for the neuropeptide NPY. GPR14 and GPR24, both distant members of the somatostatin/opioid receptor family were subsequently found to be the receptors for urotensin II (11–13) and melanin-concentrating hormone (MCH) (14–18), respectively. Interestingly, the similarities among the receptors are also reflected at the level of the structures of their ligands. From these parallels derives the nearly invariant rule among GPCRs that similar receptors have similar ligands (9). This principle is so far the most solid criterion in the identification of the natural ligands for orphan GPCRs.

4 Expression of orphan GPCRs

4.1 Vectors for use of expression in mammalian cells

Plasmid expression vectors harbouring viral promoters to ensure high levels of mRNA production are the most commonly used vectors to express heterologous GPCRs in eukaryotic cells. Typical viral promoters are derived from cytomegalovirus (CMV) or Rous sarcoma virus (RSV). The plasmid vectors also contain artificial signals for polyadenylation of the mRNA and antibiotic resistance genes, which allow for selection of positively transfected clones and are usually driven by an independent promoter. Recently, mammalian bicistronic expression vectors

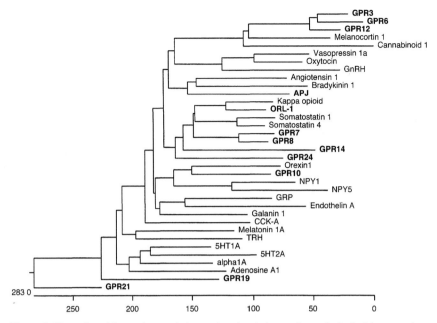

Figure 2 Clustering of G protein-coupled receptors by phylogenetic analysis. In this example, amino acid sequences of representative members of the GPCR class A family were selected and aligned using the CLUSTAL method (LaserGene, DNASTAR, Inc.). The *x*-axis represents a virtual biological clock with nodes assuming the time at which two genes separated. The length of the branches is inversely proportional to the similarities of the sequences; shorter branches mean higher similarity. Orphan GPCRs (including de-orphaned ones) are printed in bold.

that contain the encephalomyocarditis virus internal ribosome entry site (19, 20), which permits the translation of two open reading frames from one mRNA, have become commercially available (Invitrogen). Translation of both genes of interest and the selection marker from the same mRNA results in nearly all surviving colonies expressing the gene of interest after selection with antibiotics. Commonly used resistance markers include neomycin, puromycin, hygromycin, and zeocin.

4.2 Transient versus stable expression of GPCRs

The decision to use cells expressing orphan GPCRs either stably or transiently can be crucial to the outcome of the screening. Originally, successful screening for natural ligands was carried out using stable transfected cell lines (2, 10, 12, 21, 22). However, it has also become evident that single cell clones of stably transfected cells often respond differently to complex tissue libraries, making comparative screenings a difficult task. Most of the differences can be attributed to the lack of responsiveness to individual tissue fractions or weaker signal strength to known or unknown exogenous stimuli. The changes in response patterns among different clones might be explained by differences in the levels of receptor expression or by changes in growth properties of individual clones due to the variable transgene insertion sites into the cell's genome. Transiently transfected cells have the advantage of being easier and faster to produce and of representing an environment that permits comparative analyses. This is of special importance when working with receptors for which the coupling mechanism is unknown. In the transient expression system, the receptor can also be co-expressed with chimeric G proteins (see Section 4.4) which might force the receptor to couple to a desired signalling pathway (*Protocol 1*). In our laboratory we are using the transient expression system for initial screens and usually switch to stable cell lines once a reproducible response is found. In this case the response found with transiently expressing cells can be verified in stably expressing cells and the best responding clone can be chosen. This increases the signal sensitivity and the speed of the subsequent purification process.

Protocol 1

Transient transfection[a] of orphan receptors and chimeric G protein into CHO cells

Equipment and reagents

- 100 mm tissue culture Petri dish (Falcon)
- LipofectAMINE PLUS reagent (Life Technologies)
- OPTI-MEM I serum-free medium (Life Technologies)
- Alpha-MEM (Life Technologies) medium containing nucleotides and 5% fetal calf serum (Hyclone)
- Expression plasmid of cloned orphan receptors and chimeric G proteins

Method

1 One day before transfection, seed $0.6-1.2 \times 10^6$ cells in 10 ml of growth medium per 100 mm dish and incubate overnight in a CO_2 incubator until the cells are 50-80% confluent.

2 Mix 3 μg of the orphan receptor and 3 μg of the chimeric $G\alpha_{qi3}$ plasmid in a sterile 1.5 ml microcentrifuge tube with 350 μl of OPTI-MEM I. Add 20 μl of PLUS reagent and mix again. Let stand for 15 min at room temperature.

3 In a separate microcentrifuge tube dilute 30 μl of LipofectAMINE reagent with 350 μl of OPTI-MEM I medium.

4 Combine the pre-complexed DNA from step 2 with the diluted LipofectAMINE reagent. Gently mix and incubate for 15 min at room temperature.

5 During incubation aspirate growth medium and exchange with 5 ml pre-warmed, serum-free OPTI-MEM I.

6 At the end of the 15 min incubation period, add the DNA-lipid complex drop by drop to the cell culture. Incubate the cells for 3 h in a CO_2 incubator.

7 After 3 h incubation, replace the transfection medium with 10 ml of pre-warmed growth medium.

Protocol 1 continued

8 After 14 h incubation transfected cells are ready to be re-seeded into microtitre plates and grown for an additional 24 h. Afterwards cells can be used for subsequent assay protocols (see *Protocols 2–5*).

[a] The same protocol can be used for the HEK 293 cell line and to generate stably expressing cell lines. In the latter case, cells are re-plated into five 150 mm tissue culture dishes containing growth medium supplemented with the appropriate antibiotic.

4.3 Confirmation of receptor expression

Cloning a GPCR gene into a mammalian expression vector does not ensure its proper targeting to the plasma membrane. Several reasons for lack of membrane expression have been discussed. One striking example is the CGRP/adrenomedullin receptor, a former orphan receptor known as CRLR (calcitonin receptor-like receptor). This receptor is expressed at the cell surface only in the presence of receptor accessory membrane proteins called RAMPS (23). Note that the class B receptor family is so far the only family for which such accessory proteins are necessary. Another reason to verify receptor expression is the possibility that sequence data derived from genomic sources may be misinterpreted and lead to improperly folded receptor proteins. In order to prevent faulty expression of a particular GPCR due to misinterpretation of the receptor's genomic organization, it is advisable to first ensure proper protein expression before launching the ligand purification process. The following strategies can be used to test for effective expression.

4.3.1 Receptor tagging

Because production of antibodies directed against orphan GPCR sequences is a difficult, time-consuming, and often unsuccessful procedure, it is not advisable as a general approach for determining whether an orphan receptor is expressed at the membrane. Receptor tagging is the technique of choice for monitoring receptor expression in heterologous cells. This is a technique which has the advantage of speed and versatility in selecting the desired tagging site. Amino and carboxy terminal tags can be constructed simultaneously and tested in transient transfection assays. Tag-specific antibodies are commercially available and generally applicable to any orphan receptor. Carboxy terminal tags enable the detection of membrane targeting under conditions similar to those used during ligand screening. Once proper membrane expression is confirmed, the ligand screening will usually be done with the native receptor since the importance of the amino and carboxy terminal ends on ligand binding and/or signal transduction cannot be predicted. We generally use the FLAG sequence for amino terminal and the myc tag for C-terminal tagging (see *Table 1*). For expression analysis the tagged receptors are transiently expressed in the cell line used later for screening. *Protocol 2* outlines the procedure and *Figure 3* shows the visual result of a FLAG tagged receptor expressed at the plasma membrane.

Table 1 Peptide and protein sequences used for tagging GPCRs

Name	Amino acid sequence	Antibody supplier
Peptide tags		
c-myc epitope	Glu-Gln-Lys-Leu-Ile-Ser-Glu-Glu-Asp-Leu	Invitrogen/Roche
FLAG epitope	Asp-Tyr-Lys-Asp-Asp-Asp-Asp-Lys	Sigma
Haemagglutinin (HA)	Tyr-Pro-Tyr-Asp-Val-Pro-Asp-Tyr-Ala	Roche
Polyhistidine (6 × His)	His-His-His-His-His-His	Invitrogen
V5 epitope	Glyα-Lys-Pro-Ile-Pro-Asn-Pro-Leu-Leu-Gly	Invitrogen
VSV-G epitope	Tyr-Thr-Asp-Ile-Glu-Met-Asn-Arg-Leu-Gly-Lys	Roche
Xpress	Asp-Leu-Tyr-Asp-Asp-Asp-Lys	Invitrogen
Protein tags		
EGFP	Enhanced green fluorescent protein	Clontech or direct fluorescence

Protocol 2

Analysis of receptor expression in HEK 293T cells by immunofluorescence microscopy

Equipment and reagents

- Fluorescence microscope (e.g. Nikon)
- VectaShield mounting media (Vector Laboratories)
- Poly-D-lysine coated coverslips
- Phosphate-buffered saline (PBS): 50 mM sodium phosphate, 1.5 M NaCl pH 7.5

After obtaining a plasmid encoding the epitope tagged protein sequence it should be transiently expressed according to *Protocol 1*. Proper controls for immunostaining should contain separate background controls for primary and secondary antibodies.

Method

1 Prepare coated sterile glass coverslips by treating a stock of coverslips with 3 ml poly-D-lysine (100 μg/ml) in 35 mm tissue culture dish. Let stand for 30 min at room temperature.

2 Remove the poly-D-lysine solution and wash the coverslips twice with sterile distilled water. Let the slides dry under a sterile workbench and store them at room temperature for future use.

3 Seed the HEK 293T cells 24 h after transfection (see *Protocol 1*) at a density of 1×10^5 cells/well into a 35 mm tissue culture dish or 6-well plate with sterile coated glass coverslips on the bottom of the well.

4 Incubate the cells at 37 °C overnight in DMEM, 10% FCS under a 5% CO_2 atmosphere.

5 The next day, examine the glass coverslips for cell density and choose the ones that show the most dense cell lawn. Remove the medium and gently wash the coverslips twice with PBS.

Protocol 2 continued

6 For fixation, incubate the coverslips for 30 min on ice with freshly prepared PBS containing 3% paraformaldehyde. Afterwards rinse the cells three times for 10 min with PBS.

7 Permeabilize the cell membrane by treating the cells with 0.05% Triton X-100 in PBS for 10–15 min at room temperature. Subsequently wash three times for 10 min with PBS.

8 Block non-specific antibody binding sites with blocking solution (5% bovine serum albumin, fraction V or 20% goat serum in PBS) for 30 min at room temperature.

9 Replace blocking solution with the first (anti-epitope specific) antibody prepared in blocking solution at a concentration of 1–10 μg/ml. Note: actual antibody concentrations may vary from batch to batch; manufacturers recommended dilutions should always be preferential! Incubate for 1–3 h at room temperature.

10 Wash coverslips three times with PBS 10 min each step.

11 Incubate the coverslips with secondary antibody anti-mouse IgG (H&L)[a] F(ab')$_2$-fluorescein (1:10–1:300 in blocking solution; Vector) or anti-mouse IgG (H&L) F(ab')$_2$-rhodamine (1:20–1:500 in blocking solution) for 1 h at room temperature. Repeat step 10.

12 Finally add a small drop of mounting medium (VectaShield or 50% glycerol in PBS) on a microscopic slide and mount the coverslip with the cells facing down. At this point the slide might be either stored at 4°C or −20°C in the dark or examined immediately under a fluorescence microscope using a ×60 magnification and fitted with appropriate filters. For samples stained with fluorescence-conjugated antibodies maximum emission wavelength is 523 nm, and for rhodamine conjugates 570 nm.

[a] H&L: heavy and light chain antibodies.

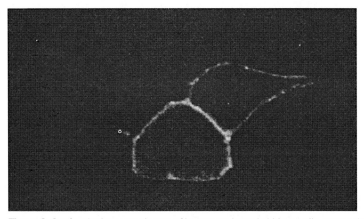

Figure 3 Confocal microscopy image of human embryonic kidney cells expressing an orphan GPCR bearing an amino terminal FLAG tag. Staining of the cells was done as described in *Protocol 2.*

4.4 Co-expression of promiscuous and chimeric G protein α-subunits

Different GPCRs couple to different Gα-subunits depending on the intrinsic structural properties of both molecules. The family of Gα-subunits can be divided into four subgroups based on their sequence homologies. The $Gα_s$-family mediates stimulatory regulation of adenylyl cyclase isoforms that lead to an increase in intracellular cAMP whereas the $Gα_i$ family inhibits adenylyl cyclases thus decreasing cAMP levels in the cell. The $Gα_q$ family of alpha subunits stimulates the activity of phospholipase Cβ (PLC) that finally leads to a rise in intracellular Ca^{2+} levels. The last subgroup comprises $Gα_{12}$ and $Gα_{13}$ and is functionally less understood, but recent studies indicate a role in the regulation of Na^+-H^+ exchange. Whereas the structural requirements for the specificity of G protein coupling in the receptor remain obscure, the C-terminal tail of the Gα-subunits has been unambiguously defined as a contact point. The importance of this region for interaction of GPCRs with the α-subunits has been demonstrated (for review see ref. 24) and mutations of the C-terminal tail yield Gα-chimeras with novel coupling properties (25). A whole series of G protein chimeras has been engineered and used to force the coupling of different GPCRs to a desired effector system (for review, see refs 26 and 27). The usefulness of chimeras is demonstrated in *Figure 4*. The orphanin FQ receptor (ORL-1/LC-132) does not normally couple to the induction of PLC when expressed in a mammalian cell line. However co-expression together with a chimeric $Gα_q$ subunit containing the substituted five amino acid C-terminus of $Gα_i$ enables the receptor to mobilize intracellular Ca^{2+}. Besides chimeric Gα, the $Gα_{16}$ subunit also couples promiscuously to ORL-1 and other GPCRs and thus can be a useful tool for screening ligands of orphan GPCRs (28).

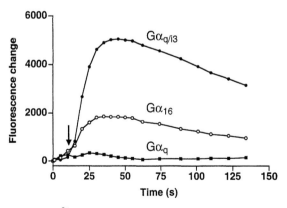

Figure 4 Ca^{2+} mobilization response induced by the agonist orphanin FQ (OFQ) in the presence of chimeric or promiscuous G protein α-subunits. A Chinese hamster ovary cell line stably expressing the G_i-linked OFQ receptor was transiently transfected with equal amounts of the indicated Gα-subunits. The cells were subsequently loaded with Fluo-4 AM and OFQ (30 nM) was applied to the cells at the indicated time point (arrow). The change in fluorescence of Fluo-4 caused by the release of intracellular Ca^{2+} was monitored using a fluorometric imaging plate reader (FLIPR™).

5 Functional assays

Receptor activation by its cognate ligand causes changes in intracellular second messengers. These can be measured by specific second messenger assays to follow purification of a ligand or to quantify the amount of a purified compound. The two major second messengers in every animal cell are cyclic AMP (cAMP) and Ca^{2+}. Under resting conditions the levels of cytosolic cAMP or Ca^{2+} are rather low but both increase upon stimulation: cAMP is produced by adenylyl cyclases through conversion of ATP while Ca^{2+} is released from storage organelles upon activation of the $InsP_3$ receptors.

5.1 cAMP assays

A number of different methods have been developed to detect and quantify intracellular cAMP. Most of the older techniques rely on using radioactively labelled ATP, followed by the extraction and counting of newly formed labelled cAMP or the disappearance of the precursor. However, the development of specific and highly sensitive radioimmunoassays and fluorescence-based immunoassays has recently largely replaced these. The immunological assays have the advantages of high sensitivity, good reproducibility, increased speed, and higher sample throughput. A general protocol for extraction and quantification of cAMP from cells is given in *Protocol 3*.

Protocol 3

Extraction and quantification of cAMP from mammalian cells using immunoassays

Equipment and reagents

- Forskolin (Sigma)
- Phosphodiesterase inhibitor: IBMX or Rolipram (Sigma)
- Colourless DMEM (without phenol red) (Life Technologies)
- cAMP radioimmunoassay kit (e.g. Amersham, NEN)

Method[a]

1   Incubate cells or tissue with the appropriate drugs or HPLC fractions in a total volume of 250 μl colourless DMEM containing 1 mM phosphodiesterase inhibitor for 15 min (for details see sections on detection of cAMP increase or decrease, respectively).

2   Place the plate on ice and quickly remove the medium by aspiration without disturbing the cells or tissue.

3   To each well, add 500 μl of ice-cold 66% ethanol to stop the reaction.

4   Freeze the plate at −80°C for at least 1 h then thaw again.

5   Remove an aliquot of the supernatant and dry down in a Speed Vac. (Usually 50 μl is a good starting point, but the amount of supernatant necessary has to be determined empirically. It is dependent on the sensitivity and dynamic range of the specific immunoassay used for quantification.)

6   Resuspend in assay buffer according to the manufacturer's instructions.

7   Use standards of known cAMP concentration to set up a calibration curve. Perform assay following the manufacturer's instructions.

8   Calculate sample cAMP content using the calibration curve.

^{a} Volumes are given for incubations in 24-well tissue culture plates, but can be adjusted accordingly to other formats.

5.1.1 Detection of cAMP increase

Stimulation of cAMP accumulation is easily detected using the method outlined in *Protocol 3*. The following protocol (*Protocol 4*) describes a method for preparing CHO or HEK 293 cells and performing the incubations with HPLC fractions or drugs of interest.

Protocol 4

Detection of cellular cAMP increase in CHO and HEK 293 cells

Equipment and reagents

• See *Protocol 1* and *Protocol 3*

Method

In the case of transiently transfected GPCRs:

1   Three days before the actual experiment, plate cells in one 100 mm tissue culture dish per 24-well or 96-well plate used later in the assay. Cells should reach 70–80% confluence 24 h after plating. Follow the instructions described in *Protocol 1*, beginning with step 2 for the transient expression of GPCRs.

The following steps are the same for both transiently or stably expressing cells:

2   Plate cells at 250 000 cells/well (24-well plate) or 50 000 cells/well (96-well plate) the day before the experiment.

3   On the day of the experiment, remove the culture medium by aspiration and replace by 200 μl (24-well plate) or 100 μl (96-well plate) colourless DMEM or equivalent medium. Place plates on ice.

4   Dry down HPLC fraction aliquots and resuspend in 50 μl of colourless DMEM or PBS. Resuspend drugs at known concentrations in the same medium and volumes.

Protocol 4 continued

5 Add samples to the cells on ice, transfer the plate to a 37°C water-bath and keep it floating for 15 min.

6 Follow the instructions of *Protocol 3* for extraction and quantification of cAMP.

5.1.2 Detection of cAMP decrease

A decrease in cellular cAMP concentrations is more difficult to determine than an increase, mainly because of the low basal levels of cAMP. Therefore, a reduction in cAMP production can be clearly detected only when adenylyl cyclases are pre-activated. The toxin, forskolin, potently stimulates most types of adenylyl cyclases and has been widely used in assays measuring inhibition of cAMP accumulation. Alternatively, endogenous receptors that coupled to $G\alpha_s$ can be used to stimulate the basal cAMP formation. In HEK 293 and in CHO cells, β-adrenergic and calcitonin receptors can be used for this purpose. In contrast to assaying cAMP stimulation, decreases in cAMP accumulation are detected in stably transfected cells because of the necessary pre-stimulation that will raise cAMP levels in all cells of which only 20–30% may express the receptor in the transient transfection format. Thus, a 50% decrease in cAMP levels in transfected cells, will amount to a mere 10% decrease in the overall cAMP levels of the total cell population, making it practically indistinguishable.

As an example of the difficulty in monitoring cAMP inhibition, *Figure 5* shows the second step of purification of orphanin FQ (OFQ) from pig hypothalamus using CHO cells stably transfected with the rat OFQ receptor (ORL-1/LC132) measuring inhibition of forskolin-stimulated cAMP accumulation. A protocol that we use to measure inhibition of cAMP accumulation after forskolin pre-stimulation is detailed below (*Protocol 5*).

Protocol 5

Detection of cellular cAMP decrease in stably transfected CHO cells

Equipment and reagents

• See *Protocol 1* and *Protocol 3*

Method

For use with stably expressing CHO cells prepared as described in *Protocol 1*.

1 To the medium used for dissolving drugs or HPLC fraction aliquots, add forskolin at x times the final concentration, where x is the ratio of total incubation volume to the volume of the sample added. Usually a final concentration of 1 μM forskolin is sufficient to obtain a strong stimulation of adenylyl cyclase, but this has to be determined empirically for each cell line.

2 Incubate for 15 min and proceed with the assay as described in *Protocol 3*.

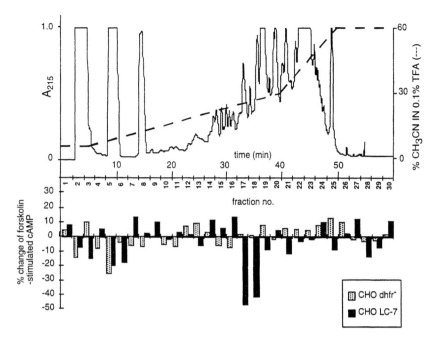

Figure 5 Purification of orphanin FQ from porcine hypothalamic extracts; second step. A stable CHO clone expressing the rat orphanin FQ receptor (LC-132) was used to measure inhibition of forskolin-stimulated cAMP accumulation after incubation with aliquots from HPLC fractions. Fractions 17 and 18 contained active material (bottom). The UV absorbance at 215 nm and the gradient profile are shown in the upper half. Column: Superspher RP 60 select B, 4 × 250 mm (Merck, Darmstadt, Germany).

5.2 Ca^{2+} assay

The concentration of cytosolic free Ca^{2+} is approximately 100 nM in unstimulated cells, whereas it is 1.5–2.0 mM in the extracellular space. Ca^{2+} has a dual role as a carrier of electrical current and as a second messenger. Many signalling pathways directly or indirectly increase cytosolic Ca^{2+}. The source of elevated Ca^{2+} can be either the endoplasmic reticulum (ER) or the extracellular space. Mobilization of ER Ca^{2+} is mediated by inositol triphosphate (InsP$_3$) derived from phospholipase Cβ (PLCβ) activation through G proteins or from PLCγ activation by receptor tyrosine kinases (e.g. EGF-R). InsP$_3$ activates the InsP$_3$ receptor, a ligand-gated Ca^{2+} channel located in the ER, which stimulates Ca^{2+} release from the intracellular stores as well as Ca^{2+} re-entry driven by the enormous concentration gradient. Nowadays, measuring intracellular Ca^{2+} is a preferential assay because it can be monitored by non-invasive fluorometric methods. It is based on the use of highly sensitive fluorescent Ca^{2+} indicator dyes, Fluo-3 and Fluo-4, which are non-fluorescent unless bound to Ca^{2+}. When bound, enhancement of fluorescence is usually at least 100-fold, and may exceed 200-fold. Between normal resting cytosolic free Ca^{2+} concentrations and indicator saturation, the enhancement is generally between five- to tenfold, making Fluo-3 particularly useful for measuring the kinetics of Ca^{2+} transients. The cells are loaded with the membrane permeable acetoxymethyl (AM) esters of these dyes which are cleaved into the cell membrane impermeable free acid by intracellular esterases. The assay described in *Protocol 6* is customized for the fluorometric imaging plate reader (FLIPR™) manufactured by Molecular Devices; this technique is described fully in Chapter 8. The key advantage of the system is that it simultaneously stimulates and optically reads all 96 wells of a microplate. The assay could also be adapted to a standard fluorescent plate reader. The activity profiles of two different GPCRs stimulated with consecutive fractions of a HPLC-fractionated rat brain extract using the FLIPR™ as a readout are shown in *Figure 6*.

Protocol 6

Intracellular Ca^{2+} measurement using the FLIPR™ (fluorometric imaging plate reader) and GPCR transfected CHO dhfr⁻

Equipment and reagents

- Black wall, clear-bottom, tissue culture treated 96-well microtitre plates (Corning)
- Access to a fluorometric imaging plate reader (Molecular Devices)
- Fluo-4 AM (Molecular Probes), 2 mM in DMSO containing 20% pluronic acid (Molecular Probes)
- 96-well plate cell washer (Labsystems)
- Probenecid (Sigma)
- Alpha MEM (Life Technologies) medium containing nucleotides and 5% fetal calf serum (Hyclone)
- 10 × Hank's buffer (Life Technologies)

Method

1 14–20 h after transfection, harvest and re-plate the cells at a density of 4×10^4 cells/well in 96-well plates. Depending on the transfected plasmid, one 10 cm tissue culture dish usually yields one full 96-well plate. Grow the cells for another 12–24 h.

2 The next day, carefully aspirate the growth medium from each well of the 96-well plate and replace with 50 µl freshly prepared Hank's buffer containing 2 µM Fluo-4 AM and 2.5 µM probenecid. Incubate the cells for 1 h at 37 °C.

3 Wash the cells three times with Hank's buffer containing 2.5 µM probenecid. Adjust the cell washer appropriately so that it leaves exactly 100 µl of wash buffer in each well after the last washing step.

4 During dye loading prepare the test samples in a separate polypropylene microtitre plate. Test samples can be fractions of tissue extracts, chemical libraries, or libraries of known ligands. Positive controls targeting GPCRs endogenous to the cell line should be included. For CHO cells we usually use UTP that yields a robust Ca^{2+} signal at 10 µM final concentration. All compounds to be tested are prepared at a 2 × concentration in 100 µl washing buffer.

5 Place the 96-well plate containing the Fluo-4 loaded cells and the drug plate into the pre-warmed FLIPR instrument and start the experiment (29). For the first minute acquire data every second, for the next two minutes reduce the sample interval to 3 sec. Add the compounds 10 sec after the start of the experiment.

6 Analyse the data by exporting the maximal fluorescence values over background.

5.3 Reporter gene assay

Besides many short-term effects, changes in the levels of intracellular second messengers also lead to altered transcription of specific genes. These genes contain responsive elements in their promoter regions which can, upon binding of specific transcription factors, activate or repress that gene expression. In the case of the cAMP system, the signalling cascade involves activation of PKA by increasing cAMP levels, which in turn phosphorylates a cAMP response element-binding protein (CREB). Phosphorylated CREB then triggers the transcription of genes containing a cAMP response element (CRE) in their promoter region. This cascade of events can be used to monitor receptor activation. If, instead of a cellular gene, a reporter gene is inserted downstream of a CRE-containing pro-moter, changes in cellular cAMP concentrations will regulate the amount of reporter gene product. This can serve as a parameter to monitor the activation of a $G\alpha_s$ or $G\alpha_i$ coupled GPCR (30). A widely used reporter gene is the firefly luciferase which, in the presence of ATP and its substrate luciferin, will produce an easily detected bioluminescence. Other response elements and combinations thereof can also be cloned in front of a luciferase reporter gene (31), making this system a versatile tool for detection of activation of various GPCRs, including those coupling to $G\alpha_q$.

Some caveats of the reporter assay have to be taken into consideration: First, transcription of the reporter gene is a downstream event that is far from the actual receptor activation. Besides the PKA pathway there are alternative ways that could lead to a phosphorylation of CREB. Therefore, different receptor events could finally converge in stimulating reporter gene transcription, which could in turn increase the 'noise' of the assay, especially when screening natural extracts. Secondly, reporter gene assays normally require several hours of in-cubation to allow for transcription and translation of the gene product. Most of the ligands, and especially peptides, are not stable over such an extended period or might be degraded by enzymes released from the cells. For each reporter gene construct the minimal necessary incubation time sufficient to produce a detect-able response must therefore be determined. This will influence the design and timing of the individual assay. Finally, as with every inducible gene expression, the gene product has a certain half-life and will eventually be degraded by cellular proteases. The optimal time for obtaining highest levels of reporter gene expression therefore has to be determined empirically for each system. Generally, a 5–6 h incubation has produced maximal signal levels.

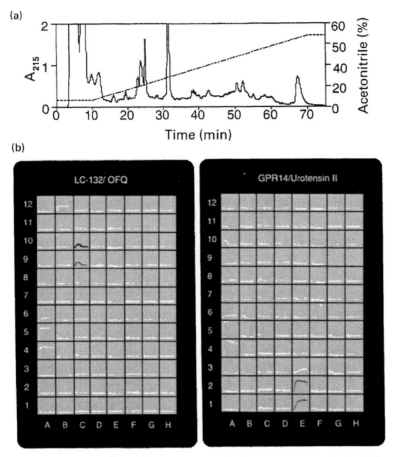

Figure 6 Screening of a fractionated bovine hypothalamic tissue extract. (a) UV absorbance of the fractionated material separated on a PrepPak-Delta-Pak C18 column (15 μm, 300 Å, 25 × 100 mm, Waters) and eluted with an acetonitrile gradient in 0.1% TFA. The eluted material was collected at 1 min intervals in individual fractions (1–72), concentrated, and tested for activity on transiently transfected cells. (b) Ca^{2+} mobilization responses elicited by the fractions collected in (a). Individual fractions (1–72, position A1 to F12) were simultaneously added to transiently transfected Chinese hamster ovary (CHO) cells expressing either the orphanin FQ/LC-132 receptor (left) or the urotensin II receptor in presence of $G\alpha_{qi3}$ chimeric G proteins. Each square corresponds to a single well in a microtitre plate. The *y*-axis in each square represents the change of intracellular Ca^{2+} levels and the *x*-axis the time. In this case, receptor-specific responses for OFQ (left; D9, D10) and urotensin II receptor (right; E1, E2), respectively, are clearly identified. The Ca^{2+} responses, seen in positions A4–A6, are caused by unknown compounds acting on receptors endogenous to CHO cells.

In the following section, an example of a CRE-dependent luciferase reporter gene assay is given.

5.3.1 Construction of the reporter gene

Mammalian expression vectors containing luciferase are commercially available. Two different strategies for inserting CRE in front of the luciferase have been described in the literature: One can use either a cloned part of a natural promoter known to contain functional CRE sequences, such as the human vasoactive intestinal peptide promoter or the rat prodynorphin promoter. Alternatively synthetic CRE sequences can be generated by ligation of oligonucleotides which allows for the introduction of multiple copies of the response element. A technical instruction for the construction of the reporter gene plasmid is beyond the scope of this chapter but is described in great detail in the literature.

When cloned into a selectable mammalian expression vector, stable cell lines can be produced and screened by incubation with forskolin. A stable clone with a large dynamic range between stimulated and unstimulated levels of reporter gene expression is a useful tool in searching for ligands of orphan GPCRs that might be coupled to $G\alpha_s$. Furthermore, it is also possible to use the CRE-luciferase reporter gene construct in transiently transfected cells.

5.3.2 Stimulation of cells and measurement of luciferase activity

Procedures for transient and stable expression have been described before. The preparation and plating of cells for reporter gene assays is identical to the procedures described in *Protocol 4*. The luciferase reaction normally has a fast kinetic, resulting in a flash of photons released within the first few seconds. This limits its application to either handling each sample individually or the use of multichannel luminometers for higher throughput.

The commercially available kits for detection of luciferase activity differ mostly in the kinetic profile of the luciferase reaction. Newer versions of the kits produce a more extended and steady bioluminescence signal, which is especially useful when quantifying in scintillation counters. In this case, the light output is stable for more than 60 minutes so that multiple plates can be prepared and counted in parallel. The following is a protocol (*Protocol 7*) using 96-well plates and a scintillation counter able to handle the same format.

Protocol 7

Measurement of luciferase activity

Equipment and reagents

- Luminometer or scintillation counter for 96-well plates (LabSystems, Packard)
- Luciferin reagent: commercial kits (e.g. Promega)
- Lysis buffer: 1% Triton X-100, 25 mM Tris–phosphate pH 7.8, 2 mM dithiothreitol, 2 mM trans-1,2-diaminocyclohexane-*N,N,N',N'*-tetraacetic acid, 10% glycerol

Method

1. Plate cells at a density of 50 000 cells/well in a total volume of 100 μl/well the day before the experiment.
2. Dissolve aliquots of HPLC fractions or drugs in 50 μl PBS and add to the cells.
3. Incubate for 5–6 h at 37 °C in a tissue culture incubator.

Protocol 7 continued

4   Aspirate the medium and add 100 µl of lysis buffer.

5   Freeze cells at −20 °C for at least 1 h; thaw.

6   Transfer an aliquot of the lysate to a white 96-well plate and dilute in assay buffer (as recommended by the manufacturer) to obtain a final volume of 50 µl.

7   Add 50 µl of luciferase reagent and place plate in the scintillation counter using the settings for detection of bioluminescence.

8   The data can either be expressed as x-fold stimulation over basal levels or as a percentage of a maximal stimulation produced by e.g. 1 µM forskolin.

6 Ligand screening

6.1 Preparation of tissue extracts as sources for ligand screening

The physicochemical nature of the ligand might vary greatly among different orphan receptors and so will the extraction and purification methods. Lipids for example, need to be approached through a strategy different from that for biogenic amines or peptides. However, the common principle of finding a suitable tissue, tissue extraction and fractionation, and finally screening for receptor activity will remain the same. In this chapter we focus only on peptides as possible receptor ligands for receptors classified as putative peptide receptors.

One of the first difficult decisions to take in applying the orphan receptor strategy is the choice of the tissue expressing the ligand. When working with brain receptors, the tissue is chosen where the receptor mRNA levels are high. For example in the case of SLC-1 (32) based on receptor expression data, it was anticipated that the ligand would be in the brain and this turned out to be the case. However, there are also examples describing the isolation from peripheral tissues of a ligand for a brain receptor (22, 33). Since high receptor expression does not ensure high levels of ligand in the same tissue, a random screening approach may sometimes prove successful. This then becomes comparable to the screening of chemical libraries performed for decades in the pharmaceutical industry and requires the availability of high throughput assays.

As a rule, purification strategies are adapted to each orphan receptor. The common denominator for success is a good initial source of the unknown ligand. Here we focus on the purification of peptides (*Protocol 8*).

Protocol 8

Peptide extracts from total rat brains

Equipment and reagents

- Infrared hotplate (Fisher Scientific)
- Waring blender (e.g. Christison Scientific)
- Polytron type tissue homogenizer (Fisher Scientific)
- Freeze-dry system (Fisher Scientific)
- Rotary evaporator (Fisher Scientific)
- Separating funnel (Fisher Scientific)
- Freshly frozen rat brains (Pel-Freez Biologicals)
- Liquid nitrogen

Method

This protocol is suitable for 40 g of wet tissue (20–25 rat brains) but can be easily scaled up.

1 Boil 160 ml of distilled water in a glass beaker using the infrared hotplate.

2 Weigh out 40 g of frozen rat brains in a small Styrofoam box and submerge with liquid nitrogen (wear protective glasses!). Transfer the whole content into the Waring blender and blend the tissue until it turns into a homogeneous powder.

3 Cautiously pour small aliquots of deep frozen powder into the boiling water and stir. Boil the homogenate for 10 min.

4 Rapidly quench the solution on ice. With stirring add 4.7 ml of glacial acetic acid dropwise to the cold solution. Subsequently treat the tissue with a Polytron homogenizer for 3–5 min at 20 000 r.p.m.

5 Add 0.05% sodium azide and stir overnight in a cold room.

6 Centrifuge the homogenate at 6000 g for 20 min in a Sorvall GSA or similar rotor.

7 Collect the supernatant and resuspend the pellet in 160 ml of cold 1 M acetic acid. Centrifuge according to step 6. Collect the supernatant and discard the pellet.

8 Pool the supernatants of step 6 and 7 and slowly add three volumes of acetone. Incubate on ice for 30 min.

9 Centrifuge the acetone-containing tissue extract at 18 000 g for 30 min in a Beckman JLA-10.500 or similar rotor. Transfer the supernatant into a 2 litre chemical flask that fits the rotary evaporator. The pellet can be discarded.

10 Evaporate the acetone by slowly applying the vacuum and rotating at a moderate speed. At this step all acetone must be removed since it interferes with the next step.

11 Remove lipids by extracting the acetone-free tissue extract from step 10 twice with 1 volume of diethyl ether using a separating funnel. Save the aqueous phase into a round-bottom flask compatible with a freeze-dryer and freeze at $-20\,°C$ or in liquid nitrogen.

12 Freeze-dry the extract for two days or until it is completely dry.

13 Dissolve the dried peptide extract in HPLC starting buffer (usually 5% acetonitrile, 0.1% trifluoroacetic acid). At this point the extract is ready for chromatographic fractionation.

6.2 Purification and structure elucidation of endogenous ligands for orphan receptors

Once a peptide extract has been prepared, it must be fractionated for use in the initial screening of cells expressing the orphan receptor. Since the peptide extract can be compared to a huge library of individual peptides of varying sizes and concentrations, the initial purification steps are aimed at a broad separation of the individual compounds. A crude peptide extract could not be used as an immediate screening source because it would elicit second messenger responses in both transfected and non-transfected cells. Cell lines contain a number of endogenous receptors: in HEK 293 cells, for instance, several different endogenous receptors coupled to the mobilization of intracellular Ca^{2+} have been characterized. Of these receptors, not all are peptide receptors and not all are generally activated by peptide extracts but more may exist.

Differentiating responses obtained from endogenous receptors (background) from those related to the exogenous orphan receptor is a major task of the approach. In principle one might discriminate signal from noise by comparing the responses of the transfected and the non-transfected cells. This is successful only if the stimulation is caused by a single compound and not superimposed with another. The initial purification therefore aims at separating signal from noise. Separation of peptides is usually achieved by chromatographic methods. A combination of two different chromatographic methods, ion exchange and reversed-phase chromatography, is generally sufficient to solve any peptide separation problem. However, it is beyond the scope of this chapter to describe these chromatographic methods in detail but the interested reader is referred to a book that also appeared in this series (34).

It has to be emphasized that the first screening is most critical to the success of the de-orphaning of a receptor and so care should be taken in preparing high quality tissue extract. The quality of the extract can be assessed by evaluating activation of known endogenous GPCRs in the transfected cell lines (*Figure 6*). Depending on these results, it might be necessary to modify the extraction conditions. For example a cold extraction procedure might prove superior to a hot one, depending on the physicochemical properties of the peptide ligand. Once a specific active fraction has been identified the peptidergic nature of the active entity should be tested, for example by peptidase treatment. Destruction of activity by specific enzymes can provide valuable structural information about the unknown peptide which can be used as a parameter in the optimization of the purification scheme. Ultimately, after several purification steps, a peptide will have been purified to near homogeneity and will await structural analysis. Depending on the amount available, mass spectroscopic methods or classical protein sequencing through Edman degradation can be applied. These techniques require equipment that is not generally found in a biochemistry laboratory and are often carried out in collaboration with specialized structural analysis groups. Once a peptide sequence has been found, an identical peptide will be synthesized and tested in the orphan receptor assay.

Acknowledgements

We greatly appreciate the contributions of all lab members, especially Drs Yumiko Saito (presently at Tokyo, Metropolitan University) and Zhiwei Wang in the development of these techniques. We are indebted to Dr Jason Bermak for providing help with confocal microscopy.

References

1. Meunier, J. C., Mollereau, C., Toll, L., Suaudeau, C., Moisand, C., Alvinerie, P., *et al.* (1995). *Nature*, **377**, 532.
2. Reinscheid, R. K., Nothacker, H. P., Bourson, A., Ardati, A., Henningsen, R. A., Bunzow, J. R., *et al.* (1995). *Science*, **270**, 792.
3. Civelli, O. (1998). *FEBS Lett.*, **430**, 55.
4. Hinuma, S., Onda, H., and Fujino, M. (1999). *J. Mol. Med.*, **77**, 495.
5. Marchese, A., George, S. R., and O'Dowd, B. F. (1998). In *Identification and expression of G-protein coupled receptors* (ed. K. R. Lynch), Vol. XII, p. 1. Wiley-Liss, New York.
6. Myers, E. W., Sutton, G. G., Delcher, A. L., Dew, I. M., Fasulo, D. P., Flanigan, M. J., *et al.* (2000). *Science*, **287**, 2196.
7. The *C. elegans* Sequencing Consortium. (1998). *Science*, **282**, 2012.
8. Lynch, K. R. (1998). In *Identification and expression of G-protein coupled receptors* (ed. K. R. Lynch), Vol. XII, p. 54. Wiley-Liss, New York.
9. Retief, J. D., Lynch, K. R., and Pearson, W. R. (1999). *Genome Res.*, **9**, 373.
10. Hinuma, S., Habata, Y., Fujii, R., Kawamata, Y., Hosoya, M., Fukusumi, S., *et al.* (1998). *Nature*, **393**, 272.
11. Nothacker, H. P., Wang, Z. H., McNeil, A. M., Saito, Y., Merten, S., O'Dowd, B., *et al.* (1999). *Nature Cell Biol.*, **1**, 383.
12. Mori, M., Sugo, T., Abe, M., Shimomura, Y., Kurihara, M., Kitada, C., *et al.* (1999). *Biochem. Biophys. Res. Commun.*, **265**, 123.
13. Ames, R. S., Sarau, H. M., Chambers, J. K., Willette, R. N., Aiyar, N. V., Romanic, A. M., *et al.* (1999). *Nature*, **401**, 282.
14. Bächner, D., Kreienkamp, H., Weise, C., Buck, F., and Richter, D. (1999). *FEBS Lett.*, **457**, 522.
15. Chambers, J., Ames, R. S., Bergsma, D., Muir, A., Fitzgerald, L. R., Hervieu, G., *et al.* (1999). *Nature*, **400**, 261.
16. Lembo, P. M., Grazzini, E., Cao, J., Hubatsch, D. A., Pelletier, M., Hoffert, C., *et al.* (1999). *Nature Cell Biol.*, **1**, 267.
17. Saito, Y., Nothacker, H. P., Wang, Z., Lin, S. H., Leslie, F., and Civelli, O. (1999). *Nature*, **400**, 265.
18. Shimomura, Y., Mori, M., Sugo, T., Ishibashi, Y., Abe, M., Kurokawa, T., *et al.* (1999). *Biochem. Biophys. Res. Commun.*, **261**, 622.
19. Rees, S., Coote, J., Stables, J., Goodson, S., Harris, S., and Lee, M. G. (1996). *Biotechniques*, **20**, 102, 106, 108.
20. Jang, S. K., Davies, M. V., Kaufman, R. J., and Wimmer, E. (1989). *J. Virol.*, **63**, 1651.
21. Sakurai, T., Amemiya, A., Ishii, M., Matsuzaki, I., Chemelli, R. M., Tanaka, H., *et al.* (1998). *Cell*, **92**, 573.
22. Tatemoto, K., Hosoya, M., Habata, Y., Fujii, R., Kakegawa, T., Zou, M. X., *et al.* (1998). *Biochem. Biophys. Res. Commun.*, **251**, 471.
23. McLatchie, L. M., Fraser, N. J., Main, M. J., Wise, A., Brown, J., Thompson, N., *et al.* (1998). *Nature*, **393**, 333.
24. Bourne, H. R. (1997). *Curr. Opin. Cell Biol.*, **9**, 134.

25. Conklin, B. R., Farfel, Z., Lustig, K. D., Julius, D., and Bourne, H. R. (1993). *Nature*, **363**, 274.
26. Wess, J. (1998). *Pharmacol. Ther.*, **80**, 231.
27. Milligan, G. and Rees, S. (1999). *Trends Pharmacol. Sci.*, **20**, 118.
28. Offermanns, S. and Simon, M. I. (1995). *J. Biol. Chem.*, **270**, 15175.
29. Schroeder, K. S. and Neagle, B. D. (1996). *J. Biomol. Screening*, **1**, 75.
30. Himmler, A., Stratowa, C., and Czernilofsky, A. P. (1993). *J. Recept. Res.*, **13**, 79.
31. Fitzgerald, L. R., Mannan, I. J., Dytko, G. M., Wu, H. L., and Nambi, P. (1999). *Anal. Biochem.*, **275**, 54.
32. Kolakowski, L. F. Jr., Jung, B. P., Nguyen, T., Johnson, M. P., Lynch, K. R., Cheng, R., *et al.* (1996). *FEBS Lett.*, **398**, 253.
33. Kojima, M., Hosoda, H., Date, Y., Nakazato, M., Matsuo, H., and Kangawa, K. (1999). *Nature*, **402**, 656.
34. Lim, C. K. (ed.) (1986). *HPLC of small molecules: a practical approach*. IRL Press, Oxford; Washington, DC.

Chapter 3

Point mutations and chimeric receptors in studies of ligand binding domains and second messenger activation of α_1-adrenoceptors

Dianne M. Perez, Michael J. Zuscik, Sean A. Ross, and David J. J. Waugh

NB5, The Department of Molecular Cardiology, The Lerner Research Institute, The Cleveland Clinic Foundation, 9500 Euclid Ave, Cleveland, Ohio 44195, USA.

1 Introduction

The creation of point mutations or the generation of chimeric receptors is a valuable model system for the evaluation of structure–function relationships of G protein-coupled receptors (GPCRs). One can essentially mutate any area in the receptor structure and then determine the functional and/or binding consequences of that mutation; thus, interpreting the importance of the original amino acid in receptor structure. This chapter will outline two basic techniques to generate mutations, how to transfect into mammalian cells, then to analyse the effects of the mutations by either binding or function.

2 Applications

2.1 Random *versus* chimeric

The use of random mutagenesis is not as common as that of directed or point mutagenesis. It is essentially used to probe for a particular domain of the protein, such as finding what areas are involved in structural integrity or to probe for the catalytic site of an enzyme. It is generated by use of either a chemical mutagen, radiation, or polymerase chain reaction (PCR) approach and is usually coupled to a selection mechanism or assay, such as colonies growing only on a certain agar plate composition which is tied to a certain metabolic or catalytic function. This type of mutagenesis has not been applied to GPCRs because of our vast knowledge of the secondary structure and the conservation that exists even among remote family members. Rather, to probe for an area of binding or function, the chimeric approach is more useful. This utilizes hybrids between two different GPCRs, the most informative being between subtypes of the same family. However, chimerics between distant members can also be utilized since this will conserve secondary structure.

2.2 Directed *versus* chimeric

In general, when one wants to scan receptor structure for gross areas of binding or function, the chimeric receptor approach is appropriate. Between subtypes of a family, this approach is strong since the overall binding pocket and structural integrity of the receptor will be conserved. The chimeras are likely to fold properly and to be expressed at workable levels. They are also likely to conserve binding of non-selective agonists and antagonists, so that the affinity or mode of binding of the radiolabel used in binding studies will not be altered. It is a powerful method to find regions of the receptor that are involved in subtype selectivity of receptor binding or efficacy. It also has the advantage of two modes of analysis, one being the loss of affinity or function and, the other, the gain of affinity or function by creating the opposite chimera. If members of a subtype family are not available, one can still generate a chimera with the closest relative. However, in most cases, the amount of switching needed to see either a gain or loss of function will be great because several helices are generally involved in binding or function. In general, the closer the homology, the fewer the number of switches needed to find any differences. Chimeras between related, but still distant family members of α_1-adrenoceptors and β-adrenoceptors have been generated to explore binding issues (1). Between more dissimilar GPCRs, the regions switched are located in the intracellular loop areas to explore G protein-coupling specificity (2).

Once the areas that are involved in binding or function by the chimeric approach have been located, one can generate point mutations to locate specifically the exact amino acid involved in the particular function. Sometimes, due to the conservation between GPCRs, or deduction about the secondary structure, due to previous mutagenesis or molecular modelling, one can predict small areas of receptor structure that are likely to be involved in function. In these cases, the point mutations can be made directly and bypass the chimeric scan. In general, the type of amino acid substituted in place of the original is varied. If one predicts that charge is important, an amino acid is replaced that neutralizes the charge but is close to the size or van der Waals radius of the original amino acid (i.e. Asp to Asn) to avoid folding artefacts. Hydrophobicity is tested by an Ile to Ala substitution, aromaticity by a Tyr to Ile substitution. If one cannot predict the type of substitution to make, create both a conservation of size and an Ala substitution. For GPCRs, if one is inducing a mutation in a transmembrane domain, substitutions that might disrupt the alpha helix, such as a glycine or proline substitution, need to be avoided.

3 Methods for creating chimeric receptors

3.1 Cassette replacement

Although restriction enzyme sites may be strategically located in the regions of the receptor it is desired to switch, in most cases they are not. The method below is the simplest procedure for generating chimeras which does not utilize a PCR approach. The model system used here will be the creation of chimeras between the α_{1a}- and α_{1d}-adrenergic receptor subtypes.

Suppose one wants to create a chimera where the switch point is located at the extracellular surface of transmembrane domain three (*Figure 1*). A unique restriction site to each cDNA is found near the beginning of the junction and after the junction. In the α_{1a}-adrenoceptor this would be a *Xho*II site while in the α_{1d}-adrenoceptor this would be an *Afl*III site. To make a chimera, the sense and antisense oligo in between these two sites can be synthesized with the two correct overhangs for *Xho*II and *Afl*III. The area in between these two sites can then be designed and synthesized to start the switch from one subtype to the other anywhere in the sequence. To complete the cDNA, an *EcoRI–Xho*II fragment

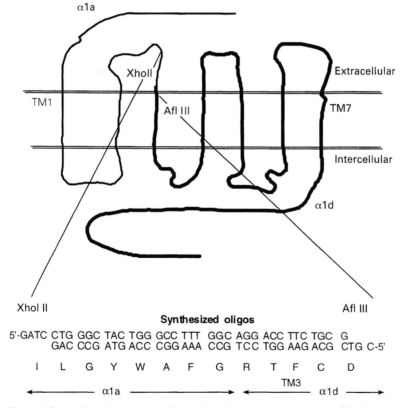

Figure 1 Generation of an α_{1a}/α_{1d} chimera through cassette replacement. Thin line represents the α_{1a}-subtype while the heavy line represents the α_{1d}-subtype. The chimeric overlap region is enhanced. TM, transmembrane domains; *Xho*II, *Afl*III, restriction sites.

of the wild-type (WT) α_{1a}-adrenoceptor is ligated with the synthesized cassette along with an *AflIII–NotI* fragment of the WT α_{1d}-adrenoceptor.

The two restriction sites can be located anywhere before or after the switch. However, there is a limitation of the distance which is constrained to the length of the two oligomers to be synthesized (generally 80–100 bases). This limitation can be circumvented by designing a cassette of four oligomers (two for the sense and two for the antisense) and overlapping the junction point by ten bases. This allows a good annealing surface between sense and antisense.

Once the oligonucleotides have been designed, they must be phosphorylated and annealed (*Protocol 1*) together prior to ligation (*Protocol 5*) with the WT fragments. Synthetic DNA is supplied with both a 5′ and 3′ hydroxyl, unlike natural DNA which has a 5′ phosphate. It is better to phosphorylate the single-stranded oligos before annealing as phosphorylation of double-stranded DNA is not as efficient. Slow cooling for the annealing ensures a perfect matching of base pairs. Once the oligos are annealed, they should have the correct overhangs for ligation with the same overhangs generated on the WT fragments by restriction digestion.

Protocol 1

Oligonucleotide phosphorylation and annealing

Equipment and reagents

- Temperature block
- Heating plate
- 100 mM ATP (Sigma), pH to 7.0
- T4 polynucleotide kinase; supplied with 10 × buffer (Amersham)
- 10 × ligase buffer (Amersham)

Method

1. Set up the kinase reaction as follows, one tube for each oligo (sense and antisense): approx. 50 pmoles of oligonucleotide, 500 nmol of ATP which is 5 μl of 100 mM stock, 5 μl of 10 × kinase buffer, and sterile water to 50 μl final volume. Incubate mixture at 37 °C for 1 h.

2. Combine the contents of each kinased oligo (100 μl total). Add ligase buffer to 1 × and heat sample to boiling by placing tubes in a beaker filled with water on the heating plate.

3. After 3 min of boiling, remove the entire beaker with the samples onto a benchtop and allow to cool slowly to room temperature.

4 Methods for creating point mutant receptors

4.1 PCR mutagenesis

Although there are many different variations of the PCR method to generate point mutations, we find the method discussed below the easiest (*Protocol 2*). A diagram of the mutational protocol is shown in *Figure 2*. Essentially, the primers are first designed and the PCR amplification carried out. The fragment is then digested with the appropriate restriction enzymes and ligated with adjoining WT fragments to a mammalian vector for expression. The requirement of this particular PCR approach is a restriction site located near to the desired mutational site (within 50 bases). Preferably, this site should be unique to the cDNA but multiple sites can be accommodated. The design of the primer begins by writing out the amino acid sequence and codons used in the cDNA of interest.

Extracellular loop and TM5 region, α_{1a} amino acid 162–193:

L–F–G–W–R–Q–P–A–P–E–D–E–T–I–C–Q–I–N–E–E–P–G–Y–V–L–**F**–S–A–L–G–S–F

This corresponds to the sense DNA sequence of:

*Nae*I site

5'-CTG TTC GGC TGG AGG CA**G CCG GC**T CCA GAG GAT GAG ACC ATC TGC CAG ATC AAT GAG GAG CCG GGC TAC GTG CTG **TTC** TCA GCG CTG GGC TCT TTC-3'

The codon of interest (Phe187) as well as the restriction site (*Nae*I), are highlighted in bold. Suppose we want to create a Phe187 to Cys mutation (*Figure 2*); the codon will be changed from TTC to TGC. We will incorporate this change in the sense primer (designated by an X) that we design as well as incorporating the *Nae*I restriction site. General considerations in this type of primer design are that the mutation site is not located near the ends of the primer and that at least ten bases are 5' of the restriction site to ensure efficiency of subsequent enzymatic digestion. The sense primer can be written as:

*Nae*I site

5'-GGC TGG AGG CA**G CCG GC**T CCA GAG GAT GAG ACC ATC TGC CAG ATC AAT GAG GAG CCG GGC TAC GTG CTG **TGC** TCA GCG CTG-3'

The length of the primer is usually not problematic, but it is limited by the ability to synthesize long oligomers. Long primers do well in PCR. Short primers of 25 bases or less should not be considered due to specificity problems.

The antisense primer is made at the end of the coding region or cDNA (*Figure 2*). Usually genes are subcloned into vectors that would incorporate a restriction site after the stop codon or 3' untranslated region, so a convenient restriction site is usually located here. In the design of the antisense primer, use 20–25 bases of the gene of interest and incorporate the unique restriction site (*Not*I). Make sure that the antisense primer has at least ten bases 5' of that restriction. The extra ten bases will then be vector sequences. The PCR fragment is then generated (*Protocol 2*) (size depends on the base pairs between the two primers) and ligated to the other pieces of WT cDNA to make the full gene product. Obviously, the opposite scenario of *Figure 2* is also valid, one can make an antisense primer to incorporate the mutational site and then make a sense primer at the start site of the cDNA.

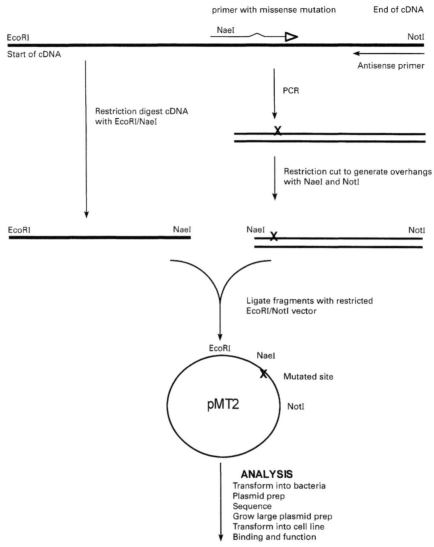

Figure 2 PCR mutagenesis flow chart. Heavy single line represents cDNA. Double thick line represents PCR product. Single thin line represents vector. X, mutated site; *Eco*RI, *Not*I, *Nae*I, restriction sites.

Protocol 2

PCR amplification of mutant DNA

Equipment and reagents

- PCR machine
- 500 μl microcentrifuge tubes
- UV transilluminator (Fisher)
- Agarose gel electrophoresis equipment with power supply (Fisher)
- Geneclean kit (BIO 101)

Protocol 2 continued

- Eppendorf or microcentrifuge tubes (1.5 ml; Fisher)
- PCR buffer, supplied by manufacturer of the *Taq*
- *Taq* polymerase or various versions (Boehringer Mannheim)
- Deoxynucleotides phosphates (dATP, TTP, dGTP, dCTP) molecular biology grade, usually supplied as 100 mM stocks
- Sense and antisense primer, synthesized by various companies (Gibco BRL)
- Mineral oil, sterile

- Agarose, low electric-endo-osmosis (EEO), molecular biology grade (Fisher, Allegience)
- 50 × TAE buffer: (per litre) 242 g Tris base, 57.1 ml glacial acetic acid, and 100 ml of 0.5 M EDTA, pH 8.0
- 6 × agarose loading buffer: 0.05% bromophenol blue, 0.05% xylene cyanol, 30% sterilized glycerol in sterile water
- Ethidium bromide solution (10 mg/ml in water) (Sigma)
- TE buffer: 10 mM Tris pH 8.0, 1 mM EDTA pH 8.0, sterilized by autoclaving

Method

1 Set up the PCR reaction as follows: 1 μl of cDNA template (wild-type receptor contained in a plasmid), 200 pmoles each primer, 10 μl of 10 × PCR buffer, dNTPs (all nucleotides) at 1.25 mM each final, 0.5 μl of *Taq* polymerase, and sterile water to 100 μl final. Overlay with 50 μl of mineral oil. You will have two tubes: one with 1 μl of DNA template and one control PCR that contains everything except the DNA template.

2 PCR conditions:

(a) One cycle: denature at 95 °C for 5 min, anneal at 45 °C for 5 min, extend at 72 °C for 45 min.

(b) PCR regular cycle for 35 cycles at 94 °C for 1 min, 45 °C for 2 min, and 72 °C for 3 min.

3 Analyse PCR products on a 1.2% agarose gel. Weigh out 1.0 g agarose (low EEO) and place into 100 ml of 1 × TAE buffer. Microwave or heat sample until all of the gel fragments have solubilized. Add 5 μl of an ethidium bromide solution for visualization of the DNA, mix, and pour into a casting tray with comb. Allow to solidify for about 1 h, then remove comb. Place hardened gel in running chamber and just cover with 1 × TAE. Add loading buffer to each sample. Run a molecular weight marker and 20 μl of the PCR sample in different lanes.

4 Run gel at an amperage that will not overheat the gel. Run gel until the bromophenol blue in the loading buffer has migrated to the end of the gel. Localize DNA products by placing gel on a UV transilluminator and take a Polaroid picture for documentation. A band at the predicted size based upon the primers should be apparent. You can isolate the DNA insert by gel extraction using a Geneclean (BIO 101) or similar product according to manufacturer's specifications. Resuspend recovered PCR fragment in 20 μl of TE buffer.

5 Manipulation of DNA for expression

5.1 Subcloning the PCR or chimeric fragments

The next step is to restriction digest the PCR or chimeric fragment (*Protocol 3*) to generate the correct sticky ends for insertion with the adjoining WT fragments into an expression plasmid. Most restriction sites do not cut well near the ends of DNA. This is why we placed at least ten bases before the restriction site in the PCR primers to ensure the attachment of the restriction enzyme. These digestions should be carried out overnight since simple 1–2 h digestions are not efficient. While preparing the PCR fragment and vector, the rest of the cDNA that was not used in the PCR amplification and, in our case is represented by a *EcoRI-NaeI* WT fragment, should be prepared at this time. A stock of the WT plasmid can be restriction digested with *EcoRI/NaeI* and the WT fragments run and separated on 1.2% agarose gel and purified by Geneclean. This protocol can also be used to generate the WT fragments for the chimeric receptor in Section 3 (i.e. *EcoRI-XhoII* and *AflIII-NotI*) to be ligated with the cassette.

Protocol 3

Restriction digestion of DNA

Equipment and reagents

- Heating block, oven, or water-bath
- Eppendorf or microcentrifuge tubes (1.5 ml; Fisher)
- Geneclean kit (BIO 101)
- TE buffer: 10 mM Tris pH 8.0, 1 mM EDTA pH 8.0, sterilized by autoclaving
- Restriction enzymes (various vendors; New England Biolabs, Boehringer Mannheim)

Method

1   Set up the digestion reaction according to the restriction enzyme as follows: 20 µl of PCR DNA or plasmids used in the chimerics in TE buffer, 2.5 µl of 10 × digestion buffer (supplied by manufacturer of the enzyme), and 2.5 µl of restriction enzyme. Incubate at 37°C (or required temperature) overnight in a heat block (16–18 h).

2   After digestion, purify the PCR fragment via the Geneclean kit according to manufacturer's instructions. This is needed to remove the small pieces of DNA released by the digestion and to remove the protein and buffers. Resuspend DNA into TE buffer (20 µl).

3   If WT DNA was used in the digestion and a fragment of DNA needs to be isolated, as in the chimeric approach, the DNA is applied to a 1.2% agarose gel and the correct fragment isolated according to *Protocol 2*, step 3.

4   Approximate the concentration of DNA by spectral analysis. Take 1 or 2 µl of the DNA and add to a quartz cuvette with 1 ml of water. Take the absorbance at 260 nm. 1 optical density (OD) is equal to 50 µg/ml of DNA.

A general rule of restriction digestions is never to add restriction enzyme(s) to more than 10% of the total reaction volume. This method prevents 'star activity' in which the high glycerol (used in enzyme stock solutions) content of the reaction mixture (> 10%) alters the specificity of the recognition sequence.

The next step is to prepare the expression vector for the ligation reaction (*Protocol 4*). Most mammalian expression vectors have a simian virus (SV40) origin of replication for reproduction in mammalian cells as well as a mammalian-type promoter (SV40, CMV). Bacterial reproduction can be achieved in the same vector by the incorporation of ColE1 origin and the β-lactamase gene for ampicillin resistance. Most mammalian expression plasmids also have the ability to reproduce (but not express) in bacteria. The plasmid needs to be digested with the same two restriction sites that are at the start and end of the WT cDNA. After digestion, the plasmid is dephosphorylated at the 5' ends to reduce the ability of the vector to ligate to itself during the ligation with the cDNA fragments. The ligation will work without the vector being dephosphorylated but a high level of vector alone (with no cDNA inserted) will be obtained after ligation. In essence, this means a more rigorous searching of the colonies to obtain the correct construct. For transfections into COS-1 cells, we use the plasmid pMT2' which contains unique *Eco*RI and *Not*I restriction sites for subcloning cDNAs, but most commercially available vectors have unique multiple cloning sites for gene insertion.

Protocol 4

Preparation of expression vector for subcloning

Equipment and reagents

- Heating block, oven, or water-bath
- Eppendorf or microcentrifuge tubes (1.5 ml; Fisher)
- Geneclean kit (BIO 101)
- TE buffer: 10 mM Tris pH 8.0, 1 mM EDTA pH 8.0, sterilized by autoclaving
- Restriction enzymes (various vendors; New England Biolabs, Boehringer Mannheim)
- Equilibrated phenol, chloroform (Amersham)
- Calf intestine phosphatase (CIP) (Amersham)

Method

1 Take 10 μg of vector stock such as pMT2'. Add 2.5 μl of 10 × digestion buffer (supplied by the manufacturer of the enzyme) and 2.5 μl total of the restriction enzymes *Eco*RI and *Not*I (or appropriate enzymes for your vector). Bring total volume to 25 μl with sterile water. Mix well by pipetting and incubate at 37 °C (or required temperature) for 2 h.

2 Isolate cut vector DNA by the Geneclean kit. Resuspend DNA in 20 μl TE.

3 To 20 μl of *Eco*RI/*Not*I cut vector, add 2.5 μl of calf intestine phosphatase (CIP) 10 × buffer (supplied by manufacturer) and 2.5 μl of the enzyme CIP. Incubate at 37 °C for

Protocol 4 continued

2–4 h. Alternatively, some restriction enzymes cut very well in the CIP buffer, so both the restriction enzyme and CIP can be digested together. However, separate digestion is always more efficient.

4   Isolate the CIP-treated vector as in step 2. Resuspend the DNA in 30–50 μl of TE. Approximate the DNA concentration as in *Protocol 3*, step 4. This is your cut vector stock.

The next step in the subcloning process is the ligation reaction (*Protocol 5*). For information regarding ligation theory, see the Maniatis cloning manual (3). The basic recipe is to use vector and insert DNA at a molar ratio of 1:3. Therefore, you have more moles of the insert than vector which allows a higher efficiency of ligation. You usually start with 0.1 pmoles of the CIP-treated vector, but this is usually established empirically depending upon how well the vector was digested with the restriction enzyme and dephosphorylated with the CIP. We recommend a test ligation of the vector alone to determine background activity (the amount of self-ligated vector with no insert).

Test ligation: Take 1, 3, and 5 μl of the CIP-treated vector stock and transform into bacteria according to *Protocol 6*. Use the same aliquot in the ligation reaction that gives you some colonies, but under ten, as background. If you have too many colonies, 30 or more with the lowest aliquot, the restriction digest and/or CIP treatment was inefficient and it is best to redo the vector preparation and cut with longer incubation times. If you get no colonies, you probably lost the DNA in the manipulations. This can be verified by running a small aliquot on an agarose gel. Once a good vector stock is made, isolate enough of the PCR fragment to achieve a sufficient stock for ligation at a 1:3 molar ratio.

Protocol 5

Ligation of PCR fragments into an expression plasmid

Equipment and reagents

- Water-bath placed in a cold room or other mechanism to achieve 16 °C
- T4 DNA ligase (Amersham)

- Eppendorf or microcentrifuge tubes (1.5 ml; Fisher)

Method

1   Set up a ligation reaction for each mutated PCR fragment along with one control that contains everything except for the insert DNA to determine the vector background colonies after transformation.

2   To 1 μl or 0.1 pmoles of cut vector stock (or amount that gives you good back-

Protocol 5 continued

ground), add 5–10 μl of digested WT DNA fragment and digested PCR fragment or chimeric oligo cassette (to approximate 1:3 molar ratio), 2 μl of 10 × ligation buffer (supplied with enzyme), 2 μl of T4 DNA ligase, and sterile water to 20 μl. Volume can be adjusted accordingly.

3 Mix and incubate overnight at 16°C (obtained by placing a water-bath in a cold room).

5.2 Identification and propagation of mutants in bacteria

The next step is to transform the ligation products into bacteria for identification and propagation (*Protocol 6*). A good all-around bacteria is DH5α, which accepts large pieces of DNA and does not undergo recombination of inserts. The key steps of transformation are to buy or make good competent cells with a high efficiency and to be gentle in the use of the cells. Always keep them on ice and never vortex. The 42°C heat shock is timed and the timing varies with the type of cells. Do not use a heat block for the heat shock treatment as the heat transfer to plastic tubes is not efficient. The best technique is to use a water-bath or a heated beaker of water.

Protocol 6

Transformation into bacteria

Equipment and reagents

- 15 ml conical sterile centrifuge tubes (polypropylene) (Fisher)
- 42°C water-bath
- Shaking culture incubator
- 37°C incubator
- Table-top centrifuge (T6000, Beckman)
- Glass spreader (bend a glass pipette under flame to achieve an L shape)

- DH5α competent cells (subcloning efficiency or higher; Gibco BRL)
- NZCYM media (Gibco BRL): 22 g/litre of water, sterilized by autoclaving
- Ampicillin plates: NZCYM media plus 16 g of Bacto agar per litre of media, sterilized by autoclaving, and let cool to 55°C. Add 100 mg of ampicillin (Sigma) per litre. Mix and pour into Petri dishes.

Method

1 For each ligation, you need 100 μl of competent cells. Thaw a vial on ice, then aliquot 100 μl into ice-cold 15 ml sterile polypropylene centrifuge tubes, also on ice, for each separate ligation reaction. Keep the cap on to maintain sterility.

2 Add no more than 10 μl of the ligation mixture to the centrifuge tubes. To use up all of the mixture, you can perform two separate transformations on one ligation mixture. Mix slowly by moving the pipette tip back and forth. Do not vortex or pipette up and down. Incubate on ice for 30 min or more.

Protocol 6 continued

3 Heat shock by placing the centrifuge tubes in a 42°C water-bath for 1 min (timed). Do not use a heat block as heat transfer is compromised. Return to the ice for another 2 min.

4 Add 2 ml of sterile NZCYM media (no antibiotic). Incubate with shaking for 1 h at 37°C to allow the cells to develop antibiotic resistance.

5 Concentrate cells by centrifuging tubes for 10 min at top speed in a table-top centrifuge. Pipette off and discard all of the media except for about 200 µl. Resuspend the cell pellet in the 200 µl and spread on ampicillin plates until moisture is absorbed. Invert plates and place in 37°C incubator overnight.

Colonies should appear on the plates with overnight incubation. Usually 10–50 colonies are present, depending upon the efficiency of the competent cells. The background ligation with vector alone should have fewer colonies. The next step is to grow the bacteria from isolated colonies and to purify the plasmid DNA for sequencing to identify the mutant (*Protocol 7*).

Protocol 7

Plasmid preparation for double-stranded DNA sequencing

Equipment and reagents

- 15 ml conical sterile centrifuge tubes (polypropylene) (Fisher)
- Shaking culture incubator
- Sterile toothpicks (place toothpicks in a glass beaker, cover with foil, and autoclave)
- 37°C incubator
- Table-top centrifuge (T6000, Beckman)
- Eppendorf or microcentrifuge tubes (1.5 ml; Fisher)
- Solution 1: 50 mM glucose, 10 mM EDTA, 25 mM Tris–HCl pH 8.0, sterilized or make from sterile components

- Solution 2: 0.2 M NaOH, 1% SDS, no need to sterilize
- Solution 3: 60 ml of 5 M potassium acetate, 11.5 ml glacial acetic acid, and 28.5 ml H_2O
- Microcentrifuge (Eppendorf)
- NZCYM media: see *Protocol 6*
- Ampicillin plates: see *Protocol 6*
- Equilibrated phenol, chloroform (Amersham)
- NZCYM with ampicillin: NZCYM media, sterile, and add 100 mg/litre of ampicillin
- 100% ethanol (Fisher)

Method

1 Pick individual colonies with a sterile toothpick. Touch the toothpick onto an NZCYM ampicillin master plate (gridded and labelled for cataloguing of numerous colonies), and drop the toothpick into a 15 ml centrifuge or culture tube containing 2.5 ml sterile NZCYM with ampicillin.

2 Shake tubes vigorously in a 37°C incubator overnight. Also, incubate the master plate in a 37°C incubator overnight.

3 The next morning, centrifuge culture tubes at 2000 g for 10 min to pellet bacteria. Also, store the master plate at 4 °C for later reference.

4 Aspirate medium and resuspend bacterial cell pellets in 100 μl of solution 1. Transfer resuspended pellets to 1.5 ml Eppendorf tubes.

5 Incubate at room temperature for 5 min and then add 200 μl of solution 2. Mix by inversion—do not vortex.

6 Incubate on ice for 5 min and then add 300 μl of ice-cold solution 3. Vortex gently for 10 sec.

7 Incubate on ice for 5 min and spin for 5 min at maximum speed in a micro-centrifuge. Transfer supernatants to fresh Eppendorf tubes.

8 Phenol/chloroform extract by adding 100 μl each of equilibrated phenol and chloro-form. Vortex samples to completely mix and spin for 3 min at maximum speed in a microcentrifuge. Transfer aqueous phase (upper phase) to fresh tubes.

9 Add 2 vol. of room temperature 100% ethanol to recovered aqueous phase, vortex, incubate at room temperature for 10 min, and spin at maximum speed for 10 min in a microcentrifuge.

10 Decant supernatants and wash pellets with 70% ethanol (200 μl). Drain all fluid from tubes and air dry DNA pellets for 15–30 min at room temperature. Resuspend pellets in 40 μl TE buffer.

Once the plasmid is isolated, the easiest method for sequencing is to use a Center of Research (CORE) or centralized facility since each primer can read 500 base pairs. It is by far the most efficient method in both time and money. However, if a CORE facility is not available at your institute, refer to the Maniatis manual (3) for protocols on sequencing plasmid DNA. Design a sequencing primer about 20 bases downstream of the mutation and also design primers to read the entire coding region of the cDNA since this could have altered during the PCR manipulations and culturing in bacteria. After verification, the plasmid is grown in larger volumes of media (scale up *Protocol 7* to 1 litre) to obtain a large stock of plasmid for subsequent transient transfections. This stock is further purified by using a kit such as Maxiprep from Fisher to remove small amounts of protein and RNA which can interfere with the efficiency of transfections.

6 Expression of mutant receptors

6.1 Culturing of cells

For expression in mammalian cells, we use COS-1 which is a green monkey kidney fibroblast cell line. This cell line has been transformed with the large T antigen of SV40 which also immortalizes the cell line. When transfected with plasmids that carry the SV40 origin of replication, high levels of expression can be achieved transiently. Primary cell lines which are made from the tissue itself, or from other cell lines which are not transformed, can also be transfected. However, their efficiency is very low and they should be transfected with cationic lipid reagents or viruses. The following procedure (*Protocol 8*) describes the growing of COS-1 cells to be used in a transient transfection. Many cell lines can be grown in this manner, but sometimes the media requirements vary from cell line to cell line. Information on the culturing of cells can be obtained from American Type Culture Collection (ATCC).

Protocol 8

Culturing of COS-1 cells

Equipment and reagents

- Tissue culture fume hood
- 20 μm sterilizing filters (Zapcaps; Schleter and Schuell)
- Sterile plastic pipettes (Fisher)
- Tissue culture flasks (T25, T125; Fisher)
- 5% humidified CO_2 incubator
- Water-bath, 37 °C
- Tissue culture microscope (Olympus)
- –80 °C freezer and liquid nitrogen storage Dewar
- Dulbecco's modified Eagle's medium (DMEM) with glutamine (Biowhitaker)
- Nunc tubes for freezing (Fisher)
- HBSS (Hank's balanced salt solution; Biowhitaker)
- Trypsin (Biowhitaker)
- Fetal bovine serum (Biowhitaker)
- Penicillin/streptomycin: 100 × stock (Biowhitaker)
- COS-1 or COS-7 cells, frozen stock (American Type Tissue Culture Collection; ATCC)
- Dimethylsulfoxide (DMSO); tissue culture grade (Sigma)

Method

1. Prepare complete culture medium by filtering the serum into the DMEM (with a Zapcap) to make a final concentration of 10% fetal bovine serum and 1 × of the penicillin/streptomycin solution. This must be done in the tissue culture hood and this is termed complete medium. When solution is to be used, warm media to 37 °C in a water-bath; placing cold media onto cells can shock them.

2. In a tissue culture fume hood, thaw the frozen stock of COS from ATCC by rolling between your fingers. Score and break glass vial using ethanol-coated paper towels. With a sterile pipette, remove liquid and place into a T25 flask which contains enough of the complete medium to cover the bottom by 0.5 cm. Place in a humidified 5% CO_2 incubator.

3. Each day examine the flask in a tissue culture microscope. Cells should be attached to the bottom of the flask and appear 'boxy' in appearance. Cells that stay floating after several days are dead. Every three or four days remove the old medium in the hood and replace with fresh complete medium until the cells become confluent.

Protocol 8 continued

4 To split the cells, remove old medium and add 5–10 ml of HBSS. Swirl to wash off the serum and remove. Add 1–2 ml of the warmed trypsin solution and place flask in the CO_2 incubator for 10 min. Remove flask and add complete medium and pipette it over the flask surface to remove cells. Divide cells into three other T25 flasks or into one T125. Add enough complete medium to cover the cells. In this way, cells are divided 1:3 until enough cells are maintained for experiments. A confluent T25 is about 10^6 cells while a T125 represents 10^7 cells.

5 To freeze cells for storage and to maintain a frozen stock, resuspend cells in a minimum of complete medium after trypsinization. To each Nunc tube, add 200 µl of DMSO, 800 µl of sterile fetal bovine serum, and the concentrated cell suspension to 1.8 ml. Mix and freeze overnight in a $-80\,^{\circ}C$ freezer. The next day, place the frozen vials into a liquid nitrogen storage Dewar.

Tissue culture medium comes with a red indicator to measure pH. When cells need the medium to be changed, the solution turns orange to yellow indicating a lower pH. Never let the medium turn yellow but change at the orange stage. During cell splitting, if the trypsin solution does not appear to be working well, a common problem is the incomplete washing of the cells with HBSS. The large amount of serum in the complete medium interferes with trypsin by occupying all of its active sites. All of the solutions needed for tissue culture can be bought or made from powder which is substantially cheaper than buying prepared media.

The key to good cell culturing is sterile technique. Wash hands prior to use and wear gloves. Maintain caps on media bottles while in use and be generous with the use of sterile pipettes. Contamination is evidenced by a cloudy suspension in the media which under microscopic examination indicates either bacteria or yeast. Any signs of contamination should be dealt with swiftly by disposing of media stocks and the cell lines. It is best to start from scratch and prepare fresh media and cell stocks instead of trying to discover the source of the contamination. Incubators should be wiped down with ethanol and other disinfectants (quatsyl) as should the culture hood before and after each use.

When adding solutions to the cells, touch the tip of the pipette to the inside rim of the tissue culture plate and release slowly. Do not add medium directly to the surface of the cells as this may remove them from the plastic bottom. When removing media, set up a vacuum-aspiration flask with a hose that can reach into the culture hood. The tip of the hose can then be inserted with a 10 ml sterile pipette or a sterile transfer pipette. Aspiration is done by placing the tip of the aspiration pipette to a bottom corner of the culture plate and tipping the plate to direct the medium. At all times, keep the cover on the plate and only open the cover enough to insert the pipette.

6.2 Transient transfection

In studies where a high level of expression is needed, such as in binding studies, transient transfections provide large amounts of the needed membranes. Signalling studies can also be done in a transient system, but the receptor number has to be titred to levels similar to those of a control so as to avoid artefacts due to differences in receptor number. In many cases, one can generate stable cell lines by performing a transient transfection and selecting clones based upon an ability to survive in media containing a chemical in which resistance is imparted by virtue of a protein expressed on the same plasmid as the receptor. In this way, many clones are isolated and represent different receptor expression levels. The transient system has the advantage of being fast. The stable system, although initially slower to make, has the advantage if many studies are to be performed with the mutant cell line and saves time by avoiding a transfection with each experiment. A constant level of receptor expression can also be achieved.

Most of our studies use a transient transfection for several reasons based upon speed and receptor expression. The most important of these is that many mutations may not express in high numbers, making analysis difficult and the use of stable cell lines impractical. Most of our studies are transient and once the mutation is made and analysed, it is not used again. Perhaps in situations where an interesting mutation is found, a stable cell line is worth the investment. Detailed below is a procedure to transfect cells transiently via the diethylaminoethane (DEAE)–dextran method (*Protocol 9*). As discussed in Section 6.1, primary and non-transformed cell lines should be transfected *via* cationic lipids.

Protocol 9

DEAE–dextran transfection

Equipment and reagents

- Tissue culture fume hood
- Tissue culture flasks and plates (Fisher)
- HBSS (Hank's balanced salt solution; Biowhitaker)
- DMEM complete media without serum (*Protocol 8*)
- Diethylaminoethane(DEAE)–dextran: 2.5 mg/ml of DEAE–dextran (Amersham) in DMEM (no serum) and sterile filtered

- DMEM complete media (*Protocol 8*)
- 1 M Tris pH 7.3, sterile filtered or autoclaved
- 1 mM chloroquine stock, in water and sterile filtered, store in foil-covered bottle at 4 °C
- 10% dimethylsulfoxide (DMSO) made in HBSS, sterile filtered

Method

1 Two days or so before the experiment, split the COS cells into 100 mm tissue culture dishes. The cells should be about 50–75% confluent at the time of transfection.

2 While solutions are being warmed to 37 °C, a master mix of DNA cocktail is prepared by mixing in a sterile centrifuge tube the following in order:

Protocol 9 continued

    (a) DNA (8 μg/6-7 × 10^6 cells).

    (b) 4.8 ml of 1 M Tris pH 7.3.

    (c) 38.6 ml of DMEM (without serum).

    (d) 4.8 ml of DEAE-dextran in DMEM. This amount is for a 12 × 100 mm plate transfection.

3 Remove medium from cells. Add 5 ml of HBSS, swirl, and aspirate. Repeat.

4 Add 4 ml of the DNA cocktail per plate. Incubate for 5-6 h. Swirl the plates every couple of hours.

5 Prepare DMEM complete with 0.1 mM chloroquine solution. Dilute the chloroquine from the 1 mM stock. 5 ml are needed per transfection plate.

6 After incubation, aspirate the DNA cocktail. Rinse plates with DMEM (no serum), once only.

7 DMSO shock: add 2 ml of 10% DMSO in HBSS for 3 min at room temperature. Time this reaction and perform on the plates sequentially. Aspirate the DMSO.

8 Add medium from step 5 at this point. Incubate for 2 h.

9 Aspirate medium. Rinse plates twice with 7 ml of DMEM (no serum).

10 Add 10 ml of complete media.

11 Harvest after 48-72 h. If medium starts to turn orange during this time, feed with complete medium.

The science behind transfection involves DNA which is negatively charged due to its phosphate backbone and its binding to the DEAE-dextran which is positively charged. This is why the complex is formed without the presence of serum because of its interference in complex formation. This complex then interacts with the cell membranes which is also electronegative. The DMSO shock allows the cell to internalize the complex and the chloroquine inhibits DNA breakdown which can occur in the lysosomes. The chloroquine inhibits microtubule development which is needed in lysosome vesicle formation. The DMSO shock is also toxic to the cells, this is why it must be timed: a certain proportion of the cells will die.

The transient transfection must be harvested for membranes, or used in cell signalling studies, within the 48-72 h. During cell division, the plasmid is lost to daughter cells since there is no selection pressure to keep the plasmid. Other common problems encountered in transfections are the use of serum in the DNA cocktail or the use of water in the DMSO stock. At all times, cell solutions should be buffered and osmotically balanced or the cells are disrupted. This procedure can be scaled up to transfect multiple 150 mm plates. Receptor number can be varied by altering the amount of DNA added in the cocktail. However, this variation is not linear or proportional and must be determined empirically.

7 Ligand binding

After the cells have been transfected, they are either analysed for their binding properties and/or their ability to signal is measured. It is hard to be comprehensive about all of the techniques and the theory behind ligand binding. It is recommended that the reader reviews the methodology in *Current protocols in pharmacology* (4). However, the protocol below provides a brief experimental outline to make membranes, perform a saturation binding assay to determine receptor density as well as affinity for the radioligand, and then subsequent competitive binding to determine the affinity of a number of compounds in the analysis of binding domains.

7.1 Membrane preparation

There are many different protocols for the preparation of membranes. Ours as outlined below (*Protocol 10*) is a little more comprehensive by using multiple washing steps to remove DNA and nuclear debris which can interfere with binding. This provides a yield of membranes which results in low non-specific binding. This protocol can be used with any cell type but needs modification with whole tissue. Since membranes are being prepared, this procedure can be done on the benchtop with non-sterile solutions.

Protocol 10

Preparation of COS cell membranes

Equipment and reagents

- Cell scraper or lifter (Fisher)
- Glass douncer (Fisher)
- High speed centrifuge and rotor
- Table-top centrifuge (Sorvall RT6000)
- Hank's balanced salt solution (HBBS), non-sterile
- 0.25 M sucrose

- Protease inhibitors: leupeptin (1 mg/ml); PMSF (1.7 mg/ml in ethanol); bacitracin (20 mg/ml); benzamidine (20 mg/ml)
- HEM buffer: 20 mM Hepes pH 7.4, 12.5 mM $MgCl_2$, and 1.5 mM EGTA
- Bradford method of protein determination (Bio-Rad kit)

Method

1 Remove medium from culture plates. Wash with 2×5 ml of HBSS. Add 1 ml of cold HBSS and scrape cells. Collect in a 50 ml conical centrifuge tube.

2 Add an additional 1 ml of cold HBSS, re-scrape, and collect cells.

3 Centrifuge for 5 min at 300 g (1200 r.p.m. in Sorvall RT6000, 4°C, setting 4). Resuspend pellet in 5 ml of cold 0.25 M sucrose. Centrifuge for 5 min at 300 g, 4°C.

4 Resuspend pellet in 10 ml of water and add: 40 µl of leupeptin, 40 µl of PMSF, 20 µl of bacitracin, and 20 µl of benzamidine. Freeze at −70°C for 30 min. Thaw on ice.

5 Dounce ten times with a 'B' or tight-fit pestle. Turn the pestle as you stroke. Keep sample and pestle in ice.

6 Centrifuge at 1260 g, 4°C to precipitate nuclei. Collect supernatant and fill to 35 ml with cold HEM buffer. Centrifuge at 30 000 g for 15 min, 4°C.

7 Resuspend pellet in cold HEM and fill to 30 ml. Centrifuge at 30 000 g for 15 min.

8 Resuspend pellet in 4 ml of HEM with 10% glycerol. Determine protein concentration with the Bradford method (5). Aliquot 50–200 μl per vial and store at −70°C.

As with any protein purification protocol, the membranes and solutions should be ice-cold and centrifugation steps carried out at 4°C. Protein concentration is typically near 1 mg/ml. Wild-type receptor membranes can generally be frozen for up to one year without any loss of receptor density. However, different mutants behave differently and can be quite unstable. With constitutively active mutants, receptor density lasts only for a couple of days and the receptor degrades or is internalized. In these cases, it is best to isolate the membranes and perform the binding immediately.

7.2 Saturation binding

Before determination of affinity or signalling changes, saturation analysis should be performed to measure the extent of receptor expression and any changes in the radioligand affinity (*Protocol 11*). This needs to be performed first because affinity calculations for competing drugs (K_i) uses the K_D measurements of the radioligand. In signalling studies, the affinity of the compound also determines the range used in the dose–response curve and, if the mutations are not expressing, these studies may be in vain.

Protocol 11

Radioligand binding isotherms at α_1-adrenergic receptors

Equipment and reagents

- Polypropylene tubes (12 × 75 mm)
- Polystyrene tubes (12 × 75 mm)
- Ice bath
- Shaking water-bath with adjustable temperature and speed controls
- Brandel cell harvester and vacuum pump
- Glass fibre filters (Grade GF/C; Brandel)
- Appropriate lead shielding and dedicated radioactive disposal means for solid and liquid waste
- Data analysis software package (GraphPad Prism 3.0)
- Gamma counter
- [^{125}I]-HEAT, 2-[β-(4-hydroxyl-3-[^{125}I]-iodophenyl)ethylaminomethyl]tetralone (NEN Life Science Products)
- HEM wash buffer: 20 mM Hepes pH 7.4, 1.4 mM EGTA, 12.5 mM MgCl$_2$
- HEM incubation buffer: 20 mM Hepes pH 7.4, 1.4 mM EGTA, 12.5 mM MgCl$_2$, containing 0.1% (w/v) BSA (Sigma, Fraction V grade)
- Phentolamine (Sigma)
- Polyethylenimine (Sigma)

Protocol 11 continued

Method

1 Label polypropylene tubes accordingly and place in incubation racks.[a] Prepare an ice bath and conduct all further steps on ice; each data point or concentration will require four tubes.

2 Determine the activity of the radioligand [^{125}I]-HEAT on the day of the experiment to account for radioactive decay. Make appropriate dilution of the stock solution to generate final radioligand concentrations that range from 10-fold below to at least 100-fold above the theoretical affinity of the radioligand for the receptor.[b] All dilutions should be made in incubation buffer.

3 For determining the non-specific binding of radioligand make a stock solution of 100 μM phentolamine.

4 Thaw cell membranes expressing α_1-adrenceptors on ice. The amount of protein to be added to each tube depends on the expression level of the receptor. For membranes prepared from transfected cell lines, the amount of membrane protein used in saturation binding assays is approximately 5–10 μg/tube. Prepare a stock membrane solution such that a 25 μl volume contains 5–10 μg membrane protein.

5 Set up the incubation reaction as follows.[c] Add in succession, 25 μl of radioligand dilution to the respective total and non-specific binding tubes, 25 μl of phentolamine to all non-specific binding tubes only, and 25 μl of membrane to all tubes. Bring the final assay volume in all tubes to 250 μl by adding incubation buffer (200 μl to all total binding tubes and 175 μl to all non-specific binding tubes). A sample of a saturation set-up is illustrated below.

Tube numbers	Membranes	Phentolamine	[^{125}I]-HEAT ($K_D \approx 90$ pM)	HEM buffer
1, 2 (total binding)	25 μl	0 μl	25 μl (10 pM final)	200 μl
3, 4 (non-specific)	25 μl	25 μl	25 μl (10 pM final)	175 μl
5, 6 (total)	25 μl	0 μl	25 μl (50 pM final)	200 μl
7, 8 (non-specific)	25 μl	25 μl	25 μl (50 pM final)	175 μl

6 Vortex all tubes and transfer racks to a shaking water-bath equilibrated at 22 °C. Incubate reactions with continuous shaking at moderate speed for 1 h.

7 To determine the exact concentrations of radioligand added to the assay, pipette 25 μl of each radioligand concentration into labelled polystyrene tubes in duplicate. Dispose of the remaining volume of radioligand dilutions appropriately.

8 Prepare the cell harvester for the separation procedure by washing the lines with ice-cold HEM wash buffer. Also soak the glass fibre filter in a 0.3% (w/v) polyethylenimine solution for 2 min, then transfer filter to soak in a 0.1% BSA solution for a further 2 min.

9 At the end of the 1 h incubation, separate bound and free radioactivity by vacuum filtration through a pre-treated glass fibre filter. Wash the filters with a further

4-tube volume of ice-cold wash buffer. Remove the vacuum and transfer the glass
fibre discs to labelled polystyrene tubes.

10 Determine the residual radioactivity on the glass fibre filter discs using a gamma
counter.

11 Determine specific binding for each ligand concentration by subtracting the mean
non-specific value from the mean total binding value. The radioligand affinity (K_D)
and receptor density (B_{max}) can then be determined by analysis of the specific binding values using an appropriate data analysis software application (e.g. GraphPad
Prism 3.0).

[a] In a typical experiment nine different concentrations of radioligand will be used, and for each
concentration, measurements of both total and non-specific binding will be made in duplicate.

[b] Remember to consider the further 10-fold dilution of the radioligand concentration in the
tube when calculating the final concentration in the assay.

[c] Appropriate precautions for working with radioactivity should be implemented including use
of lead shielding and dedicated disposal of contaminated pipette tips and tubes.

Saturation binding is performed under equilibrium conditions, therefore, the
incubation must continue until this condition is fulfilled. With a different radioligand or temperature, the time may vary but is usually within 30 min to 2 h.
Generally, we use [125I]-HEAT for measuring α_1-adrenergic receptors. However,
[3H]-prazosin can also be used and is preferred by many pharmacologists because
of the long half-life of tritium. However, one must also consider the specific
activity which is at least 30 times higher with 125I. Therefore, [125I]-HEAT is more
sensitive and can generally detect receptor densities as low as 10–50 fmoles.
Using [3H]-prazosin for this level of receptor density is highly error-prone and not
recommended. For every 1 d.p.m. of specific binding with [3H]-prazosin, there
are 30 d.p.m. with [125I]-HEAT. Wild-type α_1-adrenergic receptors generally express
around 1 pmole/mg membrane protein. Mutations at the receptor vary from
wild-type to as low as 50 fmoles/mg

A common mistake in saturation binding and the subsequent Scatchard
analysis used in Graphpad prism or other data analysis programs, is the incorporation of too much membrane such that the expressed receptor binds too
much of the radioligand. An assumption of the Scatchard analysis is that most of
the radioligand remains free, commonly taken as less than 10% of the radiolabel
is bound. If more than 10% of the radiolabel is bound, then the calculation of K_D
and B_{max} is altered (conversion plot is shifted to the right) leading to inaccurate
estimates. Commonly, a pilot experiment is carried out to determine the
amount of membranes needed to be under the 10% bound rule.

7.3 Competitive binding

Saturation binding measures the receptor density and K_D of the radioligand.
To measure the affinity of non-radiolabelled compounds, competitive binding

experiments are used (*Protocol 12*). The K_i calculated by the Cheng–Prusoff method (6) needs to incorporate the true affinity of the radiolabel. Therefore, if this changes because of the mutation, the actual K_D measured should be used in this equation.

Protocol 12

Competition binding isotherms at α_1-adrenergic receptors

Equipment and reagents

- See *Protocol 11*

Method

1. Label polypropylene tubes in duplicate for total, non-specific, and at least 10–12 increasing concentrations of competing drug. Prepare an ice bath and conduct all further steps of the assay on ice.

2. Determine the activity of the radioligand [^{125}I]-HEAT on the day of the experiment to account for radioactive decay. To optimize specific binding, prepare a solution such that the final concentration of [^{125}I]-HEAT added to each assay tube is equivalent to its K_D value for the α_1-adrenoceptor subtypes (ca. 90 pM).

3. Prepare dilutions of the cold competing ligand, e.g. epinephrine. The concentration range should span at least two orders of magnitude above and below its theoretical affinity in order to effect maximal displacement of the radioligand. Remember to factor in the further dilution of the cold ligand upon addition to the incubation tube. If the affinity of the ligand is unknown, pilot experiments with only 10 dilutions but at log values apart are performed in order to determine the range of affinity.

4. Prepare a stock solution of membrane such that a 100 μl volume contains approx. 5–10 μg membrane protein.

5. Set up the incubation reaction as follows. Add 100 μl of the radioligand solution to all tubes. Add 50 μl of incubation buffer to the total binding tubes and 50 μl of phentolamine (50 μM) to the non-specific binding tubes (final conc. 10 μM). Proceeding from the lowest to highest concentration, pipette 50 μl of epinephrine in duplicate. Finally, add 100 μl of the membrane suspension to all tubes. A set-up experiment is shown below.

Tube numbers	Membranes	Phentola-mine	[^{125}I]-HEAT ($K_D \approx 90$ pM)	HEM buffer
1, 2 (total binding)	100 μl	0 μl	100 μl (90 pM final)	50 μl
3, 4 (non-specific)	100 μl	50 μl	100 μl (90 pM final)	0 μl
5, 6 (epi, 10^{-9} M, final)	100 μl of 10^{-10} M	0 μl	100 μl (90 pM final)	0 μl
7, 8 (epi, 10^{-8} M, final)	100 μl of 10^{-9} M	0 μl	100 μl (90 pM final)	0 μl

epi: epinephrine.

6 Vortex all tubes and transfer racks to a shaking water-bath equilibrated at 22 °C. Incubate reactions with continuous shaking at moderate speed for 1 h.

7 Add 100 μl of the radioligand solution in duplicate to polystyrene tubes to determine accurately the concentration of [^{125}I]-HEAT added to the assay tubes and count with the experimental tubes. Complete the competition binding experiment by repeating *Protocol 11*, steps 8–10.

8 Analyse the raw data using a data analysis software application (e.g. GraphPad Prism). Plot the specific binding of [^{125}I]-HEAT against the concentration of epinephrine. The IC$_{50}$ value determined by the competition reaction can be used to calculate the affinity of epinephrine binding to the α$_1$-adrenoceptors by using the Cheng–Prusoff equation.

8 Second messenger assays

Some mutations may affect binding, signalling, or both. In general, mutations are analysed first for binding changes then for signalling changes. α$_1$-Adrenergic receptors predominately signal through the Gq family of G proteins to stimulate phospholipase C (PLC). This in turn activates the release of diacylglycerol (DAG) and inositol (1,3,5) triphosphate (IP$_3$). DAG activates the protein kinase C pathway (PKC) and IP$_3$ activates the Ca^{2+} release pathway. Although there are many more signalling pathways involved, some of them are known not to be independent from each another and may derive from the PKC or Ca^{2+} messengers. Since IP$_3$ is the predominate pathway which activates all three α$_1$-adrenoceptor subtypes, this assay is described below and can be used in any cell type. It should be noted that signalling studies are highly dependent on receptor number. Therefore, controls and mutated receptors should be titred in the transfection to produce equal receptor density by the Scatchard analysis. This is accomplished by altering the amount of cDNA added in the transfection protocol.

8.1 Total inositol phosphates

Total inositol phosphates (IP) are a combination of three different forms of IPs. After stimulation of PLC and the formation of IP$_3$, IP$_3$ is broken down into IP$_2$ and IP (monophosphate form) through phosphatases, with the rate-limiting step at the IP conversion inhibited by LiCl. Therefore, the concentration of all three forms increases as the metabolic pathway is inhibited and this makes their detection easier. In order to measure only the IP$_3$ form, we recommend the use of a radioassay kit from Dupont/NEN as the total IP assay is usually not sensitive enough. To measure the IP$_3$ assay via the total IP assay (*Protocol 13*), sequentially elute the three different forms of IP using 0.2 M ammonium formate in the formate buffer, followed by 0.4 M ammonium formate, and finally by the 1 M ammonium formate. The stronger the salt, the more phosphate-containing compounds will elute.

Protocol 13

Inositol phosphate assays in COS-1 cells

Equipment and reagents

- Table-top centrifuge (Sorvall RT6000)
- Disposable chromatography columns (Bio-Rad poly-prep)
- β-counter
- AGI-X8 resin, formate form, 100–200 dry mesh size (Bio-Rad)
- [^{3}H]inositol (aqueous; Dupont/New England Nuclear)
- DMEM complete (no serum)
- 1 M LiCl (sterile filtered)
- 0.4 M perchloric acid
- 0.72 M KOH/0.6 M KHCO$_3$
- 0.1 M formic acid
- 0.1 M formic acid/1 M ammonium formate
- Ecosint A liquid scintillation fluid (National Diagnostics)

Method

1 Divide COS-1 cells into 60 mm dishes and transfect according to *Protocol 9*. On the day before analysis (usually day 2 post-transfection), add [^{3}H]inositol at 1 μCi/ml of medium. Swirl to mix and incubate overnight.

2 Aspirate radioactive media into an appropriate container for radioactive waste. Wash cells with 3–4 ml of DMEM complete with no serum. Aspirate into radioactive waste.

3 Add 3 ml of DMEM complete with no serum and LiCl to a final concentration of 10 mM. Incubate 20 min at 37 °C.

4 Prepare dilutions of the agonists and/or antagonists to be used in the assay. If adding an antagonist, this can be added at the same time as the LiCl. Agonists should be added after the 20 min incubation. Keep in mind the final dilution factor of the drug when incorporated into the medium. Add agonist or drug to the cells and incubate for an additional 45 min at 37 °C. Agonists can be incubated longer.

5 Pour off media into radioactive waste. Add 1 ml of cold 0.4 M perchloric acid. Scrape dish with a cell lifter and transfer into a test-tube.

6 Add 0.5 ml of 0.72 M KOH/0.6 M KHCO$_3$ to each tube. You will see a colour change and bubbling. Mix and centrifuge in a table-top centrifuge at 2000 g for 10 min to precipitate the salts. These must be removed or they will interfere with the binding of the IPs to the resin column.

7 While centrifuging, prepare resin columns. Pack poly-prep columns to 1 ml with AGI-X8 resin and equilibrated with 0.1 M formic acid.

8 Transfer supernatant to the pre-packed resin columns. Wash columns by adding 4 ml of 0.1 M formic acid. If bubbling occurs, bang down the columns to promote flow through. Repeat wash.

9 Elute total inositol phosphates into scintillation vials by adding 2 ml of 0.1 M formic acid/1 M ammonium formate. Repeat elution in a second set of scintillation vials. These are called fractions 1 and 2.

10 Add 18 ml of Ecoscint A or other water soluble scintillant to each vial and count in a β-counter. Add c.p.m.s of fraction 1 and 2 together to calculate total IP release.

11 Results can be plotted as total release *versus* concentration of the drug and analysed through a graphics software package such as Graphpad Prism 3.0.

The assay can be optimized by altering the amount of tritium, increasing the amount to increase incorporation and subsequent release. The incubation times with the agonists can be altered and optimized; this is especially true if using the IP_3 kit from Dupont as we have found that a 10 min incubation was optimal. Optimizing the neutralization of the solution (the salt precipitation step 6) will increase the proportion of IPs that bind to the column. For efficient counting, make sure that the scintillant used can solubilize a high content of water.

9 Interpretation

In the analysis of binding, the mutation can affect either the affinity of the radio-label and/or the affinity of certain classes of ligands (agonists versus antagonists; or different chemical moieties). If the mutation affects the affinity of all ligands tested, the changes are probably due to global changes in conformation and misfolding. These types of results are difficult to interpret. In this particular case, the mutation can also be assessed in its signalling to see if it is also impaired. It is recommended to mutate another area of the receptor structure that is not so sensitive to mutation. On the other hand, many mutations affect just antagonist binding or agonist binding and may be particular to the chemical structure of the drug. In this scenario, one can argue that the global conformation has not changed since all drugs have not been altered and the effects are localized. Therefore, many different types of drugs should be tested with this mutation to demonstrate its specificity. Also, the mutation can be altered to enhance or inhibit the effect as additional proof, such as promoting hydrogen bonding or hydrophobic interaction.

In signalling studies, two parameters are often measured:

(a) The potency of the drug as determined by its EC_{50} or half-effective concentration response, or IC_{50}, in the case of antagonists.

(b) The efficacy of the drug or its ability to evoke a maximum response as determined by a E_{max} value.

Both are relative and are highly dependent upon the receptor number. This is why, in the signalling studies, receptor number should be titred to maintain equal receptor density between the control and the mutated receptor. The efficacy is usually calculated relative to a known full agonist such as epinephrine or norepinephrine and this is referred to as intrinsic activity (0–1.0). Relative efficacy is measured by accounting for its potency or EC_{50} since two drugs may cause the same output of message but one drug may do it at a lower concentration, therefore having a higher relative efficacy. Mutations can affect either or both the potency or efficacy. Potency changes that are proportional to changes in agonist binding are generally taken to be due to changes in binding since occupancy of the agonist correlates with the change in signalling. If they do not correlate, then arguments can be made as to their role in function. Any change in efficacy can be used to argue for a change in the signal transduction process. If the potency changes correlate with the binding changes but the efficacy is also reduced, it can be argued that these amino acids are involved in the signal transduction process.

References

1. Zhao, M.-M., Gaivin, R. J., and Perez, D. M. (1998). *Mol. Pharmacol.*, **53**, 524.
2. Wess, J., Blin, N., Yun, J., Shoneberg, T., and Liu, J. (1996). In *Structure and function of 7TM receptors* (ed. W. Schwartz, S. A. Hjorth, and J. S. Kastrup). Munksgaard, Copenhagen.
3. Sambrook, I., Fritsch, E. F., and Maniatis, T. (ed.) (1989). *Molecular cloning: a laboratory manual*, 2nd edn, pp. 1.53–1.73 and pp. 13.1–13.95. Cold Spring Harbor Laboratory, Cold Spring Harbor, NY.
4. Enna, S. J., Williams, M., Ferkany, J. W., Kenakin, T., Porsolt, R. D., and Sullivan, J. P. (ed.) (1998). *Current protocols in pharmacology*, pp. 1.0.1–1.5.17. John Wiley & Sons, Inc., NY.
5. Bradford, M. M. (1976). *Anal. Biochem.*, **72**, 248.
6. Cheng, Y. and Prusoff, W. H. (1973). *Biochem. Pharmacol.*, **22**, 3099.

Chapter 4

Determination of the subunit composition of native and cloned NMDA receptors

F. Anne Stephenson

Department of Pharmaceutical and Biological Chemistry, School of Pharmacy, University of London, 29/39 Brunswick Square, London WC1N 1AX, UK.

Lynda M. Hawkins

National Institute for Deafness and Communication Disorders (NIDCD), NIH, Building 50, Room 4146, 50 South Drive, Bethesda, MD 20892-4162, USA.

1 Introduction

N-Methyl-D-aspartate (NMDA) receptors are a subtype of the excitatory L-glutamate neurotransmitter receptor family. They are fast-acting cation channels with a high permeability for Ca^{2+} that are activated by the binding of both L-glutamate and the co-agonist, glycine. Six genes encode homologous NMDA receptor subunits. They are the NR1 subunit gene which undergoes extensive alternative splicing to yield eight variants, NR1-1a, -1b to NR1-4a, -4b; the NR2 subunit genes NR2A–NR2D, and the NR3A subunit gene. Expression of different NMDA receptor subunit combinations in mammalian cells or in *Xenopus* oocytes has shown that functional NMDA receptors are formed only when the NR1 subunit is co-expressed with an NR2 subunit and that NR1/NR2 NMDA receptor subpopulations have different pharmacological and biophysical properties (reviewed in refs 1 and 2). Although correlations can be made between the properties of cloned NMDA receptors versus native receptors, expression studies do not yield definitive information with regard to the subunit composition and subunit ratios of native NMDA receptors. Since the different NMDA receptor subunits share a high degree of amino acid sequence homology, it is necessary to generate research tools which discriminate each NMDA receptor NR1 splice variant and NR2 subunit in order to determine *in vivo* subunit combinations.

We have pioneered the use of anti-peptide antibodies to create such research reagents to determine the subunit permutations of both native NMDA and $GABA_A$ receptors (e.g. 3–6). More recently, we have developed a different strategy for the elucidation of the NR1 and NR2 subunit copy number per receptor oligomer using epitope tagged NMDA receptor clones (7). This chapter will describe the protocols in current use in the laboratory for the generation of anti-peptide antibodies which are NMDA receptor subunit-specific, their use in the determination of the subunit composition of native NMDA receptors via immunoaffinity purification and immunoprecipitation studies and, lastly, the use of epitope tagging to determine NMDA receptor subunit stoichiometry.

2 Production and characterization of amino acid sequence-directed anti-NMDA receptor subunit antibodies

For the production of amino acid sequence-directed anti-NMDA receptor subunit antibodies, the antigen is a synthetic peptide whose primary structure corresponds to a unique sequence within each NMDA receptor subunit. This peptide needs to be at least eight amino acids in length but in practice, it is usually about 20 amino acids with at least 75% purity. The peptide itself is unable to elicit an immune response, and so it is coupled to a carrier protein and polyclonal antibodies raised against the peptide–protein conjugate. Anti-carrier protein antibodies can be readily removed by the appropriate peptide affinity column purification of the resultant immune serum yielding affinity-purified anti-peptide antibodies. It is relatively easy to generate anti-peptide antibodies but there is no guarantee that these antibodies will recognize the protein from which their sequence is derived. The antigenic determinants of a protein can be predicted, thus increasing the probability that anti-peptide antibodies will recognize the protein congener. However, these regions are often inappropriate since the antigenic amino acid sequences may be conserved across all subunits. Years of experience in the production of anti-peptide antibodies directed against subunits of ligand-gated ion channels have resulted in the following check-list for increasing success in anti-peptide antibody production.

For subunit-specific antibodies:-

(a) The selected peptide, ~ 18–25 amino acids, should be from a divergent region of the deduced subunit sequence as close as possible to either the N- or the C-terminus of the subunit. Experience has shown that these antibodies have a high probability of reacting with both denatured and native conformations of the subunits thus they are probably readily accessible on the surface of the assembled receptors.

(b) The peptide should have a net charge at pH 7.4.

(c) The peptide sequence should be such that it ensures chemical coupling with the carrier protein at a single unique site which is either the N- or C-terminal amino acid. Peptides can be covalently coupled to proteins by four different methods: via primary amine groups at the N-terminus or in lysine residues, carboxyl groups at the C-terminus or in aspartate and glutamate residues, or via tyrosine, or cysteine residues. The preferred method of covalent attachment of selected peptide sequence to the carrier protein is via cysteine residues, since they occur relatively infrequently in proteins. Provided there are no internal cysteines, one is added to the end of the peptide that most closely mimics the sequence within the native NMDA receptor subunit.

(d) The peptide should not contain consensus sequences for N-glycosylation or phosphorylation.

(e) The peptide affinity columns should be synthesized to mimic the coupling between peptide and carrier protein, thus for cysteine-coupled peptides this would be via the thiol.

(f) The subunit-specificity of the antibodies should be verified. Ideally this involves the expression of each individual NMDA receptor subunit in mammalian cells followed by analysis of the expressed polypeptides by either immuno-blotting or immunocytochemical methods. For example, anti-NR2A subunit-specific antibodies should only recognize the expressed NR2A subunit with the correct molecular weight of $M_r \sim 180\,000$

Table 1 Examples of peptides which have been used for successful anti-NMDA receptor subunit-specific antibody production

Peptide sequence	Comments	Application[a]
NR1 35–53 Cys TRKHEQMFREAVNQANKRHC	An N-terminal sequence which is common to all splice forms of the NR1 subunit + C-terminal cysteine.	IB ✓ IP ✓ IC ✓
NR1 C2 LQLCSRHRES	A C-terminal sequence present in NR1-1a, 1b and NR1-2a, 2b which was attached to carrier protein via the N-terminal leucine.	IB ✓ IC ✓
NR1 C2′ Cys QYHPTDITGPLNLSDPSVS	A C-terminal sequence which is common to NR1-3a, 3b and NR1-4a, b with N-terminal glutaraldehyde coupling.	IB ✓ IP ✓ IC nt
NR1 N1 Cys SKKRNYENLDQLSYDNKRGPKC	The amino acid sequence of the N1 insert present in all b forms of NR1 subunits with a C-terminal cysteine.	IB ✓ IP nt IC nt
NR2A Lys 1381–1394 KRCPSDPYKHSLPSQ	A unique NR2A sequence close to the C-terminus with an extra lysine residue for N-terminal coupling to the carrier protein.	IB ✓ IP ✓ IC ✓
NR2B 46–60 Cys DEVAIKDAHEKDDKHC	An NR2B unique N-terminal sequence + a C-terminal cysteine.	IB ✓ IP ✓ IC ✓
Cys NR2C 20–34 CSGPLQTQARTRLTSQ	A unique NR2C N-terminal sequence coupled via an added N-terminal cysteine.	IB ✓ IP nt IC nt
NR2C 1227–1237 WRRVSSLESEV	A C-terminal NR2C sequence with homology to NR2D (see below) coupled via the N-terminus using the glutaraldehyde method.	IB ✓ IP nt IC nt
NR2D 1307–1323 LGTRRGSAHFSSLESEV	An NR2D C-terminal sequence which has high amino acid homology with NR2C; again N-terminal glutaraldehyde coupling.	IB ✓ IP nt IC ✓

[a] IB, immunoblotting; IP, immunoprecipitation; IC, immunocytochemistry; ✓ means that the antibodies recognize the receptor subunit in the appropriate experimental paradigm; nt, not tested.

The protocol is given in this chapter for the coupling of peptides to carrier proteins and affinity matrices via cysteine residues together with the method for the affinity purification of anti-peptide antibodies (*Protocols 1* and *2*). The other coupling methods may be found in previous chapters from our group (8–10). These chapters also contain more detailed discussions of the choice of peptide sequence and the use of alternative antigens such as bacterially-derived fusion proteins and multiple antigen presentation systems, where the need for a carrier protein is negated. *Table 1* summarizes examples of peptides which have been used for successful anti-NMDA receptor subunit-specific antibody production. Their characterization and use have been detailed elsewhere (5, 7, 11–17).

Protocol 1

The *m*-maleimidobenzoic acid *N*-hydroxysuccinimide ester (MBS) method for the coupling of peptides and carrier protein via the cysteine of the peptide to the primary amine of the carrier protein

Equipment and reagents

- Peristaltic pump
- Bio-gel P2 and Bio-gel P30 (Bio-Rad Laboratories)
- Thyroglobulin: 20 mg/ml in 10 mM potassium phosphate pH 7.2

- Selected peptide
- *m*-Maleimidobenzoic acid *N*-hydroxysuccinimide ester (MBS): 3 mg/ml in dimethylformamide (Sigma Chemical Company)

Method

1 Dissolve the carrier protein, thyroglobulin, at 20 mg/ml in 10 mM potassium phosphate pH 7.2. Dialyse overnight at 4°C against 10 mM phosphate buffer pH 7.2, and adjust the concentration to 16 mg/ml thyroglobulin.

2 The presence of a cysteine residue within the peptide gives the possibility of oxidation thus the peptide should be reduced before use. To reduce, dissolve the peptide at 10 mg/ml in potassium phosphate pH 7.2. Add solid dithiothreitol to a final concentration of 200 mM; dissolve and incubate for 1 h at room temperature. Desalt the peptide by gel filtration column chromatography in 10 mM potassium phosphate pH 7.2. For the desalting, we use Bio-gel P2 which has a molecular weight cut-off of 1800 daltons.

3 Take 250 μl of dialysed thyroglobulin and add slowly 85 μl MBS with mixing. Mix end over end for 30 min at room temperature.

4 Desalt the activated thyroglobulin on a 20 ml Bio-gel P30 column equilibrated with 50 mM potassium phosphate pH 6.0. Collect 1 ml fractions. The activated thyroglobulin is eluted in the exclusion volume of the column in fractions 6–8. The recovery of the carrier protein is ~ 95%.

Protocol 1 continued

5 Dissolve the peptide in 10 mM potassium phosphate pH 7.2, to a final concentration of 5 mg/ml. Add 1 ml of this solution to the activated thyroglobulin (3 ml), mix, and adjust to pH 7.4. Mix for 3 h at room temperature.

6 Add solid NaCl to a final concentration of 0.9% (w/v). The peptide conjugate (now at a concentration of 1 mg/ml) is ready for use.

Protocol 2

Purification of anti-peptide antibodies by peptide affinity chromatography

Equipment and reagents

- Sintered glass filter
- Activated thiol-Sepharose (Sigma Chemical Company)
- 100 mM citric acid
- Tris equilibration buffer: 0.1 M Tris–HCl pH 8.0, containing 0.3 M NaCl, 1 mM EDTA

A. Preparation of the peptide affinity resin

1 Swell 0.35 g activated thiol-Sepharose in H_2O (1 ml final volume) and wash on a sintered glass filter with 100 ml Tris equilibration buffer which has been degassed under vacuum. Transfer to a capped tube.

2 Add 1 ml of 5 mg/ml peptide in the Tris equilibration buffer as above in step 1 and mix end over end for 2 h at room temperature.

3 Terminate the reaction by washing the gel on a sintered glass filter with the Tris equilibration buffer (25 ml) then wash with 100 mM citric acid adjusted to pH 4.5 with 2 M KOH (10 ml). Block the unreacted thiol groups by incubating the gel with 1 mM β-mercaptoethanol in 100 mM citric acid pH 4.5 (3 ml), and mix end over end for 45 min at room temperature.

4 Terminate the blocking reaction by washing with 100 mM citric acid pH 4.5 (25 ml).

5 Pour the column and equilibrate with phosphate-buffered saline; use immediately or store in the presence of 0.02% (w/v) sodium azide at 4 °C.

Equipment and reagents

- Peristaltic pump
- Spectrophotometer
- Phosphate-buffered saline
- Elution buffer: 50 mM glycine–HCl pH 2.3
- 1 M Tris (pH quenching reagent)

B. Affinity purification of anti-peptide antibodies

1 Equilibrate the appropriate peptide affinity resin (1 ml) with phosphate-buffered saline. Add the respective immune serum (~5 ml) and re-circulate through the column at a rate of 40 ml/h for 2 h at room temperature or overnight at 4 °C.

Protocol 2 continued

2 Collect the filtrate and wash the column with 100 ml phosphate-buffered saline at 40 ml/min.

3 Elute the antibody with 10 ml elution buffer at 10 ml/h collecting 1 ml fractions. Neutralize each fraction immediately after elution by the addition of 1 M Tris (20 μl) to give pH 7.4. Antibody-containing fractions are determined by measuring the optical density (OD) at $\lambda = 280$ nm.

4 Pool the antibody-containing fractions and dialyse overnight against phosphate-buffered saline containing 0.02% (w/v) sodium azide at 4°C. Measure the OD at $\lambda = 280$ nm and calculate the antibody concentration using the Beer–Lambert law (OD = ε. c. l where ε, the extinction coefficient for IgG = 1.35, c = concentration in mg/ml, and l is the path length, usually 1 cm). Store at 4°C until use. The yield of affinity purified antibody varies between peptides but it is usually in the range 0.2–1 mg/ml from 5 ml immune serum.

3 Determination of the subunit composition of native NMDA receptors by immunoaffinity purification using anti-NMDA receptor subunit-specific antibodies

The strategy that we have used for the determination of the subunit compositions of native $GABA_A$ and NMDA receptors is their solubilization from brain membranes using non-denaturing detergents, followed by immunoaffinity purification using the appropriate immobilized affinity purified anti-receptor subunit-specific antibody (*Protocol 3*). This isolated receptor subpopulation is then analysed by immunoblotting using different specificity anti-receptor subunit antibodies to determine whether the other subunits have been co-purified. Generally, a single immunoaffinity purification step is inadequate for the definitive determination of native NMDA receptor subunit combinations because of the extensive heterogeneity that exists, thus we have used two different specificity immunoaffinity columns in series for the delineation of native subunit permutations. Using this approach, we have shown for example that NR1-1a/NR2A and NR1-1a/NR2B are major receptor subtypes; NR1-1a/NR2A/NR2B receptors are also expressed in mammalian brain but under our conditions of solubilization and purification, they are minor components compared to binary NR1/NR2 receptors (15). Further, quantitative immunoblotting of double immunopurified NMDA receptors has shown that the NR1:NR2 subunit ratio is 1:1.5 (7). The experimental strategies together with some of the results obtained are shown in *Figures 1* and *2*.

It is important to note that although useful information is obtained using this approach, there are several caveats to be aware of. For example, NMDA receptors are known to be resistant to detergent solubilization which is probably because

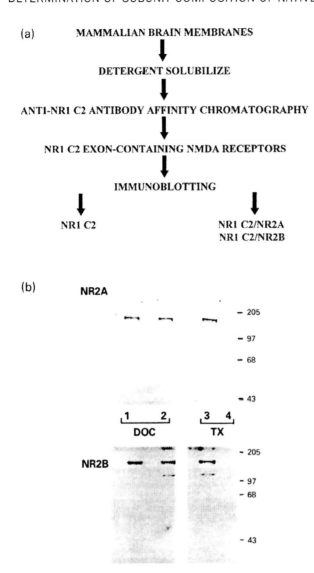

(a)

MAMMALIAN BRAIN MEMBRANES

↓

DETERGENT SOLUBILIZE

↓

ANTI-NR1 C2 ANTIBODY AFFINITY CHROMATOGRAPHY

↓

NR1 C2 EXON-CONTAINING NMDA RECEPTORS

↓

IMMUNOBLOTTING

NR1 C2   NR1 C2/NR2A
               NR1 C2/NR2B

(b)

Figure 1 Immunoprecipitation of NMDA receptors from detergent extracts of adult mammalian forebrain demonstrating the existence of a pool of unassembled C2 exon-containing NR1 subunits that are selectively solubilized by Triton X-100. (a) A schematic diagram outlining the experimental strategy. (b) An immunoblot comparing the subunit compositions of NMDA receptors purified by anti-NR1 C2 immunoaffinity chromatography after solubilization from mammalian forebrain with either Triton X-100 (TX) or sodium deoxycholate (DOC). Lanes 1 and 3, mouse forebrain membranes before solubilization; lanes 2 and 4, anti-NR1 C2 purified preparations. NR2A, probing with anti-NR2A antibodies; NR2B, probing with anti-NR2B antibodies. Note that NR2A and NR2B subunits are only detected in lane 2 thus showing that Triton X-100 selectively extracts unassembled C2 exon-containing NMDA NR1 subunits. Published with permission from the publisher (14).

(a)

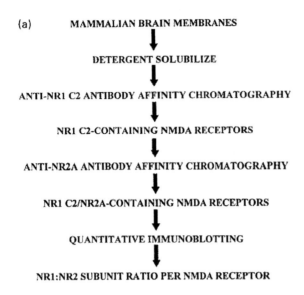

(b)

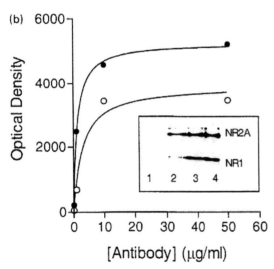

Figure 2 Double immunoaffinity purification of NMDA receptors from adult mammalian forebrain: determination of the NR1:NR2 subunit ratio. (a) A schematic diagram outlining the experimental strategy for the determination of the NR1:NR2 subunit ratio for NMDA receptors isolated from rat brain by anti-NR1 C2 and anti-NR2A antibody affinity columns in series. (b) The results of quantitative immunoblotting of the double immunopurified receptors. Briefly, NMDA receptors purified from mammalian forebrain by anti-NR1 C2 and anti-NR2A antibody affinity columns in series were analysed by immunoblotting using both anti-NR1 and anti-NR2A antibodies. The figure shows the results of a densitometric scan of an immunoblot where a fixed amount of NR1/NR2A NMDA receptors was used as antigen with increasing concentrations of anti-NR1 C2 (○) or anti-NR2A (●) antibody. The inset shows the actual immunoblot where lanes 1–4 have been probed with both anti-NR1 C2 and anti-NR2A antibodies at 0.1, 1, 10, and 50 μg/ml respectively. The figure is representative of two independent receptor double immunoaffinity purifications. The NR1/NR2 subunit ratio at 50 μg/ml antibody was 1:1.4 ± 0.2 for two independent double immunoaffinity purified preparations (three immunoblots/per preparation). Published with permission from ref. 7.

of their association with a network of scaffolding proteins within the post-synaptic density, such as the postsynaptic density-95 (PSD-95) protein (reviewed in ref. 18). One may question therefore whether the NMDA receptors solubilized under the non-denaturing conditions that we have used are actually represent-ative of synaptic NMDA receptors. Secondly, we have shown that in the brain there exists a pool of unassembled NR1 C2 exon-containing subunits which is selectively solubilized by the detergent Triton X-100 (14). It was found that NMDA receptors purified from Triton X-100 extracts of mammalian forebrain by anti-NR1 C2 antibody affinity chromatography comprised only NR1 C2 subunits with no detectable co-associating NR2A or NR2B subunits (*Figure 1*). Thus care is required in the choice of the solubilization conditions and in the interpretation of results. It is important that the findings are taken in conjunction with immunocytochemical studies which may yield insights into the localization and abundance of synaptic versus extra-synaptic receptors. Finally, it is worth pointing out that the primary structures of the GABA$_A$ receptor subunits have been known since 1987, at least four years longer than we have known the NMDA receptor subunit amino acid sequences, and although there is now general agreement about which subunit combinations occur *in vivo*, the subunit composition and subunit ratio of a single native GABA$_A$ receptor has not been determined. Information pertaining to *in vivo* subunit ratios has been extra-polated from studies of cloned receptors of predetermined subunit types (see Section 4).

Protocol 3

Immunoaffinity purification of native NMDA receptors

Equipment and reagents

- ImmunoPure IgG Orientation kit (Pierce Chemical Company)
- Equilibration buffer: 50 mM Tris–HCl pH 7.5, containing 1 M NaCl, 5 mM EDTA, and 1% (v/v) Triton X-100

- Peristaltic pump
- Elution buffer: 50 mM triethylamine pH 11.5, containing 0.05% (w/v) sodium deoxycholate

A. Synthesis of the immunoaffinity column

1 Affinity purify ~ 4 mg of the appropriate NMDA receptor anti-peptide antibody.

2 Covalently crosslink the purified IgG to protein A Sepharose (1 ml) with dimethyl pimelimidate using the ImmunoPure IgG Orientation kit according to the manu-facturer's instructions.[a,b]

3 Equilibrate the affinity column with equilibration buffer.

4 Pre-elute the affinity column with elution buffer at 10 ml/h (10 ml), and then re-equilibrate with at least 50 × column volumes of the wash buffer as above at 10 ml/h. The column is now ready for use.

Protocol 3 continued

Equipment and reagents

- Peristaltic pump
- Dialysis buffer: 10 mM Hepes pH 7.6, containing 0.1% (v/v) Triton X-100, 0.02% (w/v) sodium deoxycholate, 1 mM EDTA, 0.45 M KCl
- Affinity column wash buffer: 50 mM Tris–HCl pH 7.5, containing 1 M NaCl, 5 mM EDTA, 5 mM EGTA, and 1% (v/v) Triton X-100
- Elution buffer: 50 mM triethylamine pH 11.5, containing 0.05% (w/v) sodium deoxycholate

- Brain membrane solubilization buffer: 1% (w/v) sodium deoxycholate pH 9.0, in 50 mM Tris–HCl containing 0.15 M NaCl, 5 mM EGTA, 5 mM EDTA, 1 mM phenylmethylsulfonyl fluoride, bacitracin (5 μg/ml), soybean trypsin inhibitor (5 μg/ml), ovomucoid trypsin inhibitor (5 μg/ml), and benzamidine HCl (5 μg/ml)

B. Immunoaffinity purification of NMDA receptors from mammalian brain

1 Prepare a detergent extract of mammalian forebrain. Optimal solubilization of NMDA receptors is achieved by extracting a P2 rat forebrain membrane fraction (1.5 mg protein/ml) with brain membrane solubilization buffer for 1 h at 4°C. Collect the solubilized extract by centrifugation at 100 000 g.

2 Dialyse the solubilized extract twice against 2 litres dialysis buffer for 3 h each at 4°C.

3 Apply the dialysed, detergent solubilized brain extract to the appropriate anti-peptide antibody affinity column (e.g. anti-NR1 C2) with re-circulation at a rate of 10 ml/h for 20 h at 4°C.

4 Remove non-specifically bound proteins by washing the column with 50 ml affinity column wash buffer at a rate of 10 ml/h.

5 Elute the antibody affinity column with elution buffer, at a rate of 10 ml/h, collecting fractions of 1 ml in tubes which already contain 200 μl of 2 M glycine thus quenching to a final pH = 7.4. NMDA receptor-containing fractions are identified by immunoblotting.

6 NMDA receptor-containing fractions are pooled and applied to a second, different specificity anti-NMDA receptor antibody column, e.g. anti-NR2A Lys 1381–1394 antibody affinity column with re-circulation at 10 ml/h for 2 h at 4°C.

7 Repeat steps 4 and 5. The double immunoaffinity purified receptor preparations are then screened for immunoreactivity using different specificity anti-NMDA receptor subunit antibodies.

8 The NR1:NR2 subunit ratios are determined by quantitative immunoblotting as depicted in *Figure 2*.

[a] The use of this kit maximizes the antigen binding capacity of the immobilized antibodies.

[b] The chemical crosslinking step ensures that there is no leakage of antibody molecules during the immunoaffinity purification procedure.

4 Determination of NMDA receptor subunit stoichiometry using an epitope tagging strategy

Immunopurification and immunoprecipitation using anti-NMDA receptor subunit antibodies yields useful information with regard to which subunits co-associate *in vivo*. It has proved difficult however, to quantify subunit ratios within receptors because of the difficulty in isolating a high enough concentration of homogeneous receptors for analysis. As described above, the NR1:NR2 subunit ratio can be determined by analysing double immunoaffinity purified receptors by quantitative immunoblotting using anti-NR1 and anti-NR2 subunit antibodies but the relative avidities of the different antibodies must be determined. This requires large amounts of antibody and an assumption that the affinity purified polyclonal antibodies are recognizing a single epitope, i.e. one antibody molecule binds to one NMDA receptor subunit. In our hands, the results obtained have appreciable variation. Several different methods have been used to study subunit ratios within cloned hetero-oligomeric receptors. For NMDA receptors, this has included the co-expression of defined mixtures of wild-type and mutant NR1 or NR2 subunits in *Xenopus* oocytes. From the agonist activation profiles of the resultant receptors, the number of NR1 or NR2 subunits per receptor was deduced, although the three references cited differ in their conclusions (19–21). In an alternative approach, Im *et al.* (22) made a tandem construct of $\alpha6$-$\beta2$ GABA$_A$ receptor subunits. They found that GABA-gated chloride ion channels were formed only following the co-expression of $\alpha6$-$\beta2$ constructs with either an $\alpha6$ or a $\gamma2$ subunit indicative of $(\alpha6)_2(\beta2)_2(\gamma2)_1$ and $(\alpha6)_3(\beta2)_2$ stoichiometries. This strategy is not possible for NMDA receptors because the subunit transmembrane topology is incompatible with the use of tandem constructs. Tretter *et al.* (23) cleverly eliminated the problem of different antibody avidities by determining their relative avidities for GABA$_A$ receptor chimeric subunits where the N-terminal domain contained one epitope and the C-terminal domain, the second epitope. Using this information, they were able to deduce a subunit stoichiometry of $(\alpha1)_2(\beta3)_3$. Lastly, an elegant fluorescent energy transfer method for evaluating GABA$_A$ receptor subunit stoichiometry is described in ref. 24.

Our approach to the determination of NMDA receptor subunit stoichiometry is outlined in *Figure 3*. Thus NMDA receptor NR1 or NR2 subunits are distinguished on the basis of their antigenicity by the introduction of epitope tags, i.e. NR2A wild-type, NR2B$_{FLAG}$, and NR2B$_{c-Myc}$. Mammalian cells are transfected with NR1, NR2A, NR2B$_{FLAG}$, and NR2B$_{c-Myc}$ receptor clones, cell homogenates detergent solubilized, and receptors purified by anti-NR2A and anti-FLAG immunoaffinity columns in series. The double immunopurified material is analysed for all three specificity NR2 subunits. If all are detected and appropriate controls are carried out to ensure that non-specific receptor subunit aggregation is not a problem, this is evidence for three NR2 subunits per receptor oligomer.

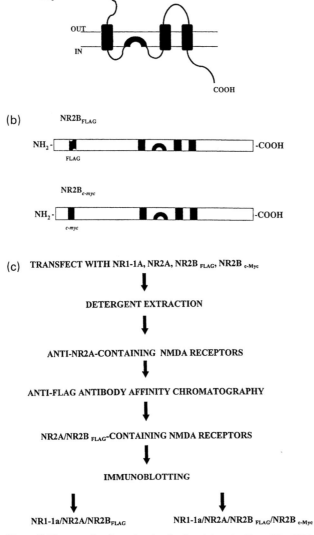

Figure 3 The use of epitope tagging for the determination of the NR1 and NR2 subunit copy number per NMDA receptor oligomer. (a) The predicted transmembrane topology of NMDA receptor subunits. (b) The positions of the epitope tags in the NMDA receptor NR2B subunit. (c) A schematic diagram outlining the experimental strategy for the determination of the copy number of NR2 subunits per NMDA receptor oligomer.

4.1 Epitope tagging of NMDA receptor subunits

Table 2 lists several commercial epitope tags. All of these are peptides of up to eight amino acids in length. The initial question is where to position the epitope tag within the NMDA subunit sequence. Obviously, introduction of the tag must not interfere with the normal function and processing of the subunits which includes post-translational modification of the subunit, its three-dimensional

Table 2 A summary of the commercially available epitope tags and their associated products[a]

Epitope tag	Associated products	Application	Source
FLAG	Monoclonal abs.	IB	Sigma; Covance
DYKDDDDK	Polyclonal ab.	IB	Sigma
	Monoclonal affinity matrix	IP	Sigma; Covance
	Biotinylated monoclonal abs.	IB	Sigma; Covance
C-Myc	Monoclonal abs.	IB, IP	Sigma; Covance
EQKLISEEDL	Polyclonal ab.	IB, IP	Affinity Bioreagents; Covance; Clontech
	Monoclonal affinity matrix	IP	Clontech
	Biotinylated monoclonal ab.	IB	Covance
	FITC-conjugated monoclonal ab.	IC	Covance
HA	Monoclonal abs.	IB, IC, IP	Covance
YPYDVPDYA	(ascites/purified)		
6-HIS	Monoclonal abs.	IB, IC	Covance
HHHHHH	Monoclonal affinity matrix	IP	Covance; Qiagen
	FITC-conjugated monoclonal ab.	IC	Covance
Glu-Glu	Monoclonal abs.	IB	Covance
EEEEYMPME			

[a] The table contains examples of some of the most commonly used epitope tags together with some of the associated products that are commercially available. The choice of tag can be dependent on the experimental paradigm in which the tagged protein is to be used; cf. IB, immunoblotting; IP, immunopurification; IC, immunocytochemistry; FITC, fluorescein isothiocyanate. Note that care should be taken with regard to the positional sensitivity of the antibody to be used. For example, the specificity of the monoclonal anti-FLAG antibody depends upon where the epitope is inserted into the target sequence, i.e. whether the tag has a free N- or C-terminus. The suppliers list is not exhaustive, examples only are given.

folding, assembly with other subunits to form functional heteromeric receptors, and also receptor–accessory protein interactions. It is difficult to predict the regions of the protein that are involved in these processes, particularly in three-dimensional folding and oligomerization, since they are not governed by amino acid consensus sequences. Thus, successful epitope tagging of subunits is a matter of an educated guess together with trial and error. We chose to introduce the FLAG and c-Myc epitope tags close to the N-termini of the NR1 and NR2 subunits resulting in the introduction of an extracellular antigenic determinant. The rationale was that this region of the subunits is removed from both the transmembrane spanning domains and amino acids implicated in the L-glutamate and glycine ligand binding sites. It is also several residues downstream of the start methionine residue thus avoiding cleavage of the tag in the processing of the respective subunit signal peptide.

There are several methods which can be used to insert the epitope tag nucleotide sequence into that encoding an NMDA receptor subunit. These are outlined in *Figure 4*. The simplest is to ligate the desired epitope tag directly into the expression vector containing the NMDA receptor subunit cDNA (*Protocol 4*). This method is, however, dependent on the presence of a unique restriction enzyme site in the appropriate position. If there is no restriction enzyme site, one may be readily generated by either site-directed mutagenesis or PCR. Using this direct strategy, we found that the pairs of oligonucleotide primers could self-ligate, resulting in the incorporation of multiple copies of the epitope tag into the NMDA receptor subunit. Whilst this is an advantage with regard to the increase of epitopes per subunit resulting in an increased sensitivity of detection in Western blots, the inclusion of the increased peptide size is a disadvantage because it is more likely to interfere with subunit folding and subunit/subunit co-association. Insertional mutagenesis is an alternative to direct ligation and has the advantage that only a single copy of the tag nucleotide sequence is inserted at the desired location. We used the Sculptor™ *in vitro* mutagenesis system from Amersham Pharmacia Biotech to carry out insertional mutagenesis in conjunction with the appropriate oligonucleotide (see *Figure 4*). For this method, more thought is required in the design of the oligonucleotide. Factors to consider are as follows:-

(a) The oligonucleotide should be designed such that there are at least 10–15 bases which are perfectly complementary to the template DNA at each side of the mutation to ensure efficient binding of the primer. Further, these 10–15 bases should have similar stabilities, i.e. they should have similar percentage guanine:cytosine (GC) contents.

(b) The GC content of the oligonucleotide as a whole should be > 50%.

(c) There should be a GC clamp at both the 5′ and 3′ ends of the oligonucleotide.

(d) The oligonucleotide should not contain a restriction site for the restriction enzyme that is to be used to digest the parental strand. (In the case of the Sculptor™ *in vitro* mutagenesis system, this would be *Nci*I.)

(e) Decrease the possibility of secondary structures in the oligonucleotide by avoiding the use of palindromic sequences which can fold to yield hairpin loops.

The major disadvantages of insertional mutagenesis are that the NMDA receptor subunit cDNAs had to be subcloned into either an M13 or phagemid vector to permit the generation of the single-stranded DNA templates and the fact that the insertion of the tag was relatively inefficient with, on average, one in ten clones containing the tag. The Sculptor™ kit has unfortunately now been discontinued although the protocol is given here for completeness (*Protocol 4B*). The Altered Sites II Mammalian Mutagenesis System from Promega is a recommended alternative.

For the NR2 subunits insertion of the tag between amino acids 53 and 54 proves a suitable position since the NR2B$_{FLAG}$ and NR2B$_{c-Myc}$ subunits behaved as wild-type. That is, NR1/NR2B wild-type and NR1/NR2B$_{FLAG}$ receptors have the same high affinity for the NR2B-selective radioligand, [^{3}H]Ro 25-6981. Also, expression of NR1/NR2B wild-type and NR1/NR2B$_{FLAG}$ receptors in mammalian cells results in the same percentage cytotoxicity post-transfection, an index of the expression of functional, i.e. assembled NMDA receptors, thus demonstrating that the introduction of the additional amino acids does not interfere with the

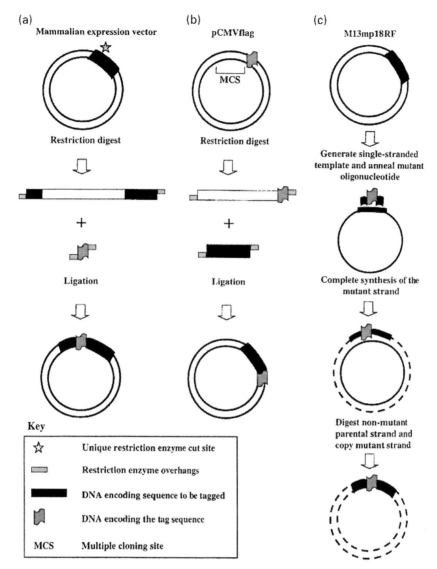

Figure 4 Schematic representation of the different methods for the introduction of an epitope tag into an NMDA receptor subunit. (a) Epitope tagging by direct ligation. (b) Epitope tagging using an epitope tagging mammalian expression vector. (c) Epitope tagging by insertional mutagenesis.

co-assembly/oligomerization of the receptor subunits (7). A further point is that the tagged subunits are expressed at approximately the same levels as the wild-type NR2B subunit, thus ensuring that the purification results are not skewed by a low level of a tagged subunit. Expression levels can be readily determined by transfection of HEK 293 cells with single subunit clones followed by comparative, quantitative immunoblotting.

Insertion of either the FLAG or c-Myc epitope tag into an analogous position of

(a)

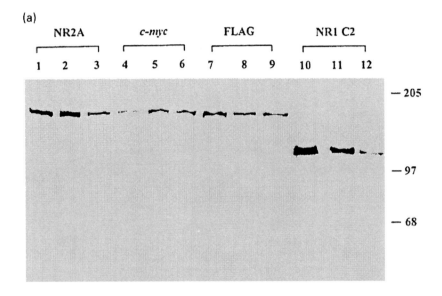

(b)

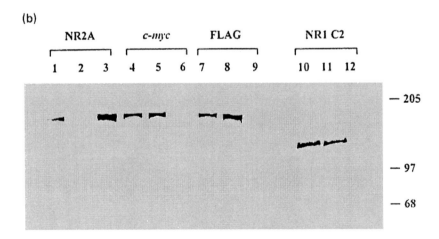

the NR1 subunit results in no cytotoxicity post-transfection. This suggests that introduction of the tag inhibits NR1/NR2 subunit association. We thus resort to an epitope tag mammalian expression vector (see below).

4.2 Epitope tagging mammalian expression vectors

An alternative approach for the epitope tagging of subunits is to use one of the epitope tagging mammalian expression vectors which are now commercially available. One example of these vectors is the pCMV-Tag series marketed by Stratagene. Each pCMV-Tag vector contains the nucleotide sequence encoding the epitope tag as part of the multiple cloning site. Subcloning of the appropriate cDNA results in the expression of a fusion protein whereby the tag is

Figure 5 Demonstration of the coexistence of three NR2 subunits in the same NMDA receptor complex. (a) An immunoblot of the anti-NR2A, anti-FLAG double immunoaffinity purified NMDA receptors. Briefly, HEK 293 cells were transfected with NR1-1a, NR2A, $NR2B_{FLAG}$, and $NR2B_{c-Myc}$ clones. Transfected cells were detergent solubilized and NMDA receptors purified by anti-NR2A and anti-FLAG antibody affinity columns in series. The receptor-containing fractions were pooled and analysed by immunoblotting. Lanes 1, 4, 7, and 10, solubilized material; lanes 2, 5, 8, and 11, anti-NR2A immunoaffinity purified material; lanes 3, 6, 9, and 12, anti-NR2A/anti-FLAG immunoaffinity purified material. Lanes 1–3 were probed with anti-NR2A, lanes 4–6 with anti-c-Myc, lanes 7–9 with anti-FLAG, and lanes 10–12, with anti-NR1 C2 antibodies. It can be seen that anti-c-Myc immunoreactivity was detected in the double purified receptor population showing that there are at least three NR2 subunits per NMDA receptor. Note that densitometric scans of lanes 3, 6, and 9 yielded ~ 1:1:1 ratios although antibody concentrations were not saturating. (b) An immunoblot of an experiment designed to show that the co-association of three NR2 subunits is not an artefact of receptor solubilization. Here, HEK 293 cells were transfected with either NR1-1a, NR2A, $NR2B_{FLAG}$, or $NR2B_{c-Myc}$ clones. The resultant single subunit transfected cells were pooled, solubilized, and receptor subunits purified by anti-NR2A immunoaffinity chromatography with the material then being analysed by immunoblotting. Lanes 1, 4, 7, and 10, solubilized, pooled transfected HEK 293 cells; lanes 2, 5, 8, and 11, anti-NR2A antibody post-column filtrate; lanes 3, 6, 9, and 12, anti-NR2A purified material. Lanes 1–3 were probed with anti-NR2A, lanes 4–6 with anti-c-Myc, lanes 7–9 with anti-FLAG, and lanes 10–12 with anti-NR1 C2 antibodies. Note that only NR2A subunits are detected in the affinity purified material demonstrating that non-specific aggregation of receptor subunits during detergent solubilization does not occur. Published with permission from ref. 7.

located at either the C-terminus or the N-terminus of the mature protein. The vectors contain a cytomegalovirus promoter to permit high levels of expression of the tagged constructs in a variety of mammalian cell lines. Further, a Kozak consensus initiation sequence is included prior to the epitope to ensure optimal expression of N-terminally tagged proteins. We have used the pCMV-Tag vector to generate a C-terminal FLAG-tagged NR1-4b subunit. Once the tagged subunit is shown to behave and, importantly, be expressed to the same level as the wild-type NR1-4b subunit (since the promoter is now different), it is used in conjunction with wild-type NR1-1a, $NR1-4b_{c-Myc}$, and NR2A subunits as outlined in *Figure 5* to determine, this time, the NR1 subunit copy number per receptor (25).

4.3 Critical assessment of the epitope tagging strategy for the determination of the number of NR1 and NR2 subunits per NMDA receptor

Overall, whilst the epitope tagging approach can yield important information on the presence of multiple copies of a particular subunit within a receptor complex, there are some caveats to be aware of. For example, we are overexpressing the various NMDA receptor subunits in a non-neuronal cell line. It is thus possible that the double immunoaffinity purified material is not representative of native, functional NMDA receptors but merely a combination of subunits that are forced to co-associate as a result of being overexpressed. Alternatively, the subunits may simply be aggregating non-specifically following detergent solubilization. *Figure 5b* shows the results of control experiments designed to show that non-specific aggregation of transfected subunits is not a problem.

The possibility that the isolated receptor is an artefact of the expression system is more difficult to address. We would argue that it is not the case here since both NR1 **and** NR2 subunits are both detected in the double immuno-affinity purified material. The problem could be investigated further by the initial isolation of cell surface receptors, rather than the collection of whole cell homogenates, which would include both precursor and functional forms of the expressed NMDA receptors. This could be carried out by cell surface biotinylation, followed by streptavidin affinity chromatography, then double immunoaffinity purification.

A final problem which relates to the determination of the NR1 subunit copy number per receptor is that when NR1 and NR2 subunits are co-expressed in mammalian cells, there is always an excess of NR1 subunits. Some evidence suggests that these NR1 subunits can co-associate to form homo-oligomers. If this is the case, the approach we have described is invalid for the determination of the NR1 subunit copy number per heteromeric receptor; resolution would now require the use of three different epitope tagged NR1 subunits or, alternatively, the separation of homomeric and heteromeric assemblies prior to anti-NR1 immunoaffinity purification. This could be achieved by an additional anti-NR2 antibody immunoaffinity chromatography step.

Protocol 4

Preparation of NMDA receptor subunit epitope tagged constructs

A. Introduction of the epitope tag into the NMDA receptor subunit sequence by direct ligation

Reagents

- T4 DNA ligase buffer (10 ×) (Roche Biochemicals)

- Two oligonucleotides[a] that have been purified by desalting and are phosphorylated at their 5' ends

Design and preparation of oligonucleotides

1   Dilute oligonucleotides 1 and 2 each to a final concentration of 50 pmol/μl in sterile water.

2   Mix 50 pmol of each oligonucleotide together with 8 μl sterile water and 1 μl of the 10 × DNA ligase buffer.

3   Heat to 95 °C for 2 min.

4   Cool the mixture from 95 °C to 55 °C at a rate of 1.5 °C/20 sec. Hold the temperature at 55 °C for 5 min.

5   Leave to cool to room temperature and store at 4 °C until ready to use.

Protocol 4 continued

Reagents

- Mammalian expression vector (1 μg/μl)
- *Ppu*MI and 10 × reaction buffer (New England Biolabs)
- T4 DNA ligase and 10 × DNA ligase buffer (Roche Biochemicals)
- 70% (v/v) ethanol (ice-cold)

- Shrimp alkaline phosphatase (United States Biochemica Corp.)
- 3 M sodium acetate pH 5.2
- Competent DH5α *E. coli* bacteria
- Luria Bertani (LB) ampicillin agar plates

Preparation of vector

1 Digest 2 μg of the mammalian expression vector with 2 μl of 10 × *Ppu*MI reaction buffer and 4 U of *Ppu*MI in a total reaction volume of 20 μl for 2 h at 37°C.

2 Add 0.1 U shrimp alkaline phosphatase and incubate at 37°C for 30 min.

3 Add a further 0.1 U shrimp alkaline phosphatase, incubate at 37°C for 30 min, then heat inactivate at 65°C for 15 min.

4 Using sterile H_2O, increase the reaction volume to 100 μl, add 100 μl phenol, and carry out a phenol/chloroform extraction of the DNA, followed by ethanol precipitation by the standard method as described (26).

Ligation of the oligonucleotides and vector

1 Mix 150 ng of the prepared vector with 1 μl of the prepared oligonucleotide, 2 μl of 10 × DNA ligase buffer, and 1 μl of T4 DNA ligase in a total reaction volume of 20 μl.

2 Incubate overnight at 16°C.

3 Transform 10 μl of the reaction mix into competent DH5α *E. coli* and plate 100 μl of the transformed bacteria onto LB ampicillin plates. Incubate the plates at 37°C overnight, collect and analyse recombinants by standard methods.

B. Introduction of the epitope tag into the NMDA receptor subunit sequence by insertional mutagenesis

Reagents

- T4 polynucleotide kinase (Amersham Pharmacia Biotech)
- Sculptor *in vitro* mutagenesis system which contains buffers A to D, dNTP mix A and B, T7 DNA polymerase, T4 DNA ligase, T5 exonuclease, *Nci*I restriction endonuclease, exonuclease III, DNA polymerase I

- 10 × T4 polynucleotide kinase buffer
- Single-stranded template DNA (1 μg/μl)
- Mutagenic oligonucleotide (1.6 pmol/μl)
- Qiaex II Gel Extraction Kit (Qiagen)
- Competent TG1 *E. coli*
- LB agar plates

Phosphorylation of oligonucleotide

1 Incubate the mutagenic oligonucleotide (30 μl) with 3 μl of 10 × kinase buffer, and 2 U of T4 polynucleotide kinase for 15 min at 37°C.

Protocol 4 continued

2 Heat inactivate at 70 °C for 10 min.

3 Store phosphorylated oligonucleotide at −15 °C to −30 °C until use.

Mutagenesis

1 Mix 2 μl of single-stranded template DNA with 1 μl of the phosphorylated oligonucleotide and 1 μl of buffer A, add sterile water to give a final reaction volume of 10 μl.

2 Incubate for 3 min at 70 °C.

3 Transfer to 55 °C and allow the mixture to cool to room temperature.

4 Add to the mix, 10 μl of dNTP mix A, 2.5 U of T4 DNA ligase, and 0.8 U of T7 DNA polymerase.

5 Incubate at room temperature for 10 min followed by 30 min at 37 °C.

6 Heat inactivate at 70 °C for 15 min.

7 Add to the mixture 50 μl of buffer B and 2000 U of T5 exonuclease. Mix.

8 Incubate at 37 °C for 30 min followed by heat inactivation at 70 °C for 15 min.

9 Add to the mix 5 μl of buffer C and 5 U of *Nci*I, incubate for 90 min at 37 °C.

10 Add to the mix 20 μl of buffer D, 160 U of exonuclease III, incubate for 30 min at 37 °C, and then 15 min at 70 °C.

11 Finally, add 20 μl of dNTP mix B, 3.5 U of DNA polymerase I and 2.5 U of T4 DNA ligase, mix, and incubate for 1 h at 37 °C.

12 Take 40 μl of the final reaction mixture and desalt and concentrate the DNA solution using the Qiaex II Gel extraction kit exactly as per the manufacturer's instructions.

13 Transform competent TG1 *E. coli* with 10 μl of the desalted and extracted solution. Incubate the plates at 37 °C overnight, collect and analyse recombinants by standard methods.

[a] An example of the oligonucleotide design is shown below. The 24 nucleotides in italics encode the eight amino acids of the FLAG epitope tag where oligonucleotide 1 is the sense strand and oligonucleotide 2 the antisense strand of the FLAG tag. The nucleotides in bold are the 5′ overhangs generated by the restriction enzyme at which site the epitope tag is to be inserted. In this example the restriction enzyme *Ppu*MI was used.

Oligonucleotide 1: 5′ **GAC**gactacaaggacgacgatgacaag 3′

Oligonucleotide 2: 3′ ctgatgttcctgctgctactgttc**CTG** 5′

5 Conclusions

In this chapter, we have described methodological approaches that are currently in use in our laboratory to determine the subunit compositions and subunit ratios of both native and cloned NMDA receptors. We have included also some of the problems, experimental and empirical, that we have encountered during the course of these studies. The techniques are applicable to any multi-subunit neurotransmitter receptor or indeed, protein. The information resulting from these studies should yield fundamental insight into the structures of these important brain proteins.

Acknowledgements

The authors would like to acknowledge the contributions of past members of the research group especially, Dr Paul L. Chazot. The work described in this chapter was supported by the Biological and Biotechnology Science Research Council (UK).

References

1. Hollmann, M. and Heinemann, S. (1994). *Annu. Rev. Neurosci.*, **17**, 31.
2. Sucher, N. J., Awobuluyi, M., Choi, Y., and Lipton, S. A. (1996). *Trends Pharmacol. Sci.*, **17**, 348.
3. Duggan, M. J. and Stephenson, F. A. (1990). *J. Biol. Chem.*, **265**, 3831.
4. Duggan, M. J., Pollard, S., and Stephenson, F. A. (1991). *J. Biol. Chem.*, **266**, 24778.
5. Chazot, P. L., Coleman, S. K., Cik, M., and Stephenson, F. A. (1994). *J. Biol. Chem.*, **29**, 24403.
6. Pollard, S., Thompson, C. L., and Stephenson, F. A. (1995). *J. Biol. Chem.*, **270**, 21285.
7. Hawkins, L. M., Chazot, P. L., and Stephenson, F. A. (1999). *J. Biol. Chem.*, **274**, 27211.
8. Stephenson, F. A. and Duggan, M. J. (1991). In *Molecular neurobiology: a practical approach* (ed. J. Chad and H. Wheal), p. 183. IRL Press, Oxford, UK.
9. Stephenson, F. A. (1995). In *Ion channels: a practical approach* (ed. R. Ashley), p. 171. IRL Press, Oxford, UK.
10. Chazot, P. L., Pollard, S., and Stephenson, F. A. (1998). In *Neuromethods* (ed. A. A. Boulton, G. B. Baker, and A. N. Bateson), Vol. 34, p. 257. Humana Press, New Jersey, USA.
11. Chazot, P. L., Cik, M., and Stephenson, F. A. (1992). *J. Neurochem.*, **59**, 1176.
12. Cik, M., Chazot, P. L., and Stephenson, F. A. (1993). *Biochem. J.*, **296**, 877.
13. Chazot, P. L., Cik, M., and Stephenson, F. A. (1995). *Mol. Membr. Biol.*, **12**, 331.
14. Chazot, P. L. and Stephenson, F. A. (1997). *J. Neurochem.*, **68**, 507.
15. Chazot, P. L. and Stephenson, F. A. (1997). *J. Neurochem.*, **69**, 2138.
16. Thompson, C. L., Drewery, D. L., Atkins, H. D., Stephenson, F. A., and Chazot, P. L. (2000). *Neurosci. Lett.*, **283**, 85.
17. Racca, C., Stephenson, F. A., Streit, P., Roberts, D. B., and Somogyi, P. (2000). *J. Neurosci.*, **20**, 2512.
18. Sheng, M. (1996). *Neuron*, **17**, 575.
19. Béhé, P., Stern, P., Wylie, D. J. A., Nassar, M., Schoepfer, R., and Colquhoun, D. (1995). *Proc. R. Soc. Lond.*, **262**, 205.
20. Laube, B., Kuhse, J., and Betz, H. (1998). *J. Neurosci.*, **18**, 2954.
21. Premkumar, L. S. and Auerbach, A. (1997). *J. Gen. Physiol.*, **110**, 485.
22. Im, W. B., Pregenzer, J. F., Binder, J. A., Dillon, G. H., and Alberts, G. L. (1995). *J. Biol. Chem.*, **270**, 26063.
23. Trettter, V., Ehya, N., Fuchs, K., and Sieghart, W. (1997). *J. Neurosci.*, **17**, 2728.

24. Farrar, S. J., Whiting, P. J., Bonnert, T. P., and McKernan, R. M. (1999). *J. Biol. Chem.*, **274**, 10100.
25. Hawkins, L. M., Chazot, P. L., and Stephenson, F. A. (2000). *Eur. J. Neurosci. Suppl.*, **12**, 37.
26. Sambrook, J., Fritsch, E. F., and Maniatis, T. (ed.) (1989). *Molecular cloning: a laboratory manual*. Cold Spring Harbor Laboratory Press, USA.

Chapter 5

Evaluating receptor stoichiometry by fluorescence resonance energy transfer

Dmitry M. Gakamsky and Israel Pecht
Department of Immunology, The Weizman Institute of Science, Rehovot 761000, Israel

Richard G. Posner
Department of Chemistry, Northern Arizona University, Flagstaff, AZ, USA.

1 Introduction

Elucidating mechanisms of molecular interactions among receptors and their ligands is a central problem in cell and molecular biology. Many different methods, including gel filtration, radiolabelling, immunoprecipitation, two-hybrid system analysis, circular dichroism, fluorescence and absorption spectroscopy, and surface plasmon resonance, have been developed to reveal the proximity of proteins in cells, their organization into molecular complexes, and to determine their stoichiometry. The present review focuses on the use of fluorescence resonance energy transfer (FRET), which is a universal, versatile, and powerful approach for evaluating/assaying distances between and within molecules. In comparison with other methods, FRET has numerous advantages. The exquisite sensitivity provided by a fluorescence detection method (single molecule detection is possible), coupled with the wide range of fluorescence-based instrumentation (e.g. microscopy, flow cytometry, steady-state fluorimetry, fluorescence polarization, fluorescence lifetime, rapid mixing devices) and a large variety of commercially available fluorescent probes, allows numerous possible approaches for studying biological structure and function. Moreover, the ability to measure the concentration of bound complexes, without the need to separate them from the unbound fraction, together with the potential for high spatial resolution, make FRET an excellent tool for the study of real time receptor–ligand interactions in individual cells (1-3). This is illustrated by the recent application of FRET to the design of high throughput screening instruments in drug design (4, 5). In this chapter we describe FRET applications in the determination of complex molecular organization and stoichiometry in solutions and in living cells.

2 Theoretical basis of fluorescence energy transfer

Fluorescence energy transfer is a process of electronic excitation transfer from one molecule, an energy donor, to another molecule, an energy acceptor. An excited donor molecule may transfer its excitation to an acceptor molecule without emitting the photon. This occurs when the resonance conditions arise from the dipole-dipole intramolecular interactions. Förster first suggested a theory for non-radiative dipole–dipole homo-energy transfer between dye molecules in 1948 (6).

2.1 Theory of FRET

The efficiency of energy transfer is determined by the ratio of photons emitted by the donor molecule, in the presence of the acceptor (F_{DA}), to the number of photons emitted in its absence (F_D):

$$E = 1 - F_{DA}/F_D \qquad [1]$$

If there are no excited state reactions, fluorescence decay of the donor molecule is a first-order process and is mono-exponential so that $F_D(t) = F_0 \exp(-t/\tau_D)$, where τ_D is the donor lifetime when there is no acceptor molecule proximal. The decay is accelerated by the transfer mechanism when an acceptor molecule is in close proximity to the donor so that:

$$F_{DA}(t) = F_0 \exp[-t(1/\tau_D + 1/\tau_T)] \qquad [2]$$

where τ_T is the characteristic decay time constant of the energy transfer. τ_T is a function of the donor's lifetime and its distance from the acceptor:

$$\tau_T = \tau_D \left(\frac{r}{R_0}\right)^6 \qquad [3]$$

where R_0 is the critical distance (Förster radius) of the energy transfer and r is the distance between the donor and acceptor:

$$R_0 = 9.79 \cdot 10^3 (J \cdot \kappa^2 \cdot n^4 \phi_D)^{1/6} \text{ (Å)} \qquad [4]$$

where κ is the orientation factor, n is the refractive index of the medium, ϕ_D is the donor quantum yield, and J is the resonance integral determined by the spectral overlap of the donor emission ($F_D(\lambda)$) and acceptor absorption ($\varepsilon_A(\lambda)$). The resonance integral J is given by the following equation:

$$J = \int_0^\infty F_D(\lambda)\varepsilon(\lambda)\,\lambda^4\,d\lambda \qquad [5]$$

The distance between donor and acceptor molecules may be calculated from the following equation:

$$E = R_0^6/(R_0^6 + r^6) \qquad [6]$$

Applying Equations 1–6 to experimental FRET measurements, one can establish the proximity of two molecules as well as determine the distance between them and their mutual orientation.

2.2 Factors affecting the efficiency of fluorescence energy transfer

The value of the resonance integral (Equation 5) is affected by several parameters, some of which are not easily determined. For example, in homogeneous systems, such as diluted solutions of dyes, the refractive index is about equal to that of the solvent refractive index whereas, for an aqueous solution of protein–dye conjugates, the refractive index of water may only serve as an approximation. However, this is not the major source of error in the calculation of the Förster radius as the uncertainty in the refractive index generally contributes only a few per cent error to the calculation. The major sources of error are the uncertainties in the geometric factor, quantum yield, and solvatochromism of the donor and acceptor electronic spectra. Absolute values for the donor's extinction coefficient, shapes and positions of emission and acceptor absorption spectra, as well as the donor's quantum yield, are the major parameters affecting the Förster radius. The maximal value of the Förster radius may be as much as 90 Å for well-overlapping spectra of donor–acceptor pairs when the acceptor has a high extinction coefficient (5). A comparative list of Förster radii for the most widely used donor–acceptor pairs is given in *Table 1*. Extensive compilations of R_0 values can be found in the literature (7).

Variations in the geometrical factor (κ^2) alter the Förster radius and result in its distribution around an average value. κ^2 is equal to ⅔ for the fast diffusion limit, which occurs when the rotational correlation times of the donor and acceptor molecules are much shorter than the donor's lifetime. Thus, whenever possible, it is preferable to conjugate donor and acceptor molecules to proteins by linkages which allow free rotation. The two parameters that affect the rotational mobility of dye molecules conjugated to proteins are the length of the linker, that attaches the dye to the protein, and the solubility of the dye, which depends on its structure. Ideally, the linker length should be long enough to allow free rotation of the fluorophore but should not be too long because

Table 1 Typical values of Forster radii, R_0

Donor	Acceptor	R_0 (Å)	Reference
Tryptophan	Dansyl	8–14	10
Fluorescein	Tetramethylrhodamine	55	16
IAEDANS	Fluorescein	46	16
EDANS	DABCYL	33	16
Fluorescein	Fluorescein	44	16
Fluorescein	QSY-7 dye	61	16
Tetramethylrhodamine	Texas Red	52	16
Eu(K)	XL665	90	5
Cy5	C5.3	69[a]	13
		83[b]	13

[a] In solution.

[b] Conjugated to a protein.

excessive spacing between fluorophore and protein can result in edge-to-edge diffusion (8) of the donor–acceptor distance which reduces the accuracy of the measurement.

Another factor that influences the resonance integral to a large extent is solvatochromism; i.e. the dependence of donor and acceptor electronic spectra on the polarity of their microenvironment (9, 10). The scale of this phenomenon is in direct proportion to the values of the dye molecule's permanent dipole moments in the ground and excited states (11, 12). Spectral shifts of fluorescence spectra are significantly larger than those of the absorption spectra. For example the fluorescence spectral shift for a tryptophan residue may be as large as 40 nm. Similar shifts are exhibited for coumarine, aminonaphthalene, and benzoxadiazole derivatives. Xanthene and cyanine derivatives exhibit smaller spectral shifts, typically less than 10 nm.

Another important factor affecting the Förster radius is the donor's quantum yield which may be altered significantly upon conjugation to proteins. Typically, the observed quantum yield of a conjugated dye molecule is higher than its value in solution. For example, upon conjugation to proteins the quantum yield of Cy5 and Cy5.5 increases from 0.27 to 0.88 and from 0.23 to 0.36, respectively. The maxima of the absorption spectra of Cy5 and Cy5.5 shift towards longer wavelength by 2 and 5 nm and, of the emission spectra, by 2 and 10 nm respectively. As a result, the Förster radius for fluorescence energy transfer between Cy5 and Cy5.5 in solution is 68.7 Å whereas it is 85.2 Å when the dyes are conjugated to proteins (13).

3 Labelling of biomolecules with fluorescent probes

In order to employ FRET for ligand–receptor binding experiments the receptors and ligands may need to be labelled with a suitable donor and acceptor pair. There are several strategies for the insertion of chromophores into such biological molecules. These have been reviewed elsewhere (14–16). Sometimes it is possible to use the protein's intrinsic chromophores (e.g. tryptophan) (17–19) or the fluorescence of their ligands (e.g. flavin) (20–22). The most extensively employed method to date has been the labelling of proteins with covalently attached suitable fluorescent groups. This may be achieved by specific reactions between fluorescent probes designed to couple to reactive groups on the protein, such as thiols or free amines.

3.1 Random labelling

The most straightforward way to label biomolecules is by performing a conjugation of chromophores to primary amines such as the surface exposed ε-amines of lysines with amine reactive fluorescent probes. In general, one obtains fairly random labelling with such reactive probes, the extent of which depends on factors such as the concentration of probe and target as well as the reaction time. One of the drawbacks to this approach is that molecules are labelled heterogeneously and so it is hard to control the number of positions of chromophores per conjugated molecule. Random conjugation yields a distribution of labelled molecules containing varying numbers of chromophores. The situation is further complicated because, when the ratio of fluorophore to protein (or oligonucleotide, etc.) is high, self-quenching may significantly reduce fluorescence quantum yield and distort the absorption spectrum. The general lack of site-specificity of amine reactive dyes limits their utility as probes of the structure of biomolecules via FRET.

Although labelling with amine reactive probes is often thought of as a 'random' process, under proper conditions some specificity can be obtained. Because amine reactive probes are generally acylating reagents that react with the free basic form of the targeted amino group, differences in basicity can sometimes be used to selectively modify amino groups and/or control the extent of the modification. For example, the ε-amino group of lysyl residues in proteins has a pKa ~ 9 while free amino groups at the N-termini typically have a pKa ~ 7. Thus, by performing the modification at neutral pH, the N-terminal region can be selectively labelled.

There are several types of commercially available amine reactive probes. These include isothiocyanates that form thioureas upon reaction with the free-base form of amino groups. The bond formed is relatively stable although it undergoes some degradation over time. Succinimidyl esters form stronger amide bonds with free amino groups than do isothiocyanates but there can be solubility problems with these reagents. Sulfonyl chlorides are highly reactive and form extremely stable sulfonamide bonds with free amino groups although the reagents are unstable in water and care must be taken in using them for conjugation.

There are a large number of readily available amino reactive dyes that belong to each of the above types. The choice of one depends on the spectral properties desired for a given application. In the preparation of fluorescently labelled antibodies, isothiocyanates of fluorescein (FITC) or tetramethylrhodamine (TRITC) are the most widely used. Amine reactive probes have been widely used to measure ligand–receptor interactions.

Protocol 1

Conjugation of amino reactive probes to proteins (16)

Equipment and reagents

- Dialysis tubes
- Protein concentrators (Amicon (Millipore), Vivascience)
- Gel filtration column (Bio-Rad)
- Anion exchange column Mono Q HR (Pharmacia)
- Dry DMSO or DMF

- An amino reactive probe (Sigma, Molecular Probes)
- Conjugation 0.1 M sodium bicarbonate buffer pH 9.0 for isothiocyanates and sulfonyl chlorides or pH 8.3 for succinimidyl esters
- Stopping reagent: 1.5 M hydroxylamine

Protocol 1 continued

Method

1 Dialyse 1 ml (1–10 mg/ml) of a protein solution against 0.1 M sodium bicarbonate buffer pH 9.0 for isothiocyanates and sulfonyl chlorides or pH 8.3 for succinimidyl esters.

2 Dissolve an amino reactive dye molecule in DMF or DMSO (DMSO should not be used with sulfonyl chlorides).

3 Slowly add 50–100 μl of the reactive dye solution to the protein solution, stirring constantly, and incubate for 1–2 h at room temperature in the dark.[a]

4 Stop the reaction by adding 0.1 ml of freshly prepared 1.5 M hydroxylamine pH 8.5.

5 Separate the modified protein from unreacted dye by dialysing against a buffer of choice containing 1 mM EDTA pH 8.0 and then pass the protein through 10 ml gel filtration column (Bio-Gel P-6DG) equilibrated with the same buffer containing 1 mM EDTA to remove the residual dye.[b]

[a] The optimal protein: probe ratio (from 2 to 10) depends upon the probe and protein reactivity and should be optimized to get the required labelling protein-to-probe ratio. Usually for fluorescein, tetramethylrhodamine, or Texas Red the labelling ratio should not excite 1 to 3 dyes per protein due to the concentration self-quenching.

[b] A subsequent protein purification and separation of homogeneously labelled protein fractions is available by using an anion exchange chromatography with Mono Q HR column (Pharmacia) in linear gradient of NaCl concentration.

3.2 Site-specific labelling

Specific labelling of a particular site is often required and reagents and protocols have been developed to accomplish this objective. This is illustrated by the conjugation of probes to free thiol groups (see *Protocol 2*). In proteins, free thiol groups occur in the amino acid cysteine. Free thiols can also be generated by the reduction of the disulfide bond in cysteine residues. The thiol functional group can be labelled with considerable selectivity with alkylating reagents such as iodoacetamides, maleimides, benzylic halides, or bromomethyl ketones. The optimal pH range for the reaction is generally 6.5–8.0 (15). Iodoacetamides react with thiols to form thioethers and are reasonably selective although they can potentially react with methionine, histidine, tyrosine, or the free-base form of amines (generally the pKa of lysine is such that modification will not occur at pH < 8 but there is a possibility of modification at the N-terminus). Maleimides react with thiols to form a thioether and are more selective as they do not react with methionine, histidine, or tyrosine.

Because the thiol functional group can be labelled with high specificity, it is often a good choice for FRET studies designed to measure inter- or intramolecular distances. For this purpose, cysteines can be introduced, by single site-directed mutagenesis, into a protein to provide a specific site for probe insertion. In employing this approach one should take care to mutate only those amino acids that do not affect the protein's structure or alter its solubility. Another precaution is that the mutation should not affect the protein–ligand interaction, i.e. the mutation should not affect the protein–ligand interface. Usually it is not difficult to satisfy all the above requirements when the protein's 3D structure has been resolved. For receptors whose structures are not known, the decision on mutation requires insight, which can sometimes be obtained from its sequence. An elegant example of this approach was employed by Baird and co-workers to examine the conformation of IgE in solution. A recombinant dansyl specific IgE class monoclonal antibody was prepared with cysteine replacing a serine near the C-terminal ends of the heavy chains. The sulfhydryl groups of the cysteines were specifically labelled with fluorescein 5-maleimide. Energy transfer between dansyl hapten residing in the Fab combining sites and the fluorescein near the C-terminus provided evidence for a bent conformation of IgE both in solution and when bound to its FcεRI receptor (23).

Organic dye molecules (16) as well as inorganic metal–ligand complexes (24) are widely used as fluorescent tags. Among organic dye molecules, aminonaph-thalene, fluorescein, rhodamine, and cyanine derivatives have found the widest applications. Perhaps the most popular donor–acceptor pairs are fluorescein-tetramethylrhodamine, iaedans-fluorescein (16), and Cy3/Cy5 (25). An exceptionally strong resonance coupling between europium cryptate (EuK) and allophycocyanin (XL665), providing a Förster radius of 90 Å, make this pair well suited to homogeneous binding assays (5, 26, 27).

A novel type of probe, fluorescent semiconductor nanocrystals, has recently been introduced into experimental practice (28, 29). Although nanocrystals are considerably larger than dye molecules, their fluorescence intensity is equivalent to that of 20–25 Rhodamine 6G molecules. In addition, highly stable, pH-insensitive nanocrystals, whose fluorescence spectra cover the entire UV–visible spectral region, are now available (28).

Protocol 2

Conjugation of thiol reactive probes with proteins (16)

Equipment and reagents

- Desalting column (Bio-Gel P-6DG, Bio-Rad)
- Gel filtration column Superdax 75 or 200 (Pharmacia)
- Anion exchange column Mono Q HR (Pharmacia)
- A thiol reactive probe
- Conjugation 20 mM Tris, 150 mM NaCl buffer pH 8.0
- Disulfide reducing reagent: DTT

Method

1 Dialyse 1 ml (1–10 mg/ml) of a protein solution against 20 mM Tris, 150 mM NaCl buffer pH 8.0.

Protocol 2 continued

2   Add 10 mM DTT and incubate for 20 min at room temperature in order to reduce the free thiol group.

3   Remove DTT by gel filtration using 10 ml Bio-Gel P-6DG column equilibrated with the same buffer without DTT.

4   Prepare 1 mM stock solution of the thiol reactive probe in dry DMSO or DMF.

5   Slowly add an aliquot of the stock dye solution to the protein solution, stirring continuously, to obtain a protein:probe ratio of 1:10.

6   Incubate for 1–2 h at room temperature in the dark.

7   Stop the reaction by adding DTT (5 mM final concentration).

8   Separate the modified protein from unreacted dye by dialysing against a buffer of choice containing 1 mM EDTA pH 8.0 and pass the protein through a 10 ml gel filtration column (Bio-Gel P-6DG) equilibrated with the same buffer containing 1 mM EDTA to remove the residual dye.

9   Purification of the labelled protein by gel filtration (Superdex 75 or Superdex 200) and, if needed, further separation of the homogeneously labelled protein fractions is available using an anion exchange chromatography with Mono Q HR column (Pharmacia) and a linear gradient of NaCl concentration.

3.3 Indirect labelling using fluorescent antibodies (see *Protocol 3*)

One of the most widely used methods for staining proteins in cells is based on mono- or polyclonal fluorescently labelled antibodies. This method has been employed in studies designed to determine the frequency of protein expression by cells as well as elements of protein organization, such as the proximity of subunits (30–33). Typically, the spatial limit that can be resolved with this approach is about 60–90 Å. One intrinsic problem in using this method is that, in most cases, a high affinity intact antibody will cluster the surface receptors on the cell (34). Thus it may be essential to use monovalent Fab fragments which, while generally exhibiting lower binding capacity, do not change the oligomerization state of the receptor. In addition, it is important to ascertain that Fab binding does not alter the receptor interactions being studied. Methods of producing Fab fragments from intact antibodies, as well as a detailed protocol for calculating the energy transfer efficiency in cell surface experiments, is described in an excellent review by Matko and Edidin (33). The efficiency of energy transfer can be measured via quenching of the donor's emission or that of the sensitized acceptor emission. Typically, one would employ three samples to evaluate the energy transfer efficiency between two proteins:

(1) Donor labelled protein and unlabelled protein.

(2) Donor labelled protein and acceptor labelled protein.

(3) Unlabelled protein and acceptor labelled protein.

Sample 1 serves as a standard for unquenched donor fluorescence, sample 2 allows for measuring energy transfer efficiency, and sample 3 accounts for the direct excitation of the acceptor without energy transfer.

Protocol 3

Indirect labelling proteins with fluorescent antibodies

Equipment and reagents

- Cell counting chamber (Hauser Scientific Company)
- 12 × 75 mm polystyrene round-bottom tube (Falcon)
- Flow cytometer (Becton Dickinson)
- Mono- or polyclonal entire antibodies or Fab fragments conjugated to an appropriate dye[a]

- Negative control monoclonal antibodies (the best of the same class) labelled with the same dye to the same dye-to-protein ratio
- PBS or Tyrode's buffers with 0.1% sodium azide
- Cell suspension

Method

1 Adjust the cell concentration to 10^6/ml with culture medium or a buffer of choice (e.g. PBS, Tyrode's) and place the cell suspension (1–2 ml) into the test-tubes.

2 Centrifuge the cells at 250 × g for 5 min. Decant the supernatant.

3 Add the appropriate amount of labelled entire monoclonal antibodies (or Fab fragment) or negative control antibodies.[b]

4 Vortex the cells and incubate for 30–60 min on ice in the dark.

5 Wash the cells three times in buffer with 0.1% sodium azide.

6 Add 1 ml of the buffer and vortex.

7 Keep the cells on ice until experimentation.[c]

[a] TRITC, FITC, PE, Cy3, Cy5, Cy-Chrome are the dyes most often employed for flow cytometry or fluorescence microscopy.

[b] The amount of antibodies should be either determined previously by titrations or supplied by the manufacturer.

[c] Cell samples should be fixed with 0.5% paraformaldehyde if it is not intended to carry out the experiment immediately after labelling (35).

3.4 Labelling with green fluroescent protein constructs

Perhaps the most elegant approach for introducing fluorescent reporter groups was developed by Prasher and co-workers (36). This was based on their work in cloning and sequencing green fluorescence protein (GFP) from the jellyfish *Aequorea Victoria* (37). One way of employing GFP is based on the creation of a protein chimera, one part of which is the 26.9 kDa GFP protein, which provides its intrinsic fluorescence. The source of this fluorescence is a p-hydroxybenzyli-dene-imidazolidinone chromophore created by cyclization and oxidation of the protein's tripeptide sequence, serine–dehydrotyrosine–glycine, at positions 65–67. By employing recombinant DNA techniques, GFP can be expressed in many organisms. Moreover, GFP mutants with a variety of spectral properties (colours) have been produced (38). GFPs have been used in FRET-based biological applications by employing pairs of mutants that have been designed in such a way as to provide good overlap of the fluorescence spectrum of the donor and the absorption spectrum of the acceptor as well as reasonable separation of the emission spectra between donor and acceptor. To date, two pairs of GFP mutants have been used for FRET-based biological applications (38). Blue fluorescent protein (BFP) is often employed as a donor and enhanced green fluorescent protein (EGFP) as an acceptor. The second pair is cyan fluorescent protein—yellow fluorescent protein (CFP-YFP) (39). As a result, GFP has become an important fluorescent label for the study of biological structure and function. Some mutants of GFP can be used for fluorescence energy transfer with conventional dye molecules such as fluorescein or tetramethylrhodamine. For example, the binding kinetics of a nuclear transcription factor–GFP chimera (p65-GFP) with tetrachlorocarboxylfluorescein labelled NF-κB-binding oligonucleotides was employed in studies in the millisecond time-domain by stopped-flow energy transfer fluorimetry (40). This analysis revealed a conformational transition of GFP-p65–DNA complex that is influenced by the transcription factor binding. This result illustrates the potential power of this method, not only to monitor a complex formation and evaluate its stoichiometry, but also to resolve the elementary steps of the complex assembly.

3.5 Specific covalent labelling via 4-cysteine's tag

Another new method for introducing a fluorescent group into recombinant proteins has been developed by Griffin and co-workers (41). It exploits the reversible covalent bond formation between organoarsenicals and pairs of thiols. Trivalent arsenic compounds bind to the paired thiol groups of proteins containing closely spaced pairs of cysteines or the cofactor lipoic acid (*Figure 1*). Such binding is completely reversed by reaction with small vicinal dithiols such as 2,3-dimercaptopropanol [British anti-Lewisite (BAL)] or 1,2-ethanedithiol (EDT), which form higher affinity compounds with the organoarsenical than do cellular dithiols. An unusually high affinity ($K_D < 1$ nM) was achieved by the design of a domain with four cysteines that had been organized to bind an organic molecule containing two appropriately spaced trivalent arsenic atoms. An important feature of this design is that the FLASH-EDT$_2$ molecule is not fluorescent (quantum yield is $\leqslant 5 \times 10^{-4}$) until it binds to its target, whereupon it becomes highly fluorescent (quantum yield is 0.49).

4 Protocols for measuring fluorescence energy transfer

In most applications FRET can be detected by the appearance of sensitized fluorescence of the acceptor or by quenching of donor fluorescence. For some dye

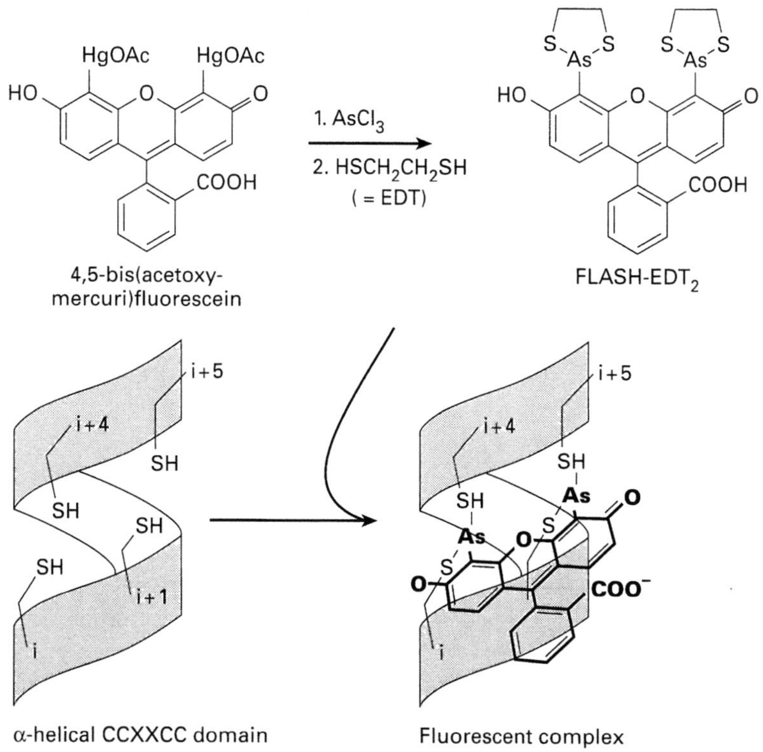

Figure 1 Synthesis of FLASH (20) and proposed structure of its adduct with a helical tetracysteine-containing peptide or protein domain. Although the structure is drawn with the i and i +4 thiols bridged by one arsenic and the i +1 and i + 5 thiols bridged by the other, one cannot rule out the isomeric compound in which one arsenic links the i and i + 1 thiols while the other links the i + 4 and i + 5 thiols (adopted from ref. 41).

molecules, with a small Stokes shift between the absorption and emission spectra, there is a high probability for homo-energy transfer (6). In such cases, FRET can be detected by the resulting fluorescence depolarization or time-resolved spectral shift of fluorescence spectra (10, 42). Non-fluorescent energy acceptors such as 4-dimethylaminoazobenzene (DABCYL) and QSY™-7 (16) dyes have the advantage of eliminating the potential problem of background fluorescence resulting from direct (i.e. non-sensitized) acceptor excitation. Fluorescence energy transfer may be used in different experimental protocols to evaluate the proximity of proteins in cells (30–33, 43) or for determining the stoichiometry of assembled protein complexes (44, 45).

4.1 Monitoring donor quenching (see *Protocol 4*)

The most popular and simple method is based on measuring steady-state fluorescence of a donor, with and without an acceptor present. Aromatic amino acid residues or externally attached probes may be used as a donor. This method is well suited to studying receptor–ligand complexes with 1:1 stoichiometry since, in general, one cannot expect that every acceptor ligand will equivalently contribute to the quenching of the donor's fluorescence.

Protocol 4

Determination of equilibrium binding constant and reaction stoichiometry by monitoring donor quenching

Equipment

- Spectrofluorimeter with a temperature control cuvette and a stirrer
- Hamilton syringe or a titration machine

Method

1 Conjugate an appropriate fluorescence quencher to the ligand using *Protocols 1* or *2*.

2 Prepare a stock solution of the labelled ligand of concentration 10- to 30-fold higher than the affinity constant.

3 Add small aliquots of the ligand stock solution to a receptor solution in a temperature controlled fluorescence cuvette with a stirrer with a Hamilton syringe.[a]

4 Plot the saturation binding curve as the donor fluorescence intensity in presence of different ligand concentrations $[I_{DA}]_i$ versus ligand concentrations.[b]

5 Evaluate the complex stoichiometry and equilibrium binding constant from the saturation binding curve.[c]

[a] (1) The receptor and ligand molecules should be purified chromatographically and well mixed before the experiment. (2) If a ligand (peptide) stock solution is prepared with organic solvents, such as DMSO or DMF, one should check that they do not affect the receptor binding and that a high concentration of the ligand does not change the pH of receptor solution. (3) The optical cuvette should be transparent in the UV spectral region (made from quartz) when the natural protein chromophores, tryptophan, tyrosine, phenylalanine, or UV-emitting organic dye molecules are used as fluorescence donors. (4) One should take care to ensure equilibrium at each step of the reaction. If the experiment is performed at a relatively high temperature (higher than 30 °C) over a long period of time (hours) the sample should be protected from evaporation. (5) Donor emission should be protected from photobleaching and (6) the temperature in the cuvette should be stabilized to an accuracy of at least $\pm$ 0.1 °C.

[b] The $\{I_{DA}\}_i$ values should be corrected for the sample dilution (C_i) and non-specific binding of the ligand (I_{DA}'): $C_i\{I_{DA} - I_{DA}'\}_i$ The dilution coefficients are calculated as $C_i = (V_0 + i^*\Delta V_i)/V_0$ where V_0 is the volume of the receptor solution, and $i^*\Delta V_i$ is the dilution value at the ith titration step. The non-specific binding of the ligand is accounted for by carrying out the same with a non-cognate ligand labelled with the same fluorescence acceptor (I_{DA}').

[c] In fact, this method evaluates stoichiometry for 1:1 complexes or detects a deviation from this simple bi-molecular reaction. Examination of more complex reactions is possible only when each ligand quenches the donor fluorescence equivalently. Generally, for ligand molecules labelled with fluorescence acceptors, only those ligands that quench the donor's fluorescence with highest efficiency will be detected. In addition, this method suffers from a low accuracy when the receptor possesses multiple fluorescent donors and an acceptor situated on a ligand molecule quenches only one of them. In such a case, the change in the receptor's fluorescence is determined by the ratio of the number of quenched donors to their total number. This value is even smaller when the energy transfer is not 100% efficient.

A similar experimental protocol may be performed in 96-well plates and the fluorescence signals determined by a fluorescence plate reader when the receptor and ligand amounts are not limited and binding is a relatively slow process.

4.2 Monitoring the acceptor-sensitized fluorescence

When the emission spectra of the donor and the acceptor molecules are well separated (and therefore fluorescence of the red slope of the donor does not contribute to the acceptor emission) it may be possible to overcome the above problems by monitoring the sensitized fluorescence of the acceptor. The protocol for this experiment is similar to that one described in the previous section with a difference that the correction of the donor–acceptor signal should also account for the acceptor direct excitation, $C_i\{I_{AD} - I_A\}_i$, where I_{AD} is the sensitized fluorescence of the acceptor and I_A is the acceptor fluorescence upon the same excitation without donor. The I_A value may be measured by carrying out the experiment with a non-cognate ligand labelled with the same acceptor molecule. This value also accounts for non-specific binding.

This protocol was employed for studying peptide binding to 'empty' class I major histocompatibility complex (MHC-I) encoded molecules. Mouse H-2K^d MHC-I contains 13 tryptophan residues, only some of which can be quenched to a certain extent by an acceptor molecule (e.g. dansyl) conjugated to a bound peptide (18). Due to the good separation between tryptophan and dansyl emission spectra (maximum at 350 nm and 510–550 nm respectively) and a significant increase in dansyl quantum yield upon dansylated peptide binding to the 'empty' MHC-I, the sensitized dansyl fluorescence enables peptide binding to be monitored with high precision and sensitivity (since the acceptor emission signal changes by 100% from zero to its maximum). This is illustrated by an equilibrium titration experiment shown in *Figure 2*.

An additional important advantage of this method is that it enables the measurement of binding and dissociation reaction time-courses as illustrated in *Figure 3*. This enables evaluation of the complex stoichiometry with greater confidence than when limited to equilibrium binding data alone. We have used such kinetic data to gain insight into the deviation from linearity of the Scatchard plot (*Figure 2* insert).

The time-course of peptide binding monitored at a 1:1 stoichiometry of heavy chain and β_2-microglobulin (β_2m), and upon addition of β_2m excess, revealed the cause for the Scatchard plot's deviation from linearity. The kinetic studies revealed that the deviations are caused by a partial dissociation of the heavy chain/β_2m heterodimer during the course of the peptide binding reaction and not due to the existence of two different peptide binding sites in class I heavy chain. Thus, the above method significantly circumvents the energy transfer 'saturation' problem and takes advantage of the large number of tryptophans present in the H-2K^d molecule, several of which are located in the vicinity of the peptide binding site and therefore participate in the energy transfer process. In addition, the tryptophan absorption band (270–300 nm) is situated in a minimum

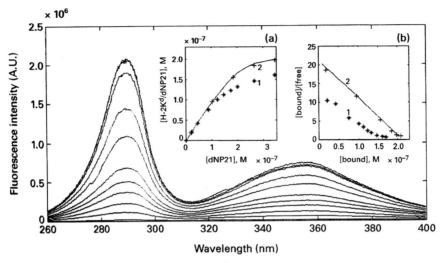

Figure 2 Equilibrium titrations of 0.22 μM H-2K^d with dNP21. A set of dansyl excitation spectra (t = 20 °C, λ_{em} = 490 nm) was recorded at increasing peptide concentrations. Inset: Effect of β_2m concentration on apparent affinity of H-2K^d heterodimer for the peptide. Equilibrium binding curves present (a) the area under the energy transfer bands as a function of peptide concentration (0.01–0.3 μM) in the presence of excess β_2m (above the equimolar ratio of heavy chain/β_2m of the heterodimer); [β_2m] = 1 μM (*, curve 1), [β_2m] = 4 μM (+, curve 2). (b) Scatchard plots of the binding data.

between the first and second absorption bands of the dansyl group. This enables the dansylated peptide binding to be monitored even in the presence of excess unbound form.

4.3 Studying stoichiometry of a complex composed of several subunits

When two or more ligands bind to the same receptor, or a complex is composed of several subunits, it is generally preferable to use an alternative protocol. If the receptor has two different binding sites for the same ligand, the latter can be labelled with different dyes which can serve as donor and acceptor, respectively. Upon incubation of the labelled ligands with the receptor, energy transfer may be observed. This method was successfully used for establishing the stoichiometry of the hetero-oligomeric γ-aminobutyric acid (GABA$_A$) receptor, a member of the major inhibitory ligand-gated ion channel family in the brain, in intact cells (*Figure 4*). Knowledge from preliminary studies indicating that the receptor is composed of α, β, and a third subunit (γ, δ, or ε), suggested an approach by which the authors modified one of the subunits, that were expressed transiently in HEK 293 cells, with an epitope tag. The maximal fluorescent signal obtained using EuK labelled mAbs to cells that exhibited the tag epitope on a single subunit demonstrated that the ratio of α_i, β_2, and γ_2 subunits is 2:2:1. This stoichiometry was confirmed by using an elegant application of FRET: Individual subunits (α_i, β_2, or γ_2) were half-saturated with EuK labelled mAbs and half-

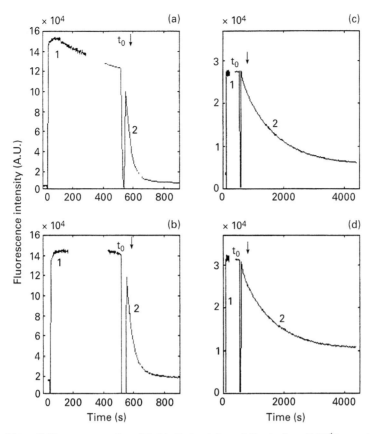

Figure 3 Time-courses of peptide binding and dissociation to/from H-2K^d heavy chain/β_2m heterodimer for different dansylated peptides as a function of β_2m concentration (t = 37 °C) (ref. 1, 18). Peptide binding reactions were started by mixing (a, b) [H-2K^d heavy chain/β_2m] = 70 nM, [TYC(Dansyl)RALV] = 2 μM or (c, d) [H-2K^d heavy chain] = 80 nM, [TYC(Dansyl)RTRALV] = 0.5 μM; 4 μM excess of β_2m above equimolar ratio of the heterodimer was added in experiments b and d. The peptide exchange reaction was started by addition of 50 μM of the unlabelled peptide TYQRTRALV to the samples.

saturated with XL665 labelled mAb. Energy transfer from EuK to XL665 occurred with both the α-, and β-labelled, but not with the γ-labelled subunits. The lack of energy transfer with γ-subunit labelled receptors indicated that there is a single γ-subunit per receptor. Thus the stoichiometry of GABA$_A$ receptors is $(\alpha_1)_2(\beta_2)_2(\gamma_2)_1$ (**44**).

FRET enables not only a complex stoichiometry to be characterized, but also the investigation of its assembly mechanism. Returning to the example of the class I MHC molecule, which is composed of a heavy chain, β_2m, and peptide, we were interested in pursuing the possible evolutionary reason for having β_2m as a non-covalent part of the complex. The 3D structure of the ternary complex has revealed that the peptide binding occurs to the heavy chain and that β_2m does not interact directly with it. Experiments based on FRET demonstrated that an

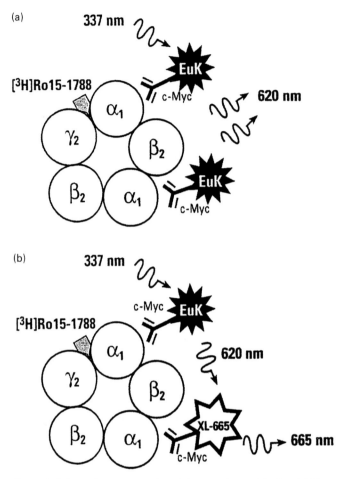

Figure 4 Schematic representation of the experimental design. (a) The number of subunits present is quantified using saturation binding of EuK-derivatized c-Myc. This is compared with the number of benzodiazepine binding sites using the radioligand [³H]Ro15-1788. (b) Fluorescence resonance energy transfer occurs only when two copies of a subunit are present. (Adopted from ref. 44.)

allosteric mechanism controls the interactions among these three components. Although peptides can bind to the 'free' heavy chain, the binding affinity increases dramatically upon association of the heavy chain with β_2m. Furthermore, the interaction between heavy chain and β_2m becomes stronger upon peptide binding to the heavy chain.

With such a mechanism in mind it is natural to ask how peptide binding stabilizes the heavy chain—β_2m interaction and whether the stabilization is a function of the peptide structure. Two different scenarios are conceivable:

(a) The heavy chain in association with β_2m can exist in two conformations and that peptide binding switches the two states in a manner that alters the heavy chain—β_2m affinity.

(b) Peptide binding changes the heavy chain conformation in a continuum of conformations which adapt it for closer interaction with β_2m.

In the first case, the extent of the affinity enhancement should not depend on peptide affinity, whereas a strong dependence on peptide affinity is expected for the second.

By labelling the ternary complex with two different probes, these hypotheses could be distinguished. One probe, a dansyl group, was attached to a side chain of an amino acid residue on the peptide that is not directly involved in the inter-action with the binding site. The other probe, Texas Red (TR), was attached to a cysteine of the β_2m C88S mutant. The mutation was made at a site of β_2m that does not interact with the heavy chain. Fluorescence energy transfer from the heavy chain tryptophans to the dansyl group (W $\rightarrow$ Dansyl) and further transfer to the TR attached to the β_2m (W $\rightarrow$ Dansyl $\rightarrow$ TR) was found to take place in the doubly-labelled ternary complexes. A ternary complex was produced by first mixing H-2K^b heavy chain with β_2m-TR, followed by addition of SIC(Dansyl) NFEKL or SIIFEC(Dansyl)KL peptide (0.5 μM final concentration). Peptide addition to the heavy chain pre-incubated with β_2m-TR further enhanced the intensity of the sensitized TR fluorescence due to the increase in affinity of β_2m and heavy chain, thereby increasing the number of the peptide complexes.

As shown in *Figure 5* the fluorescence signals (D–F) monitored at 530 nm (Dansyl) upon excitation of tryptophans at 280 nm are proportional to the con-centration of ternary complexes. Signals (A–C) monitored at 615 nm, where TR emits (upon excitation at 280 nm), are proportional to the concentration of the doubly-labelled ternary complexes. Addition of a large excess (2 μM) of unlabelled β_2m at t_1 started the exchange of labelled for unlabelled molecules at a rate that was determined by β_2m-TR dissociation from the ternary complex. At the same time, since the concentration of the labelled β_2m-TR was chosen to be lower than that of heavy chains, the addition of excess unlabelled β_2m also led to an increase in the total concentration of the ternary complexes. Unlabelled OVA (SIINFEKL) peptide (final concentration 40 μM) was then added to the sample at t_2 in order to initiate peptide exchange. The above experimental protocol allowed the independent comparison of both β_2m and peptide dissociation rates from the ternary complex under identical conditions. The experiments revealed that the β_2m exchange rate was independent of peptide affinity, suggesting that the first hypothesis, which assumes a mechanism where two discrete conforma-tions of heavy chain exist, is valid (19).

Li and co-workers (46) employed a different experimental protocol for studying the stoichiometry of phospholamban (PLB), a protein modulating the activity of the cardiac Ca^{2+} pump. Each of PLB subunits was labelled separately either with a donor or acceptor and then the labelled chains were mixed at dif-ferent donor/acceptor ratios. A theoretical simulation, assuming random mix-ing and association among protein subunits in a ring-shaped homo-oligomer enabled the authors to determine that the complex is composed of six identical subunits.

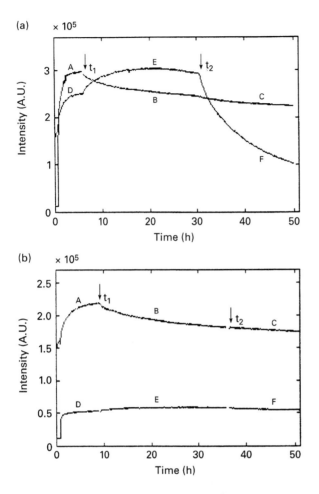

Figure 5 Studying the allosteric mechanism of heavy chain, β_2 m and peptide interactions within class I MHC ternary complex at 20 °C by fluorescence energy transfer (ref. 19). Energy transfer from the H-2K^b heavy chain intrinsic tryptophans to the peptide's dansyl group (W → Dansyl) and a cascade of further transfer to TR attached to the β_2m, (W → Dansyl → TR) was found to take place. At the first stage of the experiment ($0–t_1$) 20 μl of the heavy chain stock solution (1.8 μM) was added to 140 μM of 0.1 μM β_2m-TR followed by addition of peptide (0.5 μM final concentration): (a) SIINFEC(Dansyl)L, (b) SIC(Dansyl)NFEKL. Addition of a high concentration (2 μM) of unlabelled β_2m at t_1 started the exchange of labelled by unlabelled β_2m molecules, at a rate, which was determined by β_2m-TR dissociation from the ternary complex. Peptide exchange was initiated by addition of SIINFEKL (40 μM final concentration) at t_2. Such experimental protocol allows the independent comparison of both, β_2m and peptide dissociation rates from the ternary complex under identical conditions. The fluorescence signals A, B, and C monitored at 530 nm (Dansyl) upon excitation of tryptophans at 280 nm are proportional to the concentration of ternary complexes. Signals D, E, and F monitored at 615 nm, where TR emits upon excitation at 280 nm, are proportional to the concentration of the doubly-labelled ternary complexes.

4.4 Photobleaching FRET techniques

An interesting and useful method for measuring FRET between cell membrane components is based on monitoring changes in the rate of photobleaching of donor molecules as a function of the proximity of the acceptor molecules. The theoretical basis for this experimental method is that the photobleaching rate of the donor is affected by an additional decay path available for the excited state (i.e. FRET from donor to acceptor). When FRET occurs, the excited state lifetime of the donor is shortened and as a consequence the photobleaching time constant is lengthened. The energy transfer efficiency in this method is calculated as a ratio of the photobleaching time constants of the donor molecule in the presence or absence of a proximal acceptor molecule (47). This approach has been applied to numerous biological systems including topology of MHC class I antigen (30), the oligomerization of epidermal growth factor receptors (48), epitope mapping on human peripheral blood cells (49), and the proximity of mast cell function-associated antigen with FcεRI (43). Although this method is relatively qualitative, it still provides insights into the stoichiometry of assembled cell surface complexes. An advantage of this experimental protocol is its relative simplicity, universality, and applicability to a single living cell. The sensitivity of a fluorescence energy transfer method in combination with high spatial resolution of a near-field scanning optical microscope, allowed the extension of this method to a single donor—acceptor molecular level (2, 3). Thus, this FRET-based method has become a reliable tool for studying formation of molecular complexes on individual living cells in real time.

Protocol 5

Determination of protein proximity by photobleaching FRET (43)

Equipment and reagents

- Fluorescence microscope with a powerful excitation light source (a lamp of at least 100 W or $\geqslant 5$ mW CW laser)
- RAB buffer: 135 mM NaCl, 5 mM KCl, 1.8 mM $CaCl_2$, 1 mM $MgCl_2$, 5.6 mM glucose, 10 mM Hepes pH, 7.4, and 0.1% BSA, or a buffer of choice
- Cells, microscope glass coverslips, tissue culture plates, and tissue culture medium
- Donor and acceptor labelled fluorescence mAb specific for the molecules proximity of which is going to be ascertained
- 4% paraformaldehyde

Method

1 Grow cells on microscope glass coverslips. If the cells do not adhere, the coverslips may be coated with poly-L-lysine.

2 Wash the cells with RAB or a buffer of choice.

Protocol 5 continued

3 Incubate the cells with the donor, or donor and acceptor-labelled antibodies for 1 h at 4 °C.

4 Wash three times with buffer.

5 Fix the cells with 4% paraformaldehyde (for 5 min at 4 °C and an additional 15 min at room temperature.

6 Wash with buffer.

7 Mount the coverslips in PBS.

8 Acquire fluorescence images upon donor excitation with the light exposure for 3 sec for every image.

9 Evaluate photobleaching decay constants of the samples stained with donor, τ, and donor and acceptor labelled mAb, τ'.

10 Evaluate the energy transfer efficiency as $E = 1 - \tau/\tau'$, and a distance between the donor labelled and acceptor labelled proteins (Equation 6).

4.5 Homogeneous time-resolved FRET assay

Some experimental protocols are based on measuring the time-resolved fluorescence decay of a donor as a function of ligand concentration. Although this approach requires expensive instrumentation, it enables a more accurate measurement of the energy transfer efficiency and so the study of more complex systems. An example of this technique involved measuring energy transfer between tris-bipyridal-caged europium cryptate Eu(K), which exhibits a long lifetime emission at 620 nm, to allophycocyanin (denoted XL665). This acceptor was chosen because of its high extinction coefficient at the wavelength of Eu(K) emission. In addition, this donor–acceptor couple absorbs light at 337 nm, Eu(K), and thus is appropriate for excitation by a N_2 laser, and emits light at 665 nm (XL665). Most biological media do not affect emissions of these probes. Due to the unique combination of the donor and acceptor spectral parameters this pair possesses a long Föster radius (90 Å). Time-resolved detection of the acceptor emission provides a high dynamic range with separation of the long energy transfer signal (50–400 μsec) from a nanosecond fluorescence of background. This experimental protocol was employed in a binding assay for p53 to HDM2, the negative regulator of p53 (5).

5 Determination of a complex stoichiometry

The determination of the stoichiometry of a receptor complex is a difficult problem to which FRET can be applied. Often, one can probe the system by labelling ligands and or receptors with appropriate donor and acceptor molecules. The relationship among the spectroscopic signal change, ΔF, with concentrations of a receptor, $[M]$, and its ligands, $[L_i]$, have been reviewed in the detailed chapter by Maurice Eftink (50). In general, when a complex is an assembly of several subunits with co-operative interactions, a special experimental protocol design, similar to that mentioned above (44) is required. For some special cases, when a receptor can bind one or more ligands (equivalently or not) with or without cross-talk among them, evaluation of the reaction stoichiometry is feasible. However, special care should be taken to ascertain that the experimental protocol provides equivalent contribution to the total signal (ΔF) by each of the bound ligands. For the general case of n ligands, a reaction can be described by the following scheme:

$$M + L \overset{K_1}{\Leftrightarrow} ML_1 \overset{K_2}{\Leftrightarrow} ML_2 \ldots \overset{K_i}{\Leftrightarrow} ML_n \qquad [7]$$

$$K_i = \frac{[ML]}{[ML_{i-1}][L]} \qquad [8]$$

where K_i is the equilibrium association constant for binding of the i_{th} L molecule to the receptor, the ΔF signal is a function of n unknown parameters, and therefore evaluation of them is an ill-posed problem. In some special cases when there is only one ligand or n equivalent ligands, the binding profile is a hyperbolic function and the number of bound ligands can be calculated easily from a Scatchard plot (50). The most accurate evaluation of stoichiometry is derived from the equilibrium titration experiment in the stoichiometric limit ($K \cdot [M_0]$ is high). For more complex reactions, such as when there are multiple ligand binding sites that exhibit allosteric behaviour, the reaction parameters can be evaluated by non-linear fitting of experimental binding data (ΔF as a function of ligand concentration) in accordance with an appropriately chosen model (50, 51). However, additional experimental data, such as that obtained from complementary kinetic experiments, are generally required in order to justify the selected model.

6 Conclusions

FRET is one of the most sensitive, powerful, and promising methods for investigating the interactions of receptors with their ligands, both *in vivo* and *in vitro*. When coupled with the high spatial resolution of a confocal fluorescence microscope, FRET can be used in studies of protein interactions in living cells at the level of single molecules. Application of this method to cell biology promises to enable real time monitoring of cell signal transduction pathways, a development that should be of great benefit to the study of both fundamental and applied biology as well as research of new pharmaceuticals.

Acknowledgements

This work was supported by the M. and J. Heinemann Foundation (I. P.) and an NIH Fogarty Senior International Fellowship No. 1F06TW002308 (R. G. P.).

References

1. Kubitscheck, U., Kircheis M., Schweitzer-Stenner, R., Dreybrodt, W., Jovin, T. M., and Pecht, I. (1991). *Biophys. J.*, **60**, 307.
2. Ha, T., Ting, A. Y., Liang, J., Caldwell, W. B., Deniz, A. A., Chemla, D. S., *et al.* (1999). *Proc. Natl. Acad. Sci. USA*, **96**, 893.
3. Ha, T., Enderle, T., Ogletree, D. F., Chemla, D. S., Selvin, P. R., and Weiss, S. (1996). *Proc. Natl. Acad. Sci. USA*, **93**, 6264.
4. Stenroos, K., Eriksson, S., Hemmila, I., Blomberg, K., Linqvist, C., and Hurskainen, P. (1998). *Cytokine*, **10**, 495.
5. Kane, S. A., Fleener, C. A., Zhang, Y. S., Davis, L. J., Musselman, A. L., and Huang, P. S. (2000). *Anal. Biochem.*, **278**, 29.
6. Forster, T. (1948). *Ann. Phvsik.*, **2**, 55.
7. Van der Meer, B. W., Coker, G. III, and Chen, S.-Y. S. (1994). *Resonance energy transfer. Theory and data*, VCH, New York.
8. Haas, E., Katchalski-Katzir, E., and Steinberg, I. Z. (1975). *Biochemistry*, **17**, 5065.
9. Rubinov, A. N., Zenkevich, E. I., Nemkovich, N. A., and Tomin, V. I. (1982). *J. Luminescence*, **26**, 367.
10. Gakamsky, D. M., Haas, E., Robbins, P., Stromimger, J. L., and Pecht, I. (1995). *Immunol. Lett.*, **44**, 195.
11. Bakhshiev, N. G. (1972). *Spectroscopy of intermolecular interactions*. Nauka. Leningrad.
12. Lippert, E. (1975). In *Organic molecules photophysics* (ed. J. B. Birks), Vol. 2, pp. 1–31. Wiley-Interscience, London.
13. Schobel, U., Egelhaaf, H. J., Brecht, A., Oelkrug, D., and Gauglitz, G. (1999). *Bioconjug. Chem.*, **10**, 1107.
14. Stryer, L. (1978). *Annu. Rev. Biochem.*, **47**, 819.
15. Hermanson, G. T. (1995). *Bioconjugate techniques*. Academic Press, San Diego, USA.
16. Haugland, R. P. (1996). *Handbook of fluorescent probes and research chemicals.* Molecular Probes, Eugine, OR, USA.
17. Gakamsky, D. M. and Pecht, I. (1996). *Immunologist*, **5**, 247.
18. Gakamsky, D. M., Bjorkman, P. J., and Pecht, I. (1996). *Biochemistry*, **35**, 14841.
19. Gakamsky, D. M., Boyd, L. F., Margulies, D. H., Davis, D. M., Strominger, J. L., and Pecht, I. (1999). *Biochemistry*, **38**, 11887,
20. Shepherd, G. B. and Papadakis, N. (1976). *Biochemistry*, **15**, 2888.
21. Scouten, W. H., Graaf-Hess, A. C., De Kok, A., Grande, H. J., Visser, A. J., and Veeger, C. (1978). *Eur. J. Biochem.*, **84**, 17.
22. Koland, J. G. and Gennis, R. B. (1982). *Biochemistry*, **21**, 4438.
23. Zheng, Y., Shopes, B., Holowka, D., and Baird, B. (1991). *Biochemistry*, **30**, 9125.
24. Terpetschnig, E., Szmacinski, H., and Lakowicz, J. R. (1997). In *Methods in enzymology* (ed. L. Brand and M. L. Johnson), Vol. 278, pp. 295–321. Academic Press, New York.
25. Bastiaens, P. I., Majoul, I. V., Verveer, P. J., Soling, H. D., and Jovin, T. M. (1996). *EMBO J.*, **15**, 4246.
26. Moore, K. J., Turconi, S., Miles-Williams, A., Djaballah, H., Hurskainen, P., Harrop, J., *et al.* (1999). *J. Biomol. Screen.*, **4**, 205.
27. Earnshaw, D. L., Moore, K. J., Greenwood, C. J., Djaballah, H., Jurewicz, A. J., Murray, K. J., *et al.* (1999). *J. Biomol. Screen*, **4**, 239.
28. Bruchez, M. Jr., Moronne, M., Gin, P., Weiss, S., and Alivisatos, A. P. (1998). *Science*, **281**, 2013.
29. Chan, W. C. W. and Nie, S. (1998). *Science*, **281**, 2016.
30. Damjanovich, S., Vereb, G., Schaper, A., Jenei, A., Matko, J., Starink, J. P., *et al.* (1995). *Proc. Natl. Acad. Sci. USA*, **92**, 1122.

31. Damjanovich, S., Bene, L., Matko, J., Alileche, A., Goldman, C. K., Sharrow, S., *et al.* (1997). *Proc. Natl. Acad. Sci. USA*, **94**, 13134.

32. Damjanovich, S., Matko, J., Matyus, L., Szabo, G. Jr., Szollosi, J., Pieri, J. C., *et al.* (1998). *Cytometry*, **33**, 225.

33. Matko, J. and Edidin, M. (1997). In *Methods in enzymology* (ed. L. Brand and M. L. Johnson), Vol. 278, pp. 444–62., Academic Press, New York.

34. Mohammadi, M., Honegger, A., Sorokin, A., Ullrich, A., Schlessinger, J., and Hurwitz, D. R. (1993). *Biochemistry*, **32**, 8742.

35. Lanier, L. L. and Warner, N. L. (1981). *J. Immunol. Methods*, **47**, 25.

36. Chalfie, M., Tu, Y., Euskirchen, G., Ward, W. W., and Prasher, D. C. (1994). *Science*, **263**, 802.

37. Prasher, D. C., Eckenrode, V. K., Ward, W. W., Prendergast, F. G., and Cormier, M. J. (1992). *Gene*, **111**, 229.

38. Pollok, B. A. and Heim, R. (1999). *Trends Cell Biol.*, **9**, 57.

39. Miyawaki, A., Llopis, J., Heim, R., McCaffery, J. M., Adams, J. A., Ikura, M., *et al.* (1997). *Nature*, **388**, 882.

40. Schmid, J. A., Birbach, A., Hofer-Warbinek, R., Pengg, M., Burner, U., Furtmuller, P. G., *et al.* (2000). *J. Biol. Chem.*, **275**, 17035.

41. Griffin, B. A., Adams, S. R., and Tsien, R. Y. (1998). *Science*, **281**, 269.

42. Runnels, L. W. and Scarlata, S. F. (1995). *Biophys. J.*, **69**, 1569.

43. Jürgans, L., Arndt-Jovin, D., Pecht, I., and Jovin, T. M. (1996). *Eur. J. Immunol.*, **26**, 84.

44. Farrar, S. J., Whiting, P. J., Bonnert, T. P., and McKernan, R. M. (1999). *J. Biol. Chem.*, **274**, 10100.

45. Fernandez-Miguel, G., Alarcon, B., Iglesias, A., Bluethmann, H., Alvarez-Mon, M., Sanz, E., *et al.* (1999). *Proc. Natl. Acad. Sci. USA*, **96**, 1547.

46. Li, M., Reddy, L. G., Bennett, R., Silva, N. D. Jr., Jones, L. R., and Thomas, D. D. (1999). *Biophys. J.*, **76**, 2587.

47. Jovin, T. M., Arndt-Jovin, D. J., Marriott, G., Clegg, R. M., Robert-Nicoud, M., and Schorman, T. (1990). In *Optical microscopy in biology* (ed. B. Herman and K. Jacobson) pp. 575–602. Wiley-Liss.

48. Gadella, T. W. Jr. and Jovin, T. M. (1995). *J. Cell Biol.*, **129**, 1543.

49. Szabo, G. Jr., Weaver, J. L., Pine, P. S., Rao, P. E., and Aszalos, A. (1995). *Biophys. J.*, **68**, 1170.

50. Eftink, M. R. (1997). In *Methods in enzymology* (ed. L. Brand and M. L. Johnson), Vol. 278, pp. 221–57. Academic Press, New York.

51. Espenson, J. H. (1995). *Chemical kinetics and reaction mechanisms*. McGraw-Hill, Inc.

Chapter 6
Monitoring intracellular trafficking of receptors

Raymond Molloy and Thomas E. Hughes
Department of Ophthalmology and Visual Science, Yale School of Medicine, 330 Cedar Street, New Haven, CT 06520, USA.

Antoine Robert and James R. Howe
Department of Pharmacology, Yale School of Medicine, 330 Cedar Street, New Haven, CT 06520, USA.

1 Introduction

When appended to a membrane protein as a genetic fusion, the green fluorescent protein (GFP) can be used to mark both the presence and amount of receptor in the cellular compartments of living cells (1). The great advantage of this approach is that it affords a unique view of protein trafficking as it occurs. Fluorescent receptors can therefore serve as direct assays for the determinants of protein dynamics. Indeed, the creative ferment that has developed as researchers continue to devise new approaches is one of the main attractions of this emerging field.

It is helpful to begin by understanding why the GFP is so useful as well as some of the disadvantages. The protein comprises 238 amino acids and adopts a unique barrel-shaped structure (2) whose 11 staves in the beta sheet conformation surround the fluorophore. One drawback to using GFP as a tag, then, is simply the large size of the protein. Fortunately, its remarkably compact and self-contained structure appears, empirically, to minimize interactions with neighbouring proteins. In addition, both the N- and C-termini extend from the same end of the barrel so that, despite its size, the GFP can be placed within another coding sequence without adding much physical distance between the two parts of the receptor.

Several mutated versions of the GFP and homologous proteins from other organisms have yielded a useful set of fluorescent proteins with different spectral properties (3–5). These mutants exhibit different absorption and/or emission spectra and are often referred to colloquially by the appearance of their predominant emission peak (e.g. cyan fluorescent protein). These variants and the similar red fluorescent protein from reef coral (6, 7), make it quite easy to create different fusion proteins that can be imaged simultaneously in a living cell (see *Table 1*).

Table 1 Fusion proteins that can be visualized simultaneously in living cells

Name	Sequence	Mutations	Ex/Em maxima	Reference
GFP[a]	M62654	N/A	395/509	1
EGFP[b]	U55761	F64L; S65T	488/507	33
ECFP[c]		F64L; S65T; Y66W; N146I; M153T; V163A	433/475	4
EYFP[c]		S65G; V68L; S72A; TZ03Y	513/527	2
DsRed[d]	U62636	N/A	558/583	7
GFPuv[e]	U62636	Q80R; F99S; M153T; V173A	395/509	34
pHlourin[f]		S202H; E132D; S147E; N149L; N164I; K166Q; I167V; R168H; L220F	Varies	15

[a] Original jellyfish GFP mRNA. Useful as a reporter, but cryptic intron can be spliced in some organisms.

[b] Humanized codon usage. Excellent for confocal microscopy; can be used with DsRed for double labelling.

[c] ECFP and EYFP: suitable pair for FRET.

[d] DsRed: synthetic cDNA with humanized codon usage based on the coral protein DsFP583.

[e] GFPuv: mutant GFP in which the excitation spectrum is simplified to single peak at 395. *E. coli* codon usage.

[f] pHlourin: two excitation peaks whose amplitude varies with pH. Useful for studying trafficking through compartments of different pH.

Over the next several years, an increasing variety of fluorescent fusion proteins will continue to be made. These reagents will be appropriate tools for co-expression studies designed to measure the trafficking of a receptor in a neuron relative to different compartments and other parts of the signalling system. The more you work with the GFP the more likely you are to create multiple fusions with cyan, green, yellow, and red fluorescent proteins. This likelihood is worth considering before you begin to subclone because the inclusion of restriction sites at the junction between your receptor and the GFP can make it possible to swap different fluorescent protein coding sequences in a single step. Restriction site-independent fusion strategies such as the polymerase chain reaction (PCR) based overlap extension are less modular. (See Section 3 for protocols that discuss the creation of fusion constructs.)

A drawback germane to the new GFP user is that there is not yet much in the way of published lore on the practicalities of making a particular application operational at the bench. The aim of this chapter is to help to fill that gap by providing the new investigator with enough information to analyse the trafficking of a GFP-tagged receptor in living neurons.

The chapter is divided into five parts. In Section 2 we show how to obtain information about the likely structure of a protein and the use of such clues in the design of productive fusion constructs. An introduction to the available variants of the GFP is also included here. Section 3 provides protocols for the creation of fusion constructs. Section 4 covers transfection in continuous cell lines and in primary cultures of rodent neurons. Section 5 discusses time-lapse microscopy.

2 Planning

2.1 Getting started

In this section we describe a few of the tools that are available through the Internet for sequence analysis. Given the rapid pace at which information is updated in the electronic databases, there is really no alternative to using them. If the links described here have closed by the time you peruse these pages, it will only mean that newer, faster, and more comprehensive sites have supplanted them. While the software links may change, the evolutionary connections between sequences and the logic of their analysis will endure.

2.2 Sequence analysis

2.2.1 Goals

At this stage the task can be divided into three parts:

(a) Discover possible domains, in the receptor under study, that should not be disrupted.

(b) Decide where to append the GFP.

(c) Devise a cloning strategy that is compatible with your goals.

Before you decide where to put the GFP, you must try to define as clearly as possible what you would like to do with the construct once you have made it. It is at this stage, hypothesis generation, that electronic domain mapping is most useful. You can begin by asking where, and when, the protein moves within the cell. In addition, you may wish to ask which parts of the protein sequence are the determinants of targeting to a particular compartment. The ease with which you can make a fusion construct will vary depending on whether you plan to stick to the first question, or envisage moving on to the second.

If you will be making multiple constructs to map a targeting determinant, it is far simpler to consider at the outset all the plasmids you might wish to make, than it is to make them seriatim. The poorly considered choice of a restriction site made at one stage of a project may slow your progress unnecessarily at the next.

2.2.2 Getting the information: an example with SWISSPROT

If the receptor is known, the best first step is to look in the SwissProt protein database. In the following discussion, links on a web page will be enclosed in angled brackets (e.g. <click me>). We will use the human sequence for the glutamate-gated ion channel GLUR1 (8) as an example. Open your web browser and type in the following uniform resource locator (http://www.expasy.ch/). On the second line of the page, you will see a set of links to other mirror sites, each of which contains the same information. Click on the link to the server in the country closest to you. When the page reappears, click on the link labelled '<SWISS-PROT and TrEMBL>—Protein databases'. Click on <SRS>. Click on the button that says <Start>. As you will have noticed, on each page there are connections to a wealth of information that is worth exploring. The purpose of the Sequence Retrieval System is to help you find the information about a particular protein efficiently. On the next page, simply click the <All> button next to Sequences to indicate that you wish to search the whole protein database. Then click <continue>.

When the next page appears, type in GLUR1 into the box where the cursor is blinking. Click <Do Query>. A list of the protein records containing the term GLUR1 will be returned. Click on the link for the human GLUR1 sequence <SWISS_PROT:GLR1_HUMAN>. (The direct URL for this entry is: *http://www. expasy.ch/cgi-bin/get-full-entry?[SWISS PROT-ID:'GLR1 HUMAN']*.) At the top of the entry you will see a summary of the date the record was entered, a brief description of its function, and some relevant references. Scroll down the page to the section entitled 'Features'. Here you will find useful information about the protein and its orientation in the membrane. For example, GLUR1 contains a signal peptide and four candidate transmembrane domains. It is also alternatively spliced.

Yet more information can be found easily. Return to the SWISSPROT display for GLUR1 and scroll to the bottom of the page. You will see two adjacent links: one labelled <scan prosite> and a second that says <profilescan>. Click on <profilescan>. This command automatically compares the GLUR1 to a database of sequence patterns that are diagnostic for the presence of previously identified protein domains.

In this case, five hits are reported (see *Figure 1*). Each is shown on a separate line. The first symbol of the line indicates whether the homology is likely to be statistically significant ('!' returned) or is of marginal significance statistically ('?' returned; see arrow A in *Figure 1*). To the right of this are two numerical scores of statistical significance (labelled B). The first number is called the Nscore or normalized score; higher numbers indicate a more reliable match. The second number is a raw score and can be safely ignored. The next column (labelled C) shows the amino acid range of GLUR1 that is homologous with the given domain. For example, amino acids 532 to 638 show significant homology with the domain shown. The index number of the domain (labelled D) is shown in the

Figure 1 Organization of a ProfileScan record. The arrows point to the different parts of a record returned by the ProfileScan server. **A** The exclamation point indicates a hit or successful match between the sequence submitted and a profile stored on the server. **B** The Normalized Score and Raw Score each give estimates of how reliable statistically the match is. Higher values indicate greater reliability. **C** The amino acid range positions show the region in the sequence submitted that is similar to the profile returned. **D** The index number of the profile is also returned. **E** A brief description of the profile is shown. **F** A link to more information is also provided.

next column, followed by a vertical bar. To the right of the bar is the name of the domain (E) along with links to further information (F).

The first domain returned by the search overlaps with the putative second transmembrane domain listed in the SWISSPROT entry of GLUR1. The record therefore indicates that this region of GLUR1 may be homologous to a pore loop first noted in K^+ channels. Such protein loops enter and exit the membrane bilayer from one side without completely traversing the bilayer. When a pore loop is present, the pattern of extracellular and intracellular regions found to the carboxy side of the loop are the opposite of what one would have predicted if the putative transmembrane domain did completely span the bilayer. Several lines of evidence now indicate that glutamate-gated ion channels do in fact contain such pore loops (9). The homology search results therefore point to the possibility that the carboxy terminus of this protein might be—as it in fact is—intracellular.

Two of the remaining hits are to the same motif—a ligand-binding domain found in prokaryotic permeases. The results of structural modelling show that this region of GLUR1 is, in fact, similar to these prokaryotic domains. Subsequent experiments have shown that these regions are functionally involved in glutamate-binding in GLUR1 (10). Similarly, the atrial natriuretic factor receptor family denotes a similar domain. Thus, a region closer to the N-terminal portion of the protein contains the ligand-binding domain of the ion channel subunit.

This example is simply meant to give an indication of the kinds of data that can be gathered from the databases. The number of proteins annotated there will continue to increase as the genome sequencing efforts continue. If the protein of interest is less well characterized, it is also possible to begin with the homology searches in order to gain a similar understanding of your receptor's candidate domains.

2.2.3 The prediction of protein sorting signals search engine (PSORT)

One limitation of search engines that look for homologous domains is that they do not scan for shorter sorting sequences that play an essential role in receptor trafficking (11). The PSORTII server developed by Nakai (12) is therefore a useful adjunct for the characterization of receptor sequences. Like several other sequence analysis servers, PSORTII will scan a sequence for potential signal peptides. It is, however, the only server that will find other informative targeting sequences like endocytic sorting signals. Open a new page in your web browser. Enter the following URL: (http://psort.nibb.ac.jp/). Click on the Search <PSORTII Prediction> button and paste in your sequence in the form. Click on the <submit> button to start the search. Once the server has returned the results page, you can save the web page as a text file.

2.2.4 From protein to DNA

Now that you have gathered basic information about your protein, it is time to make use of it. Your goals at this stage are:

(a) Map the domain information onto the primary structure.

(b) Map the protein information onto the DNA.

(c) Look for appropriate restriction sites or primer sites.

(d) Decide where to place the GFP.

The easiest way to map the protein information you have gleaned is to use a computer program dedicated to manipulating and annotating DNA sequences. There are several available for this purpose, e.g. Gene Construction Kit for Mac users (www.textco.com) and the LaserGene software suite for both Mac and Windows users (www.dnastar.com). The point is to get an idea of how the potential domains cluster along the protein sequence.

With the data gathered about GLUR1, for example, it is already possible to draw several conclusions. First, to place GFP at the N-terminus in the correct translational frame, it will be necessary to insert the sequence after the signal peptide cleavage site. It is prudent to give the protease some breathing space by placing the GFP insert after the 21st or 22nd amino acid of the immature polypeptide. Secondly, one would have to be careful in building such a construct to ensure that placing the GFP near the ligand binding domain would not unduly disrupt function. Similarly, a C-terminus fusion would run the risk of disrupting potential regulatory interactions by sterically hindering the phosphorylation or dephosphorylation of the receptor. It is often the case that there will be good reasons for not putting the GFP anywhere. You should not be dismayed by this fact, but do ascertain what you will have to test in order to validate the construct functionally. In the case of GLUR1, it has been shown, surprisingly, that the presence of the GFP at the N-terminus of the polypeptide does not seem to hinder the ligand-induced gating of the channel (13). Once you know how the domains cluster, it will be a simple matter to find out whether there are useful restriction sites near domains you may wish to modify or delete.

2.2.5 GFP variants

The demonstration by Chalfie and colleagues that GFP can be expressed heterologously spurred several laboratories to search for useful variants (1). The mutations discovered, as well as those introduced rationally for the optimization of codon usage, generated a vibrant palette of proteins (see *Table 1*). Use enhanced green fluorescent protein (EGFP) if you wish to make a single fusion protein. ECFP (enhanced cyan fluorescent protein) and EYFP (enhanced yellow fluorescent protein) are mutants that can be used to follow two fusion proteins simultaneously since their respective excitation and emission spectra are well separated. In principle, one protein could also be used to fill the entire cell, while the second is employed to study trafficking of a fusion protein. The ECFP/EYFP pair has also been employed for fluorescence resonance energy transfer experiments (14). To follow two fusion proteins simultaneously via confocal microscopy— where UV excitation is not usually available—EGFP and coral-derived red fluorescent protein (DsRed) are compatible for double labelling, although it is often necessary to wait about four days after transfection for the DsRed protein to fold. Conversely, where UV excitation is a requirement, GFPuv has also been described. Green fluorescent protein with intact UV absorption peak (GFPuv) is not recommended for mammalian expression, however, since the codon usage has been optimized for bacterial expression. Finally, the pHlourin variant of GFP has been shown to change its excitation properties in a pH-dependent fashion and is useful in the study of trafficking between compartments of different pH (e.g. synaptic vesicle fusion with the plasmalemma) (15).

2.2.6 GFP placement

There are currently three possibilities for making a genetic fusion of a receptor cDNA and the GFP. The GFP can be appended at either end of the receptor open reading frame or it can be inserted between codons somewhere in the middle. To date, the great majority of published constructs have been N- or C-terminal fusions. For a first attempt, make either an N-terminal or a C-terminal fusion.

The few internal fusion constructs that have been created show that the strategy is viable, however, and have provided information that could probably not have been provided by N- or C-terminal chimeras (16). Since the GFP is constrained at two points when inserted within the coding sequence of the receptor it is less free to move and can sometimes serve as a reporter of receptor function. For example, Siegel and Isacoff placed the GFP coding sequence within that of the voltage-gated shaker K^+ channel and were able to use changes in the fluorescence of the construct to measure changes in membrane potential (17). Surprisingly, the reverse experiment—inserting a protein of interest within a modified GFP sequence—has also proved to be feasible (18). Such optical reporters are likely to play an increasingly prominent role in the genetic dissection of signal transduction pathways.

3 Building the construct

3.1 Initial considerations

The simplest way to fuse the sequence encoding a receptor to the one encoding GFP is through the judicious use of a restriction site. GFP coding sequences with elaborate multiple cloning sites that produce in-frame fusions are commercially available (e.g. Clontech K6000-1; K6001-1; www.clontech.com). These can be used to fuse the GFP coding sequence directly to a restriction site at one end of the receptor open reading frame. In rare but instructive cases, the site may be fortuitously present, but it is more likely that one will have to be introduced with site-directed mutagenesis. The most significant drawback to this approach is that extra coding region from the multiple cloning site is often introduced between the two fusion partners. The addition of this extra protein may have unintended consequences but this is not inevitable. We have successfully used a short Ala-Gly dipeptide as a linker between the C-terminus of a protein to be tagged and the start of the GFP. If a longer flexible spacer seems more appropriate, Kaech *et al.* have used a sequence derived from the vesicular stomatitis virus glycoprotein effectively (19).

A second approach is to use PCR to create the fusion protein. There are several

ways to proceed. First, one can amplify both fusion partners in separate reactions with primers that introduce unique restriction sites. The sites are then
cut and ligated to create the fusion. As novel mutants become available, the
introduced restriction sites can be used to swap in the new GFP variant. Another
PCR strategy for creating fusion constructs known as Gene Splicing through Overlap Extension (SOEing) can be used to create fusions with little or no extraneous
coding sequence (20). This approach is quicker but less modular. The largest
constraint is that it requires PCR primers that are suitable for amplification and
which are complementary to the appropriate end of the coding sequence.

3.2 Creating fusion proteins by restriction and ligation

To use a restriction site for creating a GFP fusion, one or both of the coding
sequences may have to be changed such that new sites are created that can be
used for in-frame fusions. As an example, we show how to place the GFP at the
C-terminus of the glutamate receptor subunit GluR3 (*Protocol 1*). *Figure 2* shows the

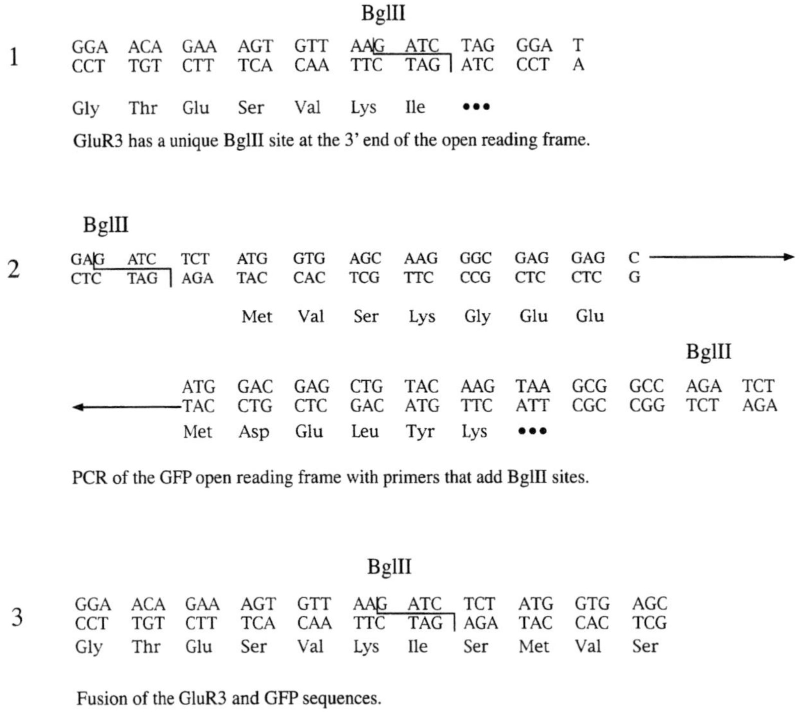

Figure 2 Creation of a fusion joint. Part 1. The presence of a *Bgl*II restriction site (AGATCT) at
the end of the open reading frame of GluR3 is shown. Part 2. An upper PCR primer sequence
that appends a *Bgl*II site at the 5′ end of the EGFP open reading frame can be used to create a
fused open reading frame between the GluR3 and EGFP coding sequences. Below this is shown
a lower primer sequence that can be used to complete the PCR reaction (see *Protocol 1*). Part
3. The result of ligating in the PCR product into the *Bgl*II site at the end of the GluR3 open
reading frame. The first seven amino acids are native to GluR3; the last three are native to
EGFP. A serine linker is present between them.

144

C-terminus portion of GluR3 which contains a unique *BglII* site just 5' of the stop codon. To fuse GFP in-frame, a *BglII* site has to be introduced into the 5' end of the GFP coding sequence in the correct frame. In *Figure 2*, two primers are illustrated that work to amplify the EGFP coding sequence. The upper primer creates a *BglII* site in the same frame as the one in GluR3. The lower primer introduces a *BglII* site 3' of the stop codon. When the GFP sequence is amplified it can either be cut and ligated directly into the unique *BglII* site in GluR3 or captured in a PCR cloning plasmid and subsequently shuttled into GluR3 (e.g. TOPO brand of cloning plasmids; Invitrogen Cat. No. K4560-01 or No. K2875-20; www.invitrogen.com). An alternative strategy is to create a GFP-bearing plasmid into which an amplified receptor sequence can be inserted. There are a few plasmids that have been designed to do this and they can greatly speed up the addition of GFP to either end of a receptor (e.g. ref. 21).

3.3 Creating a fusion protein by PCR

Protocol 1

PCR of GFP to add a restriction site in-frame

Equipment and reagents

- 0.5 ml thin-walled Eppendorf tubes (e.g. USA Scientific, 1405-4400)
- Thermal cycler (e.g. Stratagene Robocycler40, 400830)
- Primer A: ggagatctctatggtgagcaagggcgaggagc (25 μM)
- Primer B: agatctggccgctttacttgtacagctcgtccat (25 μM)

- pEGFP-N1 (Clontech, 6085-1)
- Cloned Pfu DNA polymerase (Stratagene, 600153)
- 10 × cloned Pfu polymerase buffer (supplied with enzyme)
- 25 mM deoxynucleotides mix (Roche 1969064)[a]
- Mineral oil (e.g. Leader mineral oil)

Method

1. Chill a thin-walled tube on ice for 3 min.

2. Mix the following reagents in an Eppendorf tube: 1 μl primer A, 1 μl primer B, 1 μl pEGFP-N1 50 ng/μl, 50 μl 10 × buffer 0.5 μl of 25 mM dNTP mix, 1 μl Pfu polymerase, and 40.5 μl deionized water.

3. Overlay the reaction with two drops of mineral oil.

4. Program the thermal cycler with the following parameters:
   (a) One round of 3 min at 94 °C, 1 min at annealing temperature, and 1 min at 74 °C.
   (b) 23 rounds of 1 min at 94 °C, 1 min at annealing temperature, 1 min at 74 °C.

5. Resolve 5 μl of the reaction in a 1% agarose gel made in Tris/acetate/EDTA (TAE) buffer to check the yield.

[a] 25 mM dNTPs can be made by adding an equal volume of 100 mM dATP, 100 mM dCTP, 100 mM dGTP, and 100 mM dTTP in an Eppendorf tube; the final mix will be 25 mM of each nucleotide.

Gene SOEing is simply a two-step PCR protocol that relies upon the two fusion partners priming the beginning of the second reaction (*Protocol 2* and *Figure 3*). The only difference is that the two internal primers, P2 and P3, add ~ 15 bp of the other fusion partner to each product. These regions in turn prime the creation of the full-length fusion construct in a second round of amplification which is driven solely by primers P1 and P4.

Protocol 2

Creating a fusion by SOEing

Equipment and reagents

- 0.5 ml thin-walled Eppendorf tubes (see *Protocol 1*)
- Thermal cycler (see *Protocol 1*)
- Primer 1
- Primer 2
- Primer 3
- Primer 4
- pEGFP-N1 (see *Protocol 1*) (50 ng/μl)
- Plasmid containing your receptor cDNA (50 ng/μl)
- Cloned Pfu DNA polymerase (see *Protocol 1*)
- 10 × cloned Pfu polymerase buffer (supplied with enzyme)
- 25 mM deoxynucleotides mix (see *Protocol 1*)[a]
- Mineral oil

Method

1   Amplify the initial two products as outlined in *Protocol 1* using the appropriate templates and the primer pairs P1 and P2 as well as P3 and P4.

2   Resolve 1 μg of each of these products on an agarose gel and carefully cut out the bands.

3   Purify the DNA using Geneclean (BIO 101, 1001-400; follow manufacturer's recommendations) and elute the DNA with 10 μl of water.

4   Set up another PCR reaction identical to the one described in *Protocol 1*, but in this case substitute 100 ng each of the PCR products for the template and add only the primer pair P1 and P4.[b]

[a] 25 mM dNTPs can be made by adding an equal volume of 100 mM dATP, 100 mM dCTP, 100 mM dGTP, and 100 mM dTTP in an Eppendorf tube; the final mix will be 25 mM of each nucleotide.

[b] At this point the success of the overlap extension hinges upon the concentration of the templates and the appropriate annealing temperature, so carefully laid out experimental plans that systematically explore these two variables can be critical for success.

---

4 Expression

4.1 Introduction

Once you have a construct to test, you will wish to express it in mammalian cells as efficiently as possible. There are several published methods including viral

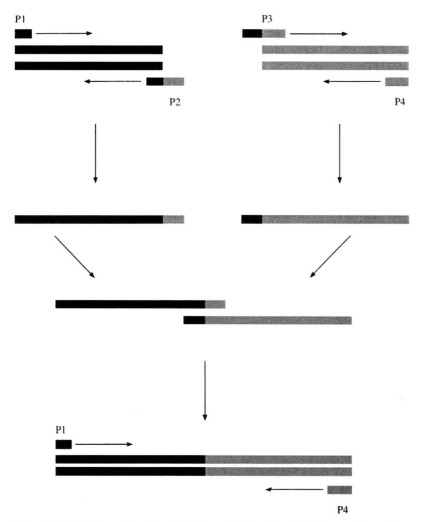

Figure 3 Splicing by overlap extension. The method developed by Horton and Pease for creating genetic fusions is shown. Two rounds of PCR are carried out. In the first round, separate PCR reactions are performed on the cDNAs to be joined. One primer in each reaction is designed to span the fusion joint to be created. Here the lower primer (P2) for the black sequence contains a region of homology to the 5′ end of the grey sequence. Similarly, the upper primer of the grey sequence (P3) contains the 3′ end of the black sequence. The products of these separate PCR reactions are shown below the vertical arrows. In the second round of PCR, the products from round one are mixed and hybridize upon melting. Successful hybrids are selectively amplified by including primers P1 and P4 in the reaction.

vectors, calcium phosphate transfection, lipid-mediated transfection, and the creation of transgenic animals (22–24). Lipid-mediated transfection is perhaps the simplest and is described below.

If you have not transfected neurons before, establish the method in your laboratory with a non-toxic plasmid (e.g. pEGFP-N1) first, before you attempt to transfect your tagged receptor. In addition, it is imperative to determine if your fusion is still functional. The assay will depend on the kind of receptor. For example, a tagged ion channel might be tested via the patch-clamp technique. If the construct is an appropriately-coupled heptahelical metabotropic receptor, one simple assay is to co-transfect with RFP tagged protein kinase C gamma (PKC; Clontech, 6929-1). If the receptor is functional, the translocation of the red-fluorescent PKC can be observed via time-lapse microscopy (see Section 5). If either cell lines, that are selectively defective for the receptor, or model genetic organisms are available, the most stringent assay is to rescue the mutant phenotype with the fusion protein.

A good plan for this stage of your project might go as follows:

(a) Transfect pEGFP-N1 in HEK 293 cells.

(b) Transfect your construct in HEK 293 cells; use pEGFP-N1 as a positive control.

(c) Determine whether your construct is functional in HEK 293 cells.

(d) Transfect pEGFP-N1 in primary neurons.

(e) Transfect your construct in primary neurons; use pEGFP-N1 as a positive control.

4.2 Protocol for HEK 293 transfection

The protocols below (*Protocol* 3 and 4) are for the transfection of HEK 293 cells and presuppose a basic knowledge of cell culture. The standard reference for cell culture techniques is Freshney (25). For transfection studies, it is particularly important to passage your cells faithfully. Avoid confluent monolayers since, once cells have entered this state, transfection efficiency often decreases permanently.

The purpose of the protocols is to ensure that your construct is visible in a living, hardy, easily transfected cell *before* you go to the trouble of neuronal expression. Since you will wish to view the living cells via fluorescence microscopy, the best results will be obtained with the cells grown on coverslips. A convenient way to do this is to use 35 mm dishes that have a hole drilled through the bottom surface. A coverslip glued over the hole becomes part of the bottom surface of the dish. Such dishes are available commercially from MatTek.

Protocol 3

Preparation of HEK 293 cells for transfection

Equipment and reagents

- MatTek dishes (part P35G-1.5-14-C)
- 15 cm tissue culture dish
- T25 flasks
- DMEM/10% FCS
- Trypsin/EDTA
- Hanks balanced salt solution
- HEK 293 cells (Microbix, PD-02-01)

Protocol 3 continued

Method

1 Grow your HEK 293 cells in a T25 flask in DMEM/10% FCS until 80% confluent.

2 Take a sleeve of MatTek dishes and open them.

3 Place eight of the dishes inside a 15 cm tissue culture dish.[a]

4 Save the remaining two dishes in the sleeve by rolling the plastic down and securing it with a clip.

5 Take the nearly confluent T25 flask from the incubator.

6 Aspirate the medium from the flask.

7 Rinse the monolayer with 1 ml of Hanks solution or trypsin solution.

8 Allow the flask to lie flat so that the entire monolayer is covered.

9 Rock the flask back and forth three times, taking care to avoid allowing any medium close to the mouth of the flask.

10 Aspirate the rinse solution.

11 Add 0.5 ml of 1 × trypsin solution and allow the flask to lie flat.

12 With the long side of the flask parallel to you, rock the flask so that the near side and then the far side alternately rise and fall.

13 Look at the bottom of the flask. As it angles toward you and away from you, observe the monolayer. When the monolayer turns white, the cells are ready to detach.

14 Rap one edge of the flask gently against the culture bench. You should do this strongly enough to help dislodge the cells, but not so strongly that the medium starts to jump. After a few tries, the monolayer should slough off the surface of the flask.

15 Add 5 ml of pre-warmed DMEM/10% FCS.

16 Pipette the medium plus sloughed-off cells up and down five times in order to disperse the cells.

17 Dilute the cells by adding 12 ml of pre-warmed DMEM/10% FCS.

18 Plate the cells by adding 2 ml to each MatTek dish.

19 Add 1 ml of the diluted cells to a new T25 flask.

20 Add 4 ml DMEM/10% FCS to the new T25.

21 Return the new T25 flask and the 15 cm dish containing the MatTek dishes to the incubator overnight.

[a] The 15 cm dish allows for more convenient handling of the MatTek dishes in the incubator.

Protocol 4

Transfection of HEK 293 cells with pEGFP

Equipment and reagents

- 1.5 ml Eppendorf tubes
- pEGFP-N1 (Clontech) 1 µg/µl
- OPTI-MEM (Life Technologies, 31985-070)
- Lipofectamine2000 (Life Technologies, 11668-019)
- HEK 293 cells from *Protocol 3*

Method

1 Place two Eppendorf tubes on the laminar flow bench.

2 Aliquot 600 µl of OPTI-MEM into each Eppendorf tube.

3 Add 36 µl of Lipofectamine2000 into one tube.

4 Mix the contents of the tube.

5 Turn on a timer set for 5 min.

6 Add 12 µg of DNA to be transfected to the second tube.

7 After 5 min, draw up the lipofectamine/OPTI-MEM mixture into a pipette.

8 Pick up the second tube that contains the DNA and put the tip of the pipette underneath the surface of the medium. Do not place the pipette tip against the wall of the tube. Rather, place it directly in the middle of the medium. Depress the plunger in a smooth motion to eject the lipofectamine/DNA complex.

9 Hold the tube between index finger and thumb. With your ring finger gently tap the bottom of the tube a few times to mix the contents efficiently.

10 Wait 20 min.

11 Swirl the mixture a few times as described.

12 Take the HEK 293 cells in the MatTek dishes out of the incubator.

13 To each MatTek dish add 200 µl of the transfection mixture.

14 Return the dishes to the incubator overnight.

15 Dishes can be observed on an inverted fluorescence microscope the next day.

4.3 Neuronal transfection (*Protocol 5*)

4.3.1 Considerations

The single most important determinant of successful neuronal transfection is the overall health of your cultures. If, for whatever reason, your cultured neurons are in less than optimal health, do not bother to transfect them. For information on the culture of these exacting cells, turn to Banker (26). Use less DNA and Lipofectamine2000 than you would with HEK 293 cells. You may also wish to consider shortening the incubation time. We generally plate hippocampal neurons at 1×10^6 neurons per poly-lysine coated MatTek dish. Neurons are concentrated on the coverslip not over the whole bottom surface of the dish.

Protocol 5

Transfecting neurons

Equipment and reagents

- 1.5 ml Eppendorf tubes
- Neurobasal medium (Life Technologies, 21103049) supplemented with 2 mM glutamine (Nb)
- pEGFP-N1 (Clontech) 1 μg/μl
- Neurobasal medium supplemented with B27 serum-free supplement (Life Technologies, 17504044) and 2 mM glutamine (Nb/B27)
- Lipofectamine2000 (see *Protocol 4*)

Method

1 Place two Eppendorf tubes on the laminar flow bench.

2 Aliquot 200 μl of Nb medium into each Eppendorf tube.

3 Add 12 μl of Lipofectamine2000 into one tube.

4 Mix the contents of the tube.

5 Turn on a timer set for 5 min.

6 Add 2 μg of DNA to be transfected to the second tube.

7 After 5 min, draw up the lipofectamine/neurobasal mixture into a pipette.

8 Pick up the second tube that contains the DNA and put the tip of the pipette underneath the surface of the medium. Do not place the pipette tip against the wall of the tube. Rather, place it directly in the middle of the medium. Depress the plunger in a smooth motion to add and mix the reagents simultaneously.

9 Hold the tube between index finger and thumb. With your ring finger gently tap the bottom of the tube a few times to mix the contents efficiently.

10 Wait 20 min.

11 Swirl the mixture a few times as described.

12 Take the neurons in the MatTek dishes out of the incubator.

13 To each MatTek dish add 50 μl of the transfection mixture.

14 Return the dishes to the incubator overnight.

15 Dishes can be observed on an inverted fluorescence microscope the next day.

While the expression is generally higher after 48 h, it is a good idea to look both on the first and second days after transfection. If the construct to be used is functional but toxic when overexpressed, you may miss green cells that have died if you look only on the second day after transfection. To ensure that the green fluorescence you are observing is in fact the GFP and not autofluorescence, you can simply compare the fluorescence observed with fluorescein and rhodamine filters. Autofluorescence will be visible with both cubes, but the GFP only with the fluorescein set.

5 Time-lapse microscopy

When it works, time-lapse microscopy is one of the great rewards of experimental life. While you are getting started, however, there is much to fix. For an introduction, consult Inoue's Video Microscopy (27). In addition, a volume devoted to the imaging of neurons has recently been published (28). For those with access to a confocal microscope, the essential reference is Pawley's 'blue bible' (29).

Time-lapse microscopy is full of trade-offs. In the case of cells transfected with GFP tagged proteins, many factors will contribute to the overall quality and number of images you can acquire. The quality of the images will depend in part, for example, on how healthy the cells are on the microscope stage, and how completely motion artefacts are suppressed. In turn, the number of images it will be possible to acquire before the cell fades will depend in part on how strongly the construct is expressed, the illumination intensity and the light collection efficiency of your microscope and detector. One way to get a basic estimate of the frame budget—how many time-lapse frames you will be able to take—is simply to image a transfected cell until it is bleached to some criterion (e.g. 50% bleach). As with transfection, the single most important determinant of success is the overall health of the neurons.

Many researchers go to the trouble of building an incubator around their microscope so that the cells can be maintained in the same CO_2-dependent media that is employed in the incubator. This is an effective approach for which the only drawback is cost. Fortunately, for neuronal cell culture, Brewer (30, 31) has developed matched media for 5% CO_2 and ambient conditions. This is the culture system that will be described here. It is also preferable to have a mechanism that will keep the cells warm on the microscope stage. Often, a hair dryer is sufficient. A better, if more expensive approach is to build or buy a resistively heated aluminium 'donought' that will surround the culture dish, and which can be coupled to a feedback controlled heater (e.g. Warner Instrument Corp TC-344B; www.warneronline.com). We have found this to be an effective method with the 35 mm dishes we use at magnifications that do not require an immersion lens. At higher magnifications, however, immersion lenses can be powerful heat sinks since they are thermally coupled to the culture dish. One solution is to maintain the nose-piece of the objective at 37°C with the same controller employed for the culture dish. The exact way in which you choose to keep your cells warm will need to be compatible with the chamber chosen. See Reider and Cole for a discussion of different perfusion chambers (32).

Once you have a method of keeping your cells warm and healthy in ambient conditions, it will be necessary to keep them in view for the period of imaging. If the cells are to be kept at, or near, 37°C, drift of the focal plane due to the expansion and contraction of the apparatus can be a significant hindrance especially at higher magnifications. It is helpful to let the dish equilibrate for ten minutes in order to lessen this problem.

Protocol 6

Time-lapse microscopy of transfected neurons

Reagents

- Hibernate E medium (Life Technologies, 10741023) supplemented with glutamine (2 mM final) and B27 supplement (Hib/B27)

Method

1   Take a single MatTek dish of transfected neurons from the incubator.

2   Place the pipette tip against the side of the MatTek dish and slowly aspirate the medium.[a]

3   Change pipettes and slowly add 2 ml of pre-warmed Hib/B27 medium to the dish. If you are adding the medium slowly enough, a disk of unchanged medium will form over the portion of the dish where the coverslip is.

4   Change pipettes and place the pipette tip by the side of the MatTek. Slowly aspirate the medium.[b] Add a second wash of Hib/B27 and aspirate as above.

5   Add a final 2 ml of pre-warmed Hibernate as above. The dish is now ready for viewing. Put it in a secondary container (e.g. a 10 or 15 cm plate) and take it to the inverted microscope for viewing.

6   Place the dish in the warming collar. Turn on the warming collar.[c]

7   Using bright-field optics, focus on the cells.

8   Switch to fluorescence optics to view your cells.[d]

[a] A residual volume will remain in the depression made by the hole in the original bottom of the dish that will leave your neurons covered. Do not remove this medium. In addition, aspiration with a vacuum attached to the house-line can disrupt the monolayer. It is less disruptive to the neurons to remove the medium with a pipettor; eject the waste into a conical tube and discard it during clean-up.

[b] This time it is better to take care to keep the pipette near the top surface of the medium as you are removing it. This will help to draw the remaining neurobasal medium to the side of the dish.

[c] It is necessary to turn on the collar after the dish has been inserted. As it heats the collar will expand and form a snug fit around the dish. If the collar is turned on beforehand, the dish will be difficult, or impossible, to insert.

[d] For time-lapse viewing, you will not be able to capture every event that may be of interest. Consider carefully the time-scale over which you wish to view the cells: for trafficking events, one frame per minute is often a good starting point, but this can easily be varied.

References

1. Chalfie, M., Tu, Y., Euskirchen, G., Ward, W. W., and Prasher, D. C. (1994). *Science*, **263**, 802.

2. Ormo, M., Cubitt, A. B., Kallio, K., Gross, L. A., Tsien, R. Y., and Remington, S. J. (1996). *Science*, **273**, 1392.

3. Heim, R., Prasher, D. C., and Tsien, R. Y. (1994). *Proc. Natl. Acad. Sci. USA*, **91**, 1250.

4. Heim, R. and Tsien, R. Y. (1996). *Curr. Biol.*, **6**, 178.

5. Brejc, K., Sixma, T. K., Kitts, P. A., Kain, S. R., Tsien, R. Y., Ormo, M., *et al.* (1997). *Proc. Natl. Acad. Sci. USA*, **94**, 2306.

6. Lukyanov, K. A., Fradkov, A. F., Gurskaya, N. G., Matz, M. V., Labas, Y. A., Savitsky, A. P., *et al.* (2000). *J. Biol. Chem.*, **275**, 25879.

7. Matz, M. V., Fradkov, A. F., Labas, Y. A., Savitsky, A. P., Zaraisky, A. G., Markelov, M. L., *et al.* (1999). *Nature Biotechnol.*, **17**, 969.

8. Hollmann, M., Rogers, S. W., O'Shea-Greenfield, A., Deneris, E. S., Hughes, T. E., Gasic, G. P., *et al.* (1990). *Cold Spring Harb. Symp. Quant. Biol.*, **55**, 41.

9. MacKinnon, R. (1995). *Neuron*, **14**, 889.

10. Stern-Bach, Y., Bettler, B., Hartley, M., Sheppard, P. O., O'Hara, P. J., and Heinemann, S. F. (1994). *Neuron*, **13**, 1345.

11. Nakai, K. (2000). *Adv. Protein Chem.*, **54**, 277.

12. Nakai, K. and Horton, P. (1999). *Trends Biochem. Sci.*, **24**, 34.

13. Shi, S. H., Hayashi, Y., Petralia, R. S., Zaman, S. H., Wenthold, R. J., Svoboda, K., *et al.* (1999). *Science*, **284**, 1811.

14. Miyawaki, A., Llopis, J., Heim, R., McCaffery, J. M., Adams, J. A., Ikura, M., *et al.* (1997). *Nature*, **388**, 882.

15. Miesenbock, G., De Angelis, D. A., and Rothman, J. E. (1998). *Nature*, **394**, 192.

16. Doi, N. and Yanagawa, H. (1999). *FEBS Lett.*, **457**, 1.

17. Siegel, M. S. and Isacoff, E. Y. (1997). *Neuron*, **19**, 735.

18. Baird, G. S., Zacharias, D. A., and Tsien, R. Y. (1999). *Proc. Natl. Acad. Sci. USA*, **96**, 11241.

19. Fischer, M., Kaech, S., Knutti, D., and Matus, A. (1998). *Neuron*, **20**, 847.

20. Horton, R. M., Ho, S. N., Pullen, J. K., Hunt, H. D., Cai, Z., and Pease, L. R. (1993). In *Methods in enzymology* (ed. R. Wu). Vol. 217, p. 270. Academic Press: San Diego.

21. Lo, W., Rodgers, W., and Hughes, T. (1998). *Biotechniques*, **25**, 94.

22. Yang, Y., Quitschke, W. W., and Brewer, G. J. (1998). *Brain Res. Mol. Brain Res.*, **60**, 40.

23. Ango, F., Albani-Torregrossa, S., Joly, C., Robbe, D., Michel, J. M., Pin, J. P., *et al.* (1999). *Neuropharmacology*, **38**, 793.

24. Xia, Z., Dudek, H., Miranti, C. K., and Greenberg, M. E. (1996). *J. Neurosci.*, **16**, 5425.

25. Freshney, R. I. (2000). *Culture of animal cells: a manual of basic techniques*, 4th edn. New York: Wiley-Liss.

26. Banker, G. and Goslin, K. (eds.) (1998). *Culturing nerve cells*, 2nd edn, Cellular and Molecular Neuroscience Series. Cambridge, MA: MIT Press.

27. Inoue, S. and Spring, K. R. (eds.) (1997). *Video microscopy: the fundamentals*, 2nd edn. New York: Plenum.

28. Yuste, R., Lanni, F., and Konnerth, A. (eds.) (2000). *Imaging neurons: a laboratory manual*. Cold Spring Harbor, NY: Cold Spring Harbor Laboratory Press.

29. Pawley, J. B. (ed.) (1995). *Handbook of biological confocal microscopy*, 2nd edn. The Language of Science. Plenum: New York.

30. Brewer, G. J., Torricelli, J. R., Evege, E. K., and Price, P. J. (1993). *J. Neurosci. Res.*, **35**, 567.

31. Brewer, G. J. (1995). *J. Neurosci. Res.*, **42**, 674.

32. Rieder, C. L. and Cole, R. W. (1998). In *Video microscopy* (ed. G. Sluder and D. E. Wolf), p. 253. Academic Press: San Diego.

33. Cormack, B. P., Valdivia, R. H., and Falkow, S. (1996). *Gene*, **173**, 33.

34. Crameri, A., Whitehorn, E. A., Tate, E., and Stemmer, W. P. (1996). *Nature Biotechnol.*, **14**, 315.

Chapter 7

Monitoring the turnover kinetics of G protein-coupled receptors

J. Michael Edwardson

Department of Pharmacology, University of Cambridge, Tennis Court Road, Cambridge CB2 1QJ, UK.

1 Introduction

G protein-coupled receptors are found predominantly at the plasma membrane in unstimulated cells (1). Following agonist stimulation, the receptors are usually efficiently endocytosed and delivered to intracellular compartments, from where they may be either recycled to the plasma membrane or transported to lysosomes for degradation. This intracellular transport is known to play an important role in the regulation of receptor function. Consequently, a complete characterization of the processes of receptor endocytosis and recycling is crucial to our understanding of cellular responses to agonists.

The quantitation of receptor trafficking is best achieved by the use of radioligand binding. In intact cells, a non-polar, membrane-permeant radioligand will detect the total population of receptors, whereas a polar, membrane-impermeant radioligand will detect only receptors present at the cell surface. The transport of receptors into and out of the plasma membrane can therefore be measured through changes in the binding of a membrane-impermeant radioligand. In response to agonist stimulation, the binding of such a radioligand typically falls rapidly, with a half-time of a few minutes, reaching a value of 40–60% of the binding to unstimulated cells after about 30 minutes (2, 3). During this 30 minute period there is no significant reduction in the total number of receptors, measured through the binding of a membrane-permeant radioligand (2, 3). If the agonist is removed after a short time, the density of receptors at the surface is restored almost to the original value, by receptor recycling (2, 3). In fact, the redistribution of receptors in response to agonist stimulation, usually termed internalization, represents a 'snap-shot' of a new steady-state, in which at any instant there are fewer receptors at the cell surface and more in internal compartments. The instantaneous situation, therefore, provides no information about the kinetics of receptor trafficking. For instance, it is impossible to determine the extent of receptor movement that underlies the internalization, or to decide whether a particular manipulation that affects the extent of internalization is doing so through an effect on endocytosis or recycling, or both. In this chapter, I describe the application of a kinetic analysis of receptor trafficking that is designed to address questions such as these. I then explain how the contribution of trafficking to changes in receptor function, such as desensitization and re-sensitization, can be assessed.

2 Mathematical modelling of receptor turnover

A complete model for intracellular receptor trafficking should consider at least four transport steps: delivery of receptors to the plasma membrane through the secretory pathway and from recycling endosomes, removal of receptors from the plasma membrane by endocytosis, and delivery of receptors from endosomes to their site of degradation in lysosomes (1) (*Figure 1*). Data for receptor trafficking can be fitted to such a model (see, for example, ref. 4), although the mathematics is rather complicated. In practice, the synthetic route can be eliminated if necessary simply by blocking protein synthesis with an agent such as cycloheximide (4–6). Further, it is now clear that receptor degradation is minimal over the time-course of internalization (typically 30 minutes) (4). Consequently, over short time periods and in the presence of cycloheximide, receptor trafficking can be adequately described by a relatively simple two-compartment model. This is the model that we have used for our recent studies.

In the two-compartment model, receptors traffic between the plasma mem-

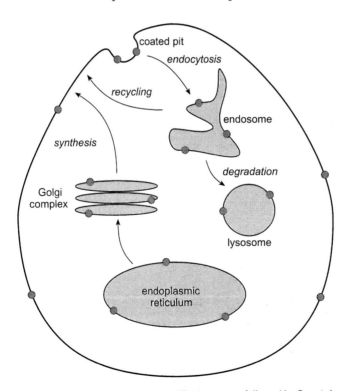

Figure 1 Illustration of the major trafficking routes followed by G protein-coupled receptors.

brane and endosomes. The rate of change of the number of receptors at the cell surface is given by the rate of arrival of receptors by recycling minus the rate of removal by endocytosis:

$$\frac{dR_s}{dt} = k_r R_e - k_e R_s \tag{1}$$

where R_s is the number of receptors at the cell surface, R_e is the number of receptors in endosomes, and k_r and k_e are the rate constants of recycling and endocytosis, respectively.

The total number of receptors, $R_s + R_e$, is equal to the number of receptors at time $t = 0$; i.e. $R_s + R_e = R_{s0} + R_{e0}$. Substituting $R_e = R_{s0} + R_{e0} - R_s$, Equation 1 can be re-written as:

$$\frac{dR_s}{dt} = k_r (R_{s0} + R_{e0}) - (k_e + k_r) R_s$$

which is now of the form $\frac{dy}{dt} = b - ay$. This can be integrated to $ay - b = Ce^{-at}$ where

$b = k_r (R_{s0} + R_{e0})$, $a = k_e + k_r$, and C is the constant of integration. Rearrangement results in an equation describing the number of receptors at the cell surface as a function of time:

$$R_s = \frac{1}{k_e + k_r} [k_r (R_{s0} + R_{e0}) + (k_e R_{s0} - k_r R_{e0}) e^{-(k_e + k_r)t}] \tag{2}$$

Receptor recycling can be described by Equation 2 with k_e set to zero. Then receptor recycling ($R_s - R_{s0}$) as a function of time is given by:

$$R_s - R_{s0} = R_{e0} (1 - e^{-k_r t}) \tag{3}$$

A plot of $R_s - R_{s0}$ against t therefore allows k_r and R_{e0}, the size of the recyclable endosomal pool of receptors, to be determined.

In some cell types it can be shown that there is no significant endosomal pool of a particular G protein-coupled receptor before the cells are stimulated with agonist (4, 5). In this case, during agonist-triggered receptor internalization, R_{e0} in Equation 2 can be set at zero, so that R_s is given by:

$$R_s = \frac{R_{s0}}{k_e + k_r} [k_r + k_e e^{-(k_e + k_r)t}] \tag{4}$$

A plot of R_s against t now allows k_e to be determined, since k_r is already known (as a result of the fitting of receptor recycling data to Equation 3). At large values of t, the exponential terms in Equation 4 approach zero, and therefore the steady-state number of receptors at the cell surface ($R_{s,ss}$) may be written as:

$$R_{s,ss} = R_{s0} \frac{k_r}{k_r + k_e} \tag{5}$$

A similar analysis will show that the steady-state number of receptors in endosomes ($R_{e,ss}$) is given by:

$$R_{e,ss} = R_{s0} \frac{k_e}{k_r + k_e} \tag{6}$$

157

It is clear from this analysis that the extent of receptor internalization in response to agonist stimulation depends not only on k_e but also on k_r (see *Figure 2*). This has important implications for the relationship between the effects of agonist stimulation on the rate of receptor endocytosis and the extent of receptor internalization, as will be described below.

My colleagues and I have applied this analysis to the intracellular trafficking of the muscarinic receptor for acetylcholine. There are three major reasons why this receptor represents an ideal model system:

(a) Excellent membrane-impermeant and membrane-permeant ligands are available.

(b) Receptors at the plasma membrane can be irreversibly alkylated using propyl-benzilylcholine mustard (PrBCM), which simplifies analysis of the delivery of receptors to the plasma membrane.

(c) There are five subtypes of receptors that are widely expressed in various cell types (7). These subtypes couple to two distinct second messenger systems (M1, M3, and M5 stimulate phosphoinositide hydrolysis, while M2 and M4 inhibit adenylyl cyclase). This provides the opportunity to look for connections between receptor trafficking and G protein-coupling/second messenger generation.

Many of the principles outlined here apply to other G protein-coupled receptors, and the trafficking of some of these, in particular the β_2-adrenoceptor, has been studied extensively (8, 9). Membrane-impermeant radioligands are available for many receptors, and these ligands can be used in straightforward assays for the movement of receptors into and out of the plasma membrane (see below). Where membrane-impermeant radioligands are not available, an alternative approach is to use a membrane-permeant radioligand to measure total receptor number and then to measure the internal receptor number by using a membrane-impermeant unlabelled ligand to displace the radioligand from receptors at the cell surface. The number of receptors at the plasma membrane can then be calculated from the difference between the two values. If this method is used, it is important to bear in mind that the binding affinity of the radioligand might be considerably affected by differences in the environment of the receptor in intracellular compartments. For instance, the lumen of the endosome is at a pH of about 5.0, which is likely to have significant effects on the binding affinities of some agonists (see, for example, ref. 4).

The use of the membrane-impermeant radioligand [^{3}H]N-methylscopolamine ([^{3}H]NMS) to measure muscarinic acetylcholine receptor density at the plasma membrane of cultured cells is described in *Protocol 1*. This protocol should work equally well for membrane-impermeant radioligands for other G protein-coupled receptors.

Protocol 1

Measurement of muscarinic acetylcholine receptor density at the plasma membrane of cultured cells (e.g. NG108-15)

Reagents

- Appropriate cell culture medium (e.g. for NG108-15 cells, Dulbecco's modification of Eagle's medium (DMEM), supplemented with 10% fetal bovine serum)
- N-[2-hydroxyethyl]piperazine-N'-[2-ethanesulfonic acid] (Hepes)-buffered DMEM pH 7.4 (made up from powder), without serum
- [^{3}H]NMS, specific activity approx. 80 Ci/mmol (Amersham), and N-methylatropine (Sigma)

Method

1. Grow cells at 37 °C in culture medium in 5% CO_2, 95% humidified air.

2. Detach cells from culture flasks by brief incubation with trypsin (0.5 mg/ml) and EDTA (0.2 mg/ml) in phosphate-buffered saline pH 7.4 (PBS).

3. Seed cells into 24-well culture plates at approx. 10^5 cells/well in 2 ml culture medium. (Depending on the extent to which the cells adhere to the culture plates, it may be necessary to pre-coat the wells with poly-D-lysine. This involves exposing the plastic to poly-D-lysine (1 mg/ml in water) for 5 min, aspirating the solution, and washing the well with water.) Use the cells after two days, when almost confluent.

4. Wash each well twice with 2 ml DMEM-Hepes, at 12 °C. This step removes the agonist, and also stops further receptor trafficking. Add 1 ml of the same medium containing radioligand ([^{3}H]NMS) at an appropriate concentration (we usually use 0.4 nM). Use three wells for each data point (i.e. carry out each determination in triplicate). Include N-methylatropine (1 μM) in one set of wells to define non-specific binding.

5. Float the plate(s) on a water-bath at 12 °C. Incubate for 2 h to allow binding of the radioligand to reach equilibrium.

6. Wash each well twice again with 2 ml DMEM-Hepes at 12 °C, and add 200 μl of 0.5% Triton X-100 in water. Leave at room temperature for 15 min to solubilize cells. Transfer samples to scintillation vials. Wash wells with a further 200 μl Triton X-100, and again transfer samples to scintillation vials.

7. Add scintillation fluid and count radioactivity in a scintillation counter.

3 Measurement of rate constants for endocytosis and recycling

To measure the rate constants for the movement of receptors into and out of the membrane it is necessary to monitor receptor recycling first, and fit the data to

Equation 3 (above) in order to determine the rate constant for recycling from endosomes, k_r. The value of k_r can then be inserted into Equation 4, which is fitted to internalization data in order to determine the rate constant for endocytosis, k_e. To determine the time-course of internalization, *Protocol 1* should be adapted to include pre-treatment of the cells with agonist for various times before the measurement of radioligand binding. For receptor recycling experiments, internalization should first be triggered by treatment of the cells with agonist for an appropriate length of time. The agonist should then be removed by washing, and recycling allowed to proceed for various times. We routinely carry out curve-fitting using GraphPad Prism software, which is able to handle user-generated equations.

3.1 Use of insurmountable receptor antagonists

Although not essential for the purpose of determining trafficking rate constants, a convenient way to monitor the delivery of receptors to the plasma membrane is to alkylate receptors already at the cell surface using an irreversible ligand, and then to follow the reappearance of radioligand binding with time. The advantage of this approach is that it reduces the value of the term R_{s0} to zero, so that Equation 3 can be simplified to:

$$R_s = R_{e0}\left(1 - e^{-krt}\right) \tag{7}$$

Muscarinic receptors can be alkylated using PrBCM (5). *Protocol 2* describes the use of this irreversible antagonist.

Protocol 2

Alkylation of muscarinic acetylcholine receptors at the surface of cultured cells using PrBCM

Reagents

- PrBCM
- PBS, and PBS containing 0.9 mM $CaCl_2$ and 0.5 mM $MgCl_2$ (PBS-Ca/Mg)

Method

1. Allow PrBCM (0.1 mM) to cyclize (i.e. to form the aziridinium ion) by incubation in PBS pH 7.4, at room temperature for 1 h. Stop the reaction by diluting the PrBCM to 10 nM in PBS-Ca/Mg.

2. Wash the cells, growing on 24-well plates, twice in 2 ml of PBS-Ca/Mg and incubate with cyclized PrBCM (10 nM) in the same buffer at 15 °C. At the appropriate time, remove the PrBCM by washing the cells twice in 2 ml of PBS-Ca/Mg. At 15 °C, receptor trafficking effectively ceases, and only receptors initially at the cell surface will be alkylated.

Protocol 2 continued

3 Monitor receptor alkylation through the reduction in specific binding of [³H]NMS. At 15°C, the half-time for alkylation of muscarinic receptors by 10 nM PrBCM is typically 6 min. After 30 min, more than 90% of the receptors will have been alkylated.

In fact, irreversible antagonists are available for only a few G protein-coupled receptors. Where no such reagent is available, receptor recycling should be measured through the recovery of radioligand binding after maximal internalization has been induced by agonist treatment. The data should then be fitted to Equation 3 to determine k_r:

$$R_s - R_{s0} = R_{e0} (1 - e^{-k_r t}) \qquad [3]$$

A good example of the basic approach outlined above is our study of the trafficking of muscarinic ACh receptors in the neuroblastoma-glioma cell line NG108-15 (4, 5). These cells express predominantly the M4 subtype of muscarinic receptor. In this early study, we fitted the data to a model that took account of the delivery of receptors to the plasma membrane along the synthetic pathway (i.e. from the Golgi complex) and the degradation of receptors in lysosomes, as well as receptor cycling between the plasma membrane and endosomes. For simplicity, we assumed that transport of receptors through the synthetic pathway and degradation of receptors in lysosomes were both linear processes. When the cells were incubated at 37°C for various times following PrBCM treatment, the recovery of [³H]NMS binding was linear and almost completely abolished by pre-treatment with cycloheximide. Hence, in unstimulated NG108-15 cells, there is no substantial intracellular (endosomal) pool of receptors (i.e. R_{e0} is effectively zero), and delivery of receptors to the plasma membrane is exclusively through the synthetic pathway. Further, the rate of recovery of [³H]NMS binding can be used to estimate the rate of synthesis of receptors by the cells (22 receptors/cell/min). When the same experiment was carried out with cells that had previously been stimulated with a maximally stimulating concentration of a muscarinic agonist (carbachol) for 30 minutes, the extent of the recovery of [³H]NMS binding was larger, and not completely sensitive to cycloheximide. The time-course of the cycloheximide-resistant component of recovery was exponential, indicating delivery of receptors from a finite source, most likely recycling endosomes. Fitting of the recovery data to Equation 7, gave a rate constant for recycling, k_r, of 0.12 min⁻¹ ($t_{1/2}$ 6 min), and an endosomal pool size of 28% of the number of receptors present at the surface of unstimulated cells. For NG108-15 cells R_{e0} was set at zero, and the data for receptor internalization were used to calculate a value for the rate constant for endocytosis, k_e, of 0.13 min⁻¹ ($t_{1/2}$ 6 min), similar to the value of k_r. The rates of receptor degradation were calculated as 60 receptors/cell/min in the presence of agonist and 22 receptors/cell/min in the absence of agonist.

Our results show that the mathematical model is reasonably successful in describing the trafficking of the muscarinic receptor. Having determined the rate constants or rates for the various steps, we used a relatively crude iterative procedure (which involved calculating how many receptors were moving minute-by-minute) to estimate the extent of receptor trafficking during and after agonist stimulation. Before stimulation, each cell had about 11 500 receptors at the plasma membrane. After 30 minutes of agonist stimulation this number fell to 5050. During the 30 minute stimulation period 22 230 receptors were endocytosed, 15 780 receptors were recycled to the plasma membrane, and 1800 were degraded in lysosomes, leaving 4650 in endosomes. In the 30 minutes following washout of the agonist, the endosomal pool of receptors was almost emptied, with about 80% of the receptors being recycled to the plasma membrane and about 20% being degraded. This example illustrates the difference between the extents of receptor internalization (11 500 − 5050 = 6450 receptors/cell in 30 minutes) and receptor endocytosis (22 230 receptors/cell in the same period). Clearly, the extent of receptor trafficking in response to agonist stimulation is much greater than might be expected from measurement of surface receptor number before and after stimulation.

3.2 Comparing rate constants for transfected and native receptors

Many studies of receptor trafficking have been carried out using cell lines into which the receptor of interest has been introduced by transfection. There are two main advantages of this approach: the behaviour of the receptor can be studied in the absence of complicating factors such as the presence of various receptor subtypes, and transfection usually results in the expression of large numbers of receptors, which makes them easier to study. This approach, however, requires two important assumptions: that the trafficking machinery for the receptor is already in place in the 'foreign' cell, and that overexpression of the receptor does not saturate or otherwise interfere with the operation of this machinery. In reality, the comparison of results for the behaviour of the same receptor in different cell types has led to considerable confusion (compare, for example, refs 10 and 11). In one study, we compared directly the behaviour of different subtypes of the muscarinic receptor in native and foreign cells (6). It is well known that the M1, M3, and M5 receptor subtypes couple to $G_{q/11}$ and stimulate phospholipase C, whereas M2 and M4 subtypes couple to $G_{i/o}$ and inhibit adenylyl cyclase. We found that M3 receptors endogenously expressed in SH-SY5Y human neuroblastoma cells and M4 receptors endogenously expressed in NG108-15 mouse neuroblastoma-glioma cells are both endocytosed efficiently in response to agonist stimulation. However, when transfected into Chinese hamster ovary (CHO) fibroblasts, M2 and M4 receptors are endocytosed efficiently, whereas M1 and M3 receptors are not (6). NG108-15 cells endogenously express about 11 500 M4 receptors per cell, whereas M4-transfected CHO cells express about 1 000 000 receptors per cell. Despite this huge difference in expression levels, the rate constants for endocytosis of receptors in response to agonist stimulation are reasonably similar: 0.13 min^{-1} in NG108-15 cells and 0.07 min^{-1} in CHO cells. Hence, there is no evidence that the endocytotic machinery is being saturated in the CHO cells. For M3 receptors in CHO cells, the rate constant for endocytosis is only 0.02 min^{-1}, which suggests that the mechanisms underlying endocytosis differ between M3 and M4 receptors. Furthermore, the fact that M1 receptors behave like M3 receptors, and M2 receptors behave like M4 receptors, suggests that the G protein/second messenger system involved contributes to this difference in endocytotic mechanism.

4 Relationship between agonist intrinsic activity and receptor endocytosis and internalization

Since agonists but not antagonists normally cause receptor internalization, it is widely assumed that it is the activated conformation of the receptor, which couples to G proteins, that is responsible for initiating receptor endocytosis. In an attempt to understand more fully the relationship between the activity of agonists and their ability to trigger internalization, we compared the ability of a number of muscarinic agonists to generate a second messenger response with their ability to increase the rate constant for receptor endocytosis (12). We used SH-SY5Y cells, which predominantly express the M3 subtype of muscarinic receptor, and measured the production of the second messenger inositol (1,4,5) trisphosphate (Ins(1,4,5)P$_3$), according to *Protocol 3*. The intrinsic activities of the various agonists (pilocarpine, arecoline, bethanechol, methacholine, and oxotremorine-M) were expressed in terms of the maximal Ins(1,4,5)P$_3$ response to the agonists relative to the maximal response to the full agonist, carbachol. We found that methacholine and oxotremorine-M were both full agonists, whereas bethanechol, arecoline, and pilocarpine were partial agonists, eliciting 52%, 26%, and 8% of the maximal response to carbachol, respectively. The full agonists all caused the removal of about 90% of the cell surface receptors to intracellular compartments. The partial agonists bethanechol, arecoline, and pilocarpine caused the internalization of 82%, 78%, and 19% of the receptors, respectively (i.e. 91%, 87%, and 21% of the internalization caused by carbachol). Hence, the ability of the partial agonists to cause receptor internalization was greater than would be expected from their ability to generate Ins(1,4,5)P$_3$, and a plot of maximal receptor internalization against maximal Ins(1,4,5)P$_3$ generation was not linear. This result is in fact predicted by the trafficking model (see Equation 5), given that the rate constants for receptor recycling following stimulation with the various agonists were all about equal (0.05–0.06 min^{-1}).

The rate constants for endocytosis in response to the agonists could be calculated using Equation 5, since the values of $R_{s,ss}$ and k_r were both known. It was found that the values of k_e were linearly related to the intrinsic activities of the agonists. This result strongly suggests that the same receptor conformation is responsible for coupling to G proteins and for stimulating endocytosis.

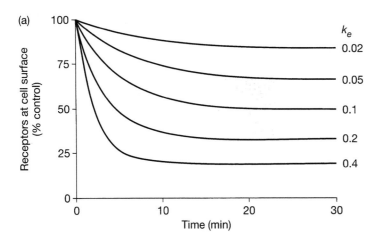

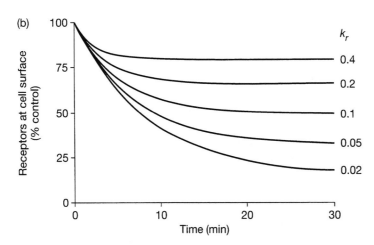

Figure 2 Theoretical predictions of the time-courses of receptor internalization calculated from Equation 4, assuming that the receptors traffic between the plasma membrane and endosomes. (a) Effect of changing the rate constant for endocytosis, k_e, while keeping the rate constant for recycling, k_r, constant at 0.1 min^{-1}. (b) Effect of changing the rate constant for recycling, k_r, while keeping the rate constant for endocytosis, k_e, constant at 0.1 min^{-1}.

Note also that, because the extent of internalization depends on both k_e and k_r (*Figure 2*), the mathematical model for trafficking predicts that weak agonists, which have only a small effect on k_e, would nevertheless be able to cause a relatively large degree of internalization. This characteristic of the model may well account for some previously puzzling observations in the literature, where it has been claimed that weak agonists, and even antagonists, are able to trigger substantial internalization (see, for example, ref. 13).

Protocol 3

Measurement of Ins(1,4,5)P$_3$ generation by receptors that couple to G$_{q/11}$

Reagents

- Ins(1,4,5)P$_3$ assay kit and [^{3}H]Ins(1,4,5)P$_3$, specific activity approx. 40 Ci/mmol (Amersham)

- Appropriate cell culture medium

Method

1   Grow cells (e.g. SH-SY5Y cells) in 24-well plates until almost confluent.

2   Wash cells twice with 250 μl of culture medium at 37 °C. Leave for 10 min at 37 °C.

3   Begin experiment by removing the wash medium and replacing 150 μl of fresh medium with or without agonist. Incubate at 37 °C for the appropriate time. (In SH-SY5Y cells, the Ins(1,4,5)P$_3$ response to agonists is biphasic, with an initial peak at 10 sec followed by a reduction over the next 120 sec to a lower, sustained phase.)

4   Terminate the incubation by adding 150 μl of ice-cold 1 M trichloroacetic acid. Place the plate(s) on ice for 15 min.

5   Add 160 μl of the solution above the cells to 40 μl of 10 mM EDTA. Add 200 μl of a 1:1 mixture of tri-n-octylamine and 1,1,2-trichlorotrifluoroethane. Vortex and leave the samples at room temperature for 15 min.

6   Centrifuge the samples at 13 000 g for 3 min. Take 100 μl of the upper aqueous phase, and add 50 μl of 25 mM NaHCO$_3$. Samples may be stored at −20 °C until the assay of Ins(1,4,5)P$_3$.

7   Measure Ins(1,4,5)P$_3$ mass using a kit from Amersham, following the manufacturer's instructions. The assay depends on the ability of cellular Ins(1,4,5)P$_3$ to displace [^{3}H]Ins(1,4,5)P$_3$ from an Ins(1,4,5)P$_3$ binding protein. If required, the Ins(1,4,5)P$_3$ binding protein supplied with the kit may be replaced with a porcine platelet membrane preparation (14).

5 Manipulating receptor turnover using inhibitors of endocytosis or recycling

5.1 Inhibitors of endocytosis

All receptor trafficking steps are almost entirely prevented by incubation of the cells at temperatures below 15 °C, or by depletion of intracellular ATP. These manoeuvres can be used to confirm that changes in cell surface receptor numbers are indeed a result of genuine trafficking events. For instance, we have found that delivery of M4 muscarinic receptors to the plasma membrane in NG108-15 cells, along both the synthetic and recycling pathways, is almost abolished by pre-treatment of the cells with 50 nM antimycin and 50 mM deoxyglucose, which block mitochondrial and glycolytic generation of ATP, respectively (5). Further, the same treatment blocked receptor-mediated internalization of agonists at the somatostatin sst_2 receptor in both Neuro2A neuroblastoma cells (15) and transfected CHO cells (16). The use of reduced incubation temperature is exploited in the techniques described above to prevent further trafficking while radioligand binding is used to assess the extent of trafficking that has occurred during an experiment.

A more selective block of endocytosis can be achieved by treatment of the cells with the lectin, concanavalin A (Con A). Con A is homo-tetrameric lectin that binds tightly to mannose residues on glycoproteins. It is therefore able to crosslink cell surface receptors and prevent them from entering the endocytotic pathway. Con A has been used successfully to block the endocytosis of the β_2-adrenoceptor in response to agonist stimulation (9, 17, 18), and this manipulation has helped to elucidate the functional significance of agonist-stimulated endocytosis of this receptor (see below). In our hands, Con A did not completely block agonist-stimulated endocytosis of the M3 muscarinic receptor in SH-SY5Y cells, but rather reduced the rate constant for endocytosis by about 50% (19). Endocytosis of proteins via clathrin-coated pits, the principal route through which G protein-coupled receptors traffic, is efficiently blocked by incubation of the cells in hyperosmotic media (usually achieved using 0.5 M sucrose). This method has been used to block the endocytosis of the β_2-adrenoceptor (9, 18), the M4 muscarinic receptor in both transfected CHO cells (20) and NG108-15 cells (21), the M3 muscarinic receptor in SH-SY5Y cells (22), and ligands of the somatostatin sst_2 receptor (15, 16). A potential advantage of the use of hyperosmolar sucrose is the fact that its effect is at least partially reversible on washout (15). Depletion of intracellular K^+ and acidification of the cytosol have also been shown to inhibit endocytosis of the M3 muscarinic receptor in SH-SY5Y cells (22). These manipulations were achieved by incubating the cells in buffers of appropriate ionic conditions containing the carboxylic ionophore nigericin, which becomes incorporated into the plasma membrane. This method has not been used widely, probably because agents such as nigericin have effects on other cellular processes (see below).

Endocytosis of a number of receptors, including the β_2-adrenoceptor (23), has been shown to be blocked by treatment of the cells with the trivalent arsenical phenylarsine oxide (PAO). This effect is apparently brought about by an interaction of PAO with vicinal sulfydryl groups on the receptor, to form a stable ring structure. It was reported that other properties of the β_2-adrenoceptor, such as ligand binding and receptor desensitization are unaffected by PAO treatment. However, in our experience (unpublished results), PAO did affect agonist activation of the muscarinic receptors, and we cannot therefore recommend its use as a specific inhibitor of endocytosis.

There is conflicting evidence as to whether activation of G proteins is necessary for agonist-stimulated receptor endocytosis. One approach to this problem, at least when studying receptors that couple to G_i/G_o, is to test the effect of pertussis toxin on internalization. Pertussis toxin ADP-ribosylates the α-subunit of G_i/G_o and prevents coupling between the receptor and the G protein. As an example, it was shown that pre-treatment of Neuro2A cells with pertussis toxin decreased sst_2 receptor-mediated internalization of the agonist ligand BIM-23027 by about 50% (15). The inhibition may reflect the removal of an amplification of internalization that normally occurs when the receptor activates the G protein. Alternatively, uncoupling of the receptor from the G protein might simply have shifted the receptor into a low-affinity state so that, at a given agonist concentration, fewer became internalized along with the receptor.

5.2 Inhibitors of recycling

The carboxylic ionophores nigericin and monensin are lipid-soluble compounds that partition into membranes of cells and dissipate transmembrane gradients of protons and monovalent cations. These compounds cause profound morphological changes in treated cells, such as vesiculation of the Golgi complex (24). They have been used to block, or slow down, protein traffic through the Golgi complex (25), and to block the recycling of G protein-coupled receptors, such as the β_2-adrenoceptor (9), from endosomes to the plasma membrane. Although effective at inhibiting receptor recycling, these compounds clearly do not have a specific effect on this transport step. Indeed, we have shown that nigericin inhibits (but does not completely block) both endocytosis and recycling of the M3 muscarinic receptor in SH-SY5Y cells (19).

6 Role of receptor trafficking in desensitization and resensitization

At first glance, it seems reasonable to propose that receptor internalization mediates functional desensitization, with a fall in the receptor density at the plasma membrane causing a reduction in cellular response. However, loss of receptors from the surface is often too slow to account for the rapid onset of desensitization, and the effects of mutations on receptor internalization and desensitization are often different, suggesting that the major role of internalization is not simply to cause desensitization. An alternative hypothesis is that receptor cycling is involved in resensitization. This hypothesis has been directly tested for the β_2-adrenoceptor. The sequence of events that follows agonist stimulation of the β_2-adrenoceptor is apparently as follows. Agonist stimulation of the receptor results in receptor phosphorylation (26), by both second messenger-regulated kinases, such as 3′,5′-cyclic adenosine monophosphate (cyclic AMP)-dependent protein kinase (27), and G protein-coupled receptor kinase 2 (GRK2), which phosphorylates only the agonist-occupied form of the receptor (28). Phosphorylation by GRK2 causes the recruitment of β-arrestin from the cytosol (29). Arrestin binding to the receptor prevents its coupling with the G protein, and leads to receptor desensitization. Arrestin also mediates an interaction between the receptor and clathrin, which leads to the targeting of phosphorylated receptors to clathrin-coated pits and subsequent receptor endocytosis (30, 31). The receptors are then delivered to endosomes, where they are efficiently dephosphorylated, probably by a specific G protein-coupled receptor phosphatase (32, 33). Finally, the receptors are recycled to the plasma membrane, where they are able to respond once again to agonist stimulation (18). If the purpose of the endocytosis and recycling of the β_2-adrenoceptor is to permit dephosphorylation and consequent resensitization, then inhibitors acting at any stage of the transport pathway should prevent receptor resensitization. It was shown that inhibition of internalization (by Con A or hyperosmolar sucrose), of dephosphorylation (by calyculin A), or of recycling (by monensin) all reduced or blocked resensitization (9), confirming the role of receptor trafficking in this case.

We have tested the roles of receptor endocytosis and recycling in the desensitization and resensitization of M3 muscarinic receptors in SH-SY5Y cells and M4 receptors in NG108-15 cells. In the SH-SY5Y cells, the intracellular response that we measured was the agonist-stimulated rise in intracellular Ca^{2+} concentration (see *Protocol 4*).

Protocol 4

Population measurements of changes in intracellular Ca^{2+} concentration receptors coupled to $G_{q/11}$

Equipment and reagents

- Fluorescence spectrophotometer (e.g. an Hitachi F-2000 model) fitted with a water jacket to maintain cuvette temperature at 37°C
- Glass coverslips (9 × 22 mm) and holders designed to hold the coverslips vertically, at 45° to the incident light beam; a diagram of the complete assembly can be found in ref. 34

- Acrylic cuvettes (capacity 3 ml)
- Krebs/Hepes buffer: 4.2 mM $NaHCO_3$, 118 mM NaCl, 4.7 mM KCl, 1.2 mM KH_2PO_4, 1.2 mM $MgSO_4$, 2 mM $CaCl_2$, 10 mM glucose, 10 mM Hepes pH 7.4
- Fura-2 acetoxymethyl ester and Ca^{2+} calibration kit (Molecular Probes)

Method

1   Grow cells (e.g. SH-SY5Y cells) until almost confluent on coverslips in 6-well plates.

2   Wash cells twice with 2 ml of Krebs/Hepes buffer.

3   Insert a coverslip into the holder, and place in a cuvette. Incubate the cells in Krebs/Hepes buffer.

4   Place the cuvette in the spectrophotometer at 37°C, and determine the autofluorescence of the sample. This should be less than 20% of the initial fura-2 fluorescence signal.

5   Load the cells with fura-2 by incubating them at 20°C in Krebs/Hepes buffer con-

taining 3.3 μM fura-2 acetoxymethyl ester, with continual stirring. After 50 min, remove the buffer and wash the cells three times with 2 ml of buffer at 37°C.

6 Replace the cuvette in the spectrophotometer. Maintain a flow of buffer through the cuvette of about 5.5 ml/min by perfusion into the bottom of the cuvette and aspiration from just above the coverslip.

7 Allow the cells to equilibrate for 10 min and then add the agonist, by changing the perfusing solution.

8 After subtraction of autofluorescence, determine the ratio of fluorescence at excitation wavelengths of 340 nm and 380 nm, with an emission wavelength of 510 nm. Calculate the free Ca^{2+} concentration by reference to a look-up table created using the Ca^{2+} calibration kit.

Carbachol caused a biphasic elevation of $[Ca^{2+}]$ in SH-SY5Y cells, consisting of an initial peak at 5–10 sec followed by a sustained plateau (19). Pre-exposure of cells to a maximally stimulating concentration of carbachol (1 mM) caused a significant reduction in the peak response to a second challenge with carbachol. The half-time of this desensitization (about 30 sec), was considerably shorter than the receptor internalization in response to agonist stimulation (about 4 min), indicating that internalization does not play a major role in desensitization. Resensitization after washout of the agonist was complete before the number of receptors at the cell surface recovered to the original level. This result suggests that there is a receptor reserve for the Ca^{2+} response to carbachol and, furthermore, that this reserve plays a role in recovery from desensitization. This possibility will be considered in further detail in Section 7.

To examine the functional roles of receptor endocytosis and recycling, we tested the effects of blockers of these processes on the desensitization and resensitization of the Ca^{2+} response to carbachol. We found that Con A reduced the rate constant for endocytosis by about 50% (see above), but had no significant effect on either the rate or the extent of receptor recycling. Con A had no effect on the extent of desensitization produced by carbachol, but did reduce the rate and extent of resensitization. Nigericin did not affect the rate constant for muscarinic receptor recycling, but did reduce the extent of recycling by about 50%. Nigericin also reduced the rate constant of endocytosis. The functional effect of nigericin was very similar to that of Con A: the extent of desensitization was unaffected, but the rate and extent of resensitization were reduced. Hence, the endocytosis and recycling of muscarinic receptors are involved in resensitization of the response, and not in desensitization.

Although the level of expression of M3 muscarinic receptors in SH-SY5Y cells is too low to permit a direct study of the phosphorylation state of the receptor under various conditions, we did find that calyculin A increased the extent of desensitization and reduced the extent of resensitization. Taken together, these results suggest that agonist-stimulated trafficking of M3 receptors in SH-SY5Y cells contributes to receptor resensitization, perhaps through an enhancement of receptor dephosphorylation by an intracellular phosphatase. This scenario is, of course, similar to that for the β_2-adrenoceptor.

In NG108-15 cells, stimulation of the M4 muscarinic receptor results in inhibition of forskolin-stimulated adenylyl cyclase activity. Intracellular cyclic AMP levels were measured as described in *Protocol 5*. (Note that this protocol can easily be modified to measure cyclic AMP generation in response to stimulation of G_s-coupling receptors: in this case the forskolin pre-stimulation is simply omitted.) We found that the time-course and extent of receptor desensitization in the NG108-15 cells were identical to the corresponding values for receptor internalization (21). Furthermore, desensitization and internalization were both increased by overexpression of GRK2. Exposure of the cells to hyperosmolar sucrose (0.6 M) almost completely blocked both internalization and desensitization. From these results, we conclude that internalization of the M4 muscarinic receptor in NG108-15 cells plays a key role in agonist-induced desensitization.

Protocol 5

Measurement of the inhibition of forskolin-stimulated adenylyl cyclase activity by receptors coupled to G_i

Equipment and reagents

- GF/B filters (Whatman)
- Appropriate cell culture medium and Hepes-buffered DMEM
- Cycloheximide and 3-isobutyl-1-methylxanthine (IBMX) (both from Sigma)
- Cyclic AMP and protein kinase A (both from Sigma)
- [5,8-^{3}H]cyclic AMP, specific activity approx. 40 Ci/mmol (Amersham)
- Cyclic AMP assay buffer: 100 mM NaCl, 5 mM EDTA, 50 mM Tris–HCl pH 7.0

Method

1 Grow cells (e.g. NG108-15 cells) on 24-well plates until almost confluent.

2 Pre-incubate cells at 37°C for 30 min in DMEM-Hepes containing the protein synthesis inhibitor cycloheximide (20 μg/ml) and the cyclic AMP phosphodiesterase inhibitor IBMX (0.5 mM).

3 Where appropriate, add the agonist and forskolin (50 μM) as pre-warmed aliquots to achieve a final volume of 300 μl. Incubate at 37°C for 5 min.

4 Terminate the incubations by adding 20 μl of 10 M HCl. Neutralize the samples by adding 20 μl of 10 M NaOH, and buffer by adding 200 μl of 1 M Tris–HCl pH 7.0.

5 Centrifuge the samples at 21 000 g for 5 min in an Eppendorf microcentrifuge to pellet cell debris.

6 Assay the cyclic AMP content of the supernatants through their ability to displace

[^{3}H]cyclic AMP from protein kinase A. Carry out all steps in the cyclic AMP assay on ice.

7 Dilute samples that have been treated with forskolin 50-fold into ice-cold DMEM-Hepes; dilute samples that have not been treated with forskolin twofold. In both cases take 100 μl for the assay.

8 Add to each sample 100 μl of [^{3}H]cyclic AMP (approx. 25 000 d.p.m.) in assay buffer, and 100 μl of protein kinase A (32 μg/ml) in the same buffer containing bovine serum albumin (2.5 mg/ml).

9 Incubate the samples for 2 h on ice and then filter through polyethyleneimine-treated GF/B filters.

10 Count the radioactivity retained on the filters by liquid scintillation counting.

11 Construct a standard curve using known concentrations of cyclic AMP and fit the data to a logistic expression using GraphPad Prism. Read cyclic AMP concentrations off this curve. Calculate cellular cyclic AMP concentrations, taking account of the volumes taken and the dilutions made during the assay.

Evidence is emerging that agonist-stimulated receptor internalization plays other roles, in addition to mediating receptor desensitization or resensitization. For instance, it has been shown recently that β-arrestin binding by the agonist-activated β$_2$-adrenoceptor causes the recruitment of the tyrosine kinase c-Src (35). This kinase in turn activates the mitogen-activated protein (MAP) kinases Erk1 and Erk2. In this way a second signalling pathway is switched on immediately after the inactivation of the G protein-mediated pathway. It was further demonstrated that β-arrestin mutants that were unable either to bind c-Src or to target the receptors to clathrin-coated pits acted as dominant negative inhibitors of β$_2$-adrenoceptor-mediated activation of the MAP kinases. Hence, receptor endocytosis is essential to the operation of this additional signalling pathway.

7 Pharmacological manipulation of the receptor reserve

One factor that is likely to be relevant to the relationship between receptor trafficking and receptor function is the number of receptors at the plasma membrane, and specifically the presence or otherwise of a receptor reserve. Intuitively, one would expect that if there is a large number of receptors at the plasma membrane, in excess of the number required to produce a maximal response, then removal of receptors by internalization would make little difference to the size of the response generated at saturating concentrations of agonist. Under such circumstances, however, a large proportion of the agonist-activated receptors might become phosphorylated, and thereby desensitized. Receptor internalization and recycling would then allow receptor dephosphorylation, and subsequent resensitization. On the other hand, if there is little or no receptor reserve, then removal of receptors from the plasma membrane might contribute to desensitization.

In SH-SY5Y cells the $[Ca^{2+}]_i$ responses to carbachol resensitize before surface receptor number recovers fully through recycling (see above) (19). This result indicates that there is a substantial receptor reserve which might contribute to the rapid resensitization of the response. We tested this possibility directly by using PrBCM to remove the receptor reserve, and examining the effect of this manipulation on receptor resensitization. Treatment of the cells with PrBCM caused a time-dependent fall in [^{3}H]NMS binding ($t_{1/2}$ 6 min). When half of the receptors were alkylated, there was no effect on the $[Ca^{2+}]_i$ response to a saturating concentration of carbachol. However, as more receptors were removed, there was a gradual reduction in the response, suggesting that the receptor reserve represents about half of the surface receptor population. When this receptor reserve was abolished by PrBCM treatment, it was found that desensitization of the response to carbachol was unaffected, but that the extent of the resensitization following removal of the agonist was significantly reduced. Hence, receptor reserve in SH-SY5Y cells does indeed play a role in resensitization, allowing the cells to respond maximally to agonist before surface receptor numbers have been restored to original levels. In NG108-15 cells, we found a contrasting picture (21). Here desensitization of the response to carbachol was enhanced by stimulation of endocytosis (e.g. by overexpression of GRK2) and blocked by inhibition of endocytosis (e.g. by hyperosmolar sucrose). We would predict that in these cells there is no receptor reserve for the response to carbachol. However, this prediction has not yet been tested.

8 Future prospects

It has been estimated that there are more than a thousand G protein-coupled receptors in the human genome. Of these, only a few have been studied in any detail. Hopefully, the type of kinetic analysis described here will prove to be useful in the study of the intracellular trafficking of other members of the G protein-coupled receptor superfamily.

Once receptors are endocytosed and delivered to endosomes, they may be either recycled to the plasma membrane or sent on to be degraded in lysosomes. Clearly, the outcome of this 'choice' at the level of the endosome has major implications for cellular function in the face of long-term agonist stimulation. For this reason, there is growing interest in the molecular mechanisms underlying this sorting event. The development of a flexible kinetic analysis of endosome-to-lysosome trafficking may be of considerable benefit in this respect.

Finally, it is clear that agonists vary greatly in the extent to which they are internalized along with their receptors. For instance, it has been shown that agonists at the somatostatin sst_2 receptor are internalized to a significant extent (15, 16), and this internalization might be important for the killing of tumour cells by radioactive somatostatin analogues. In contrast, it appears that agonists at the muscarinic receptor are not internalized to any appreciable extent. The key factor that decides the fate of the agonist is likely to be its affinity for the receptor. Put crudely, on average, low-affinity agonists dissociate before they are internalized with the receptor, whereas high-affinity agonists remain bound and undergo receptor-mediated endocytosis. The development of a suitable model should allow prediction of the extent of internalization of any agonist, provided that its binding affinity and the rate constant for endocytosis of the receptor are both known.

Acknowledgements

I am grateful to Jennifer Koenig and Philip Szekeres for their contributions to the development of the techniques described in this chapter, and to the Wellcome Trust for supporting my work on receptor trafficking.

References

1. Koenig, J. A. and Edwardson, J. M. (1997). *Trends Pharmacol. Sci.*, **18**, 276.
2. Harden, T. K., Petch, L. A., Traynelis, S. F., and Waldo, G. L. (1985). *J. Biol. Chem.*, **260**, 13060.
3. Thompson, A. K. and Fisher, S. K. (1990). *J. Pharmacol. Exp. Ther.*, **252**, 744.
4. Koenig, J. A. and Edwardson, J. M. (1994). *J. Biol. Chem.*, **269**, 17174.
5. Koenig, J. A. and Edwardson, J. M. (1994). *Br. J. Pharmacol.*, **111**, 1023.
6. Koenig, J. A. and Edwardson, J. M. (1996). *Mol. Pharmacol.*, **49**, 351.
7. Hulme, E. C., Birdsall, N. J. M., and Buckley, N. J. (1990). *Annu. Rev. Pharmacol. Toxicol.*, **30**, 633.
8. Sibley, D. R., Strasser, R. H., Benovic, J. L., Daniel, K., and Lefkowitz, R. J. (1986). *J. Biol. Chem.*, **83**, 9408.
9. Pippig, S., Andexinger, S., and Lohse, M. J. (1995). *Mol. Pharmacol.*, **47**, 666.
10. Tsuga, H., Kameyama, K., Haga, T., Kurose, H., and Nagao, T. (1994). *J. Biol. Chem.*, **269**, 32522.
11. Pals-Rylaarsdam, R., Xu, Y., Witt-Enderby, P., Benovic, J. L., and Hosey, M. M. (1995). *J. Biol. Chem.*, **270**, 29004.
12. Szekeres, P. G., Koenig, J. A., and Edwardson, J. M. (1998). *Mol. Pharmacol.*, **53**, 759.
13. Roettger, B. F., Ghanekar, D., Rao, R., Toledo, J., Yingling, J., Pinon, D., *et al.* (1997). *Mol. Pharmacol.*, **51**, 357.
14. Cullen, P. J., Dawson, A. P., and Irvine, R. F. (1995). *Biochem. J.*, **305**, 139.
15. Koenig, J. A., Edwardson, J. M., and Humphrey, P. P. A. (1997). *Br. J. Pharmacol.*, **120**, 52.
16. Koenig, J. A., Kaur, R., Dodgeon, I., Edwardson, J. M., and Humphrey, P. P. A. (1998). *Biochem. J.*, **336**, 291.
17. Lohse, M. J., Benovic, J. L., Caron, M. G., and Lefkowitz, R. J. (1990). *J. Biol. Chem.*, **265**, 3202.
18. Yu, S. S., Lefkowitz, R. J., and Hausdorff, W. P. (1993). *J. Biol. Chem.*, **268**, 337.
19. Szekeres, P. G., Koenig, J. A., and Edwardson, J. M. (1998). *J. Neurochem.*, **70**, 1694.
20. Bogatkewitsch, G. S., Lenz, W., Jakobs, K. H., and Van Koppen, C. J. (1996). *Mol. Pharmacol.*, **50**, 424.
21. Holroyd, E. W., Szekeres, P. G., Whittaker, R. D., Kelly, E., and Edwardson, J. M. (1999). *J. Neurochem.*, **73**, 1236.

22. Slowiejko, D. M., McEwen, E. L., Ernst, S. A., and Fisher, S. K. (1996). *J. Neurochem.*, **66**, 186.

23. Hertel, C., Coulter, S. J., and Perkins, J. P. (1985). *J. Biol. Chem.*, **260**, 12547.

24. Tartakoff, A. M. (1983). *Cell*, **32**, 1026.

25. Daniels, P. U. and Edwardson, J. M. (1988). *Biochem. J.*, **252**, 693.

26. Hausdorff, W. B., Caron, M. G., and Lefkowitz, R. J. (1990). *FASEB J.*, **4**, 2881.

27. Clark, R. B., Friedman, J., Dixon, R. A. F., and Strader, C. D. (1989). *Mol. Pharmacol.*, **36**, 343.

28. Benovic, J. L., Strasser, R. H., Caron, M. G., and Lefkowitz, R. J. (1986). *Proc. Natl. Acad. Sci. USA*, **83**, 2797.

29. Lohse, M. J., Benovic, J. L., Caron, M. G., and Lefkowitz, R. J. (1990). *Science*, **248**, 1547.

30. Ferguson, S. S. G., Downey, W. E. III, Colapietro, A.-M., Barak, L. S., Ménard, L., and Caron, M. G. (1996). *Science*, **271**, 363.

31. Goodman, O. B. Jr., Krupnick, J. G., Santini, F., Gurevich, V. V., Penn, R. B., Gagnon, A. W., *et al.* (1996). *Nature*, **383**, 447.

32. Krueger, K. M., Daaka, Y., Pitcher, J. A., and Lefkowitz, R. J. (1997). *J. Biol. Chem.*, **272**, 5.

33. Pitcher, J. A., Payne, E. S., Csortos, C., DePaoli-Roach, A. A., and Lefkowitz, R. J. (1995). *J. Biol. Chem.*, **92**, 8343.

34. Ohkuma, S. and Poole, B. (1978). *Proc. Natl. Acad. Sci. USA*, **75**, 3327.

35. Luttrell, L. M., Ferguson, S. S. G., Daaka, Y., Miller, W. E., Maudsley, S., Della Rocca, G. J., *et al.* (1999). *Science*, **283**, 655.

Chapter 8

Real time receptor function *in vitro*: microphysiometry and the fluorometric imaging plate reader (FLIPR)

Martyn Wood and Darren Smart

Neuroscience Research, GlaxoSmithKline, New Frontiers Science Park, Third Avenue, Harlow, Essex CM19 5AW, UK.

1 Introduction

The recent application of molecular biological techniques to the cloning of receptors has greatly increased our understanding of their function and regulation. This has enabled the study of the direct interaction of the ligand with a single receptor essentially free of many of the complications experienced with *in vitro* studies in native receptors, although the use of recombinant expression systems has introduced new problems such as overexpression and lack of, or incorrect, coupling mechanisms. However, for receptors with 7-transmembrane domains, the use of heterologous expression systems has resulted in a limited ability to study receptor function in real time. Instead of measuring contractile or secretory responses directly, receptor function is studied using biochemical techniques such as changes in the concentration of cAMP, changes in the activity of adenylyl cyclase, or accumulation of inositol phosphates. However, these changes are not measured in real time but use 'before' and 'after' or 'end-point' measurements. Here, we describe the use of two techniques which enable the real time assessment of receptor function using cloned receptors expressed in cell lines, with special reference to the family of 7-transmembrane spanning receptors.

2 Microphysiometry

2.1 Principles of microphysiometry

The Cytosensor microphysiometer (Molecular Devices) has been widely used to study both receptors and receptor signalling (1–3). The versatility of the microphysiometer is that, unlike other methods for studying receptor signalling, it is independent of the signal transduction pathway employed by the receptor. The microphysiometer quantifies cellular metabolic activity by measuring changes in the extracellular acidification rate. Physiological changes, such as receptor activation, alter the rate of energy metabolism, and this, together with associated utilization of ATP, results in the excretion of acid metabolites, such as protons, from the cell. The microphysiometer uses silicon semiconductors to detect these minute changes in the excretion of acid metabolites as a measure of changes in metabolic rate (for a detailed discussion of the principles of the sensor, see ref 4). The light-addressable potentiometric sensor (LAPS) uses a light source (LED) to induce charge movements through the silicon sensor to produce a photocurrent. At neutral pH, the silicon surface is negatively charged due to the presence of SiO-groups. Changes in the pH of the solution affect the photocurrent-induced surface potential of the sensor which is sensed electronically and enable the rate of change in the surface potential to be determined.

Experimentally cells, at a density of 100 000–300 000 per well, are grown in transwell cups. By use of a capsule insert and spacer, cells are sandwiched between two polycarbonate membranes to form a tight chamber, with a volume of only 4 μl, which is in close proximity to the sensor (*Figure 2*). A peristaltic pump controls perfusion of the cells with low buffering capacity media, typically at a flow rate of 100 μl/min, which removes acidic metabolite accumulation at the sensor surface. In a cyclical fashion, the pump is turned off (typically for 30 sec with the pump on 60–90 sec) allowing the accumulation of acid metabolites at the sensor surface. The acidification rate is determined by measuring the rate of change of the surface potential of the sensor. By means of a valve, drugs can be perfused over the cells for controlled periods allowing pharmacological, physiological, and toxicological studies. An important aspect of the microphysiometer is that it can measure decreases in metabolic activity, as seen in cell death, as well as the more normal increase in metabolic activity resulting from receptor activation. The Cytosensor microphysiometer has either four or eight sensor chambers allowing between-chamber comparisons and the determination of concentration-related effects. The versatility of the microphysiometer can be enhanced by the use of an optional autosampler (Cytosampler, Molecular Devices) allowing multiple drug or physiological challenges.

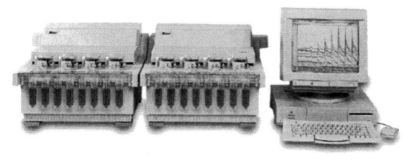

Figure 1 The Cytosensor microphysiometer.

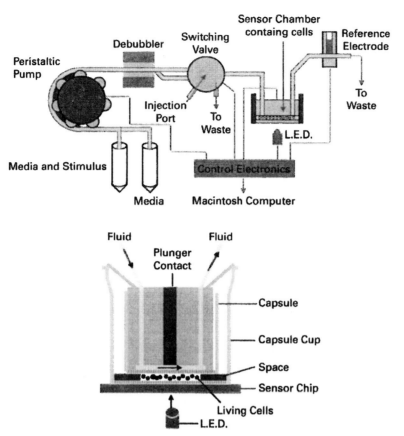

Figure 2 Schematic diagram of the Cytosensor microphysiometer and sensor chamber. Shown is the perfusion system, valve switching and detailed layout of the sensor chamber (reproduced by courtesy of Molecular Devices).

There are many advantages in the use of the microphysiometer. It is non-invasive and allows the determination of metabolic rate on a wide variety of cell types in real time. It can measure both increases and decreases in metabolic rate, but probably the greatest advantage of the microphysiometer is that, by measuring changes in metabolic rate, it is independent of the signal transduction pathway. The microphysiometer has therefore been used successfully to monitor activation of a wide variety of cell surface receptors which couple to different effectors, e.g. muscarinic M_1, M_2, M_3, M_4, and M_5 (5) and CRF (6) receptors, intracellular targets such as nuclear hormone receptors (7), and signal transduction pathways (3). The disadvantages of the microphysiometer are that it has a relatively low throughput, is labour-intensive (especially without the autosampler), and requires technical expertise.

2.2 Practical aspects

The preparation of good quality cells is critical for the generation of robust and reproducible functional responses. The method used for cell detachment must be chosen with care—we use scraping for Chinese hamster ovary (CHO) cells and trypsin for human embryonic kidney (HEK) 293 cells (being careful with exposure times as we have found that trypsin can interfere with some assays). Details are given in *Protocol 1*.

Experimental details for a typical run on the microphysiometer are given in *Protocol 2*.

Protocol 1

Cell preparation for microphysiometry using CHO cells

Equipment and reagents

- Microscope
- Cell scraper
- Centrifuge
- Pipettes
- Culture flasks

- CO_2 incubator
- Cell culture medium: MEM Alpha medium (Gibco 22571-020)
- Ribonucleosides and deoxyribonucleosides
- Fetal bovine serum (FBS) (Gibco)

Method

1 Prewarm the cell culture medium (MEM Alpha medium), plus ribo- and deoxyribo-nucleosides, 10% fetal bovine serum to 37°C.

2 Microscopically examine the culture for level of confluence, morphological appearance, and any sign of contamination.

3 Decant culture medium from cells by pipette aspiration and pipette 10 ml fresh medium into the flask.

4 Gently but firmly run the edge of a sterile scraper across the cell sheet to remove the cells.

5 Spin cells in a centrifuge 200 g at 21°C for 5 min and decant culture medium by pipette aspiration until only the pellet is left at the bottom of the tube.

6 Mix cells carefully in a small volume of culture medium by repeat pipetting through a fine-tipped glass Pasteur pipette; when no clumps are visible, make up to 10 ml in culture medium.

7 Inoculate fresh culture flasks at the correct cell number or ratio, 1:30 split is generally confluent in three to four-days, 1:60 as a seven day back-up.

8 Place inoculated flasks in a 5% CO_2 incubator at 37°C.

9 Cells will need to be subcultured every three or four days, and should be regularly checked until the operator is familiar with the cell line growth.

10 Recommended plating densities: 12-well plates: 1 ml volume giving 300 K/well (transwell plates, Molecular Devices).

Protocol 2

Typical experimental details for Cytosensor microphysiometer using human muscarinic receptors expressed in CHO cells

Equipment and reagents

- See *Protocol 1*
- Cytosensor microphysiometer
- Carbachol
- DMSO/PEG
- 4 M NaCl (Sigma S-5886)
- DMEM (w/o sodium pyruvate/NaHCO$_3$) powder (Gibco 52100-021, Life Technologies to make 5 litres)
- 200 mM glutamine (Gibco 25030-024, Life Technologies)

A. Cells

1 The cells are prepared from subconfluent/confluent flasks. Passage number between 4 and 30 for best results. Cells are plated out 24 h prior to use on the microphysiometer.

2 The cells are suspended in medium at 300 000 cells/ml and 1 ml of cell suspension added to the cup of the transwell. Outside the cup, 1 ml of medium is also added. The cells are then incubated at 37 °C (5% CO$_2$/95% O$_2$).

3 After at least 6 h, the medium is removed from the transwell and replaced with medium free of fetal calf serum (FCS) and any other supplements which may affect the cells, e.g. sodium butyrate for hM2 and hM4 cells.

4 The cells will then be incubated overnight ready to be used the next day.

B. Microphysiometer

1 The machine is run at 37 °C with the debubblers run at 6 °C above this temperature.

2 Pump cycle times vary and need to be optimized:

(a) For hM1 cells the total pump cycle time is 1:30 min with the pump on for 1 min 13 sec (50% pump speed) the pump is then stopped for 17 sec. The rate measurement is taken for 13 sec, 2 sec after the pump has stopped.

(b) For hM3 cells the total pump cycle time is 1:30 min. The pump is on for 1 min (50% pump speed), the pump is then stopped for 30 sec. The rate measurement is taken for 20 sec, 8 sec after the pump has stopped.

3 Timings of drug exposure; the timings are optimized for each receptor against carbachol. For the following experimental methods the exposure time for the receptors are as follows:

(a) hM1 CHO 32 sec (15 sec prior to pump off and during the 17 sec pump off period).

(b) hM3 CHO 1 min 30 sec (1 pump cycle).

4 Experiments use a combination of the Cytosampler and the valve switched between side one and side two. The Cytosampler is used to prime the line and then a switch

is used to feed the cells from line 2. This method ensures that the cells are exposed to the compound at an accurate time point.

5 Drug dilutions: in general all compounds are dissolved in dimethyl sulfoxide (DMSO)/polyethylene glycol (PEG) (50:50) at 10^{-2} M, they are then diluted either 1:100 or 1:1000 into running medium. If the pH of the running medium is compromised, the pH of the solution is adjusted to 7.4 using either HCl or NaOH. The pH of the solution needs to be the same as the running medium or it will produce a sham effect. (Try to adjust the pH of the solution to that of the running medium at a higher concentration than will be used on the machine as this will help to dilute any pH effects.)

C. Agonist testing

1 The cells are run with running medium for approx. 1 h before commencing the experiment as this ensures that a stable baseline has been established before the cells are stimulated. After 1 h the cells are exposed to a maximal concentration of carbachol, there is then at least a 42 min interval before the dose–response curve for the agonist can begin (the 96-well blocks can hold 2 ml of fluid, with a 1:30 pump cycle at normal speed this will last 21 min, hence 42 min).

2 Running medium is usually the same basic medium as that used to grow the cells but without HCO_3^- (this is replaced by NaCl): DMEM running medium: $1 \times$ DMEM (w/o sodium pyruvate/NaHCO$_3$) powder, 50 ml of 200 mM glutamine, 55.5 ml of 4 M NaCl (58.4 g/250 ml). Add together, make up to 5 litres and adjust pH to 7.4. Media can be stored in the fridge following sterile filtration.

3 Cells are exposed to the lowest concentration of agonist first. Each agonist exposure time is followed by a washout time of 21 min before subsequent stimulation. We usually test seven to ten concentrations of agonist to define the concentration–effect curve. Usually the cells are challenged with maximal concentrations of carbachol (acting on endogenous muscarinic receptors in CHO cells) before and after construction of the concentration–effect curve: this is to act as a reference maximal effect and to determine if there is run-down during the time of the assay.

D. Antagonist evaluation

1 The cells are allowed to settle for 1 h before stimulation. The cells are stimulated with carbachol at 21 min intervals for the appropriate exposure times (see above). There is a washout period of 21 min before a > 40 min exposure to the antagonist after which a second concentration response to carbachol is started. The antagonist is then continuously present throughout the second dose–response curve at a single concentration. The cells are again stimulated at 21 min intervals with increasing concentrations of carbachol. The concentrations are increased in half log-unit steps. The concentration of antagonist used may vary between chambers. At least one chamber is run in the absence of antagonist as a time matched control.

Protocol 2 continued

E. Data acquisition

1 Data is saved to a Macintosh file, this is then either removed from the system on floppy disk or via SyQuest/Zip disks depending on the file size. This is then transferred to a second Macintosh which is not connected to a microphysiometer. The rate data can then be exported in a form which can be analysed on a PC. This can be done on the Macintosh computer, which controls the microphysiometer, but being able to visualize the data at the same time as analysing the data is a great advantage and cannot always be achieved if the microphysiometer is already running a program.

F. Data analysis

1 Peak acidification rates are determined for agonist concentrations and the resulting concentration response curves are fitted using Robofit (8).

G. Safety issues

1 The use of chemical and biological substances described in the methodology necessitates a risk assessment, to comply with COSHH (UK) regulations.

As use of the microphysiometer does require some technical background and familiarization, some advice we have developed is given in *Protocol 3*.

Protocol 3

Microphysiometer practical advice

Temperatures, pump speeds, and pump cycle times are all suggested guidelines but can vary depending on the cell line being investigated. In this case the machine will be run at 37°C. This method also assumes the use of adherent cell lines. If the cells are non-adherent, the basic method is the same with only a variation at Step 22 where the agarose entrapment method (see Molecular Devices Tip Sheets) would be employed.

1 If the machine has not been used the previous day, run the Cytoclean programme with water the day before you use it—this will make calibrating the machine easier.

2 Running media should be 37°C so that when the liquid is added to the sensor chambers and wells, no air bubbles are produced from the heating process.

3 This method uses the Cytosampler (Molecular Devices autosampler using 96-deep well blocks) for fluid delivery.

4 Check water levels in tubes—uneven levels could indicate blockage in tubing.

5 Set the parameters to: Temperature (°C): Chambers: 37, debubblers: 6.0. The aim is to get the machine warming up to running temperature.

6 Run through warm running medium.

Protocol 3 continued

7 Align Cytosampler, thread appropriate tubing onto robot arm allowing a couple of centimetres to extend below the arm. Using the front panel . . . home . . . align.

8 Controls—Edit Pump Cycle—to suit own cells, 'Pump on' interval should be long enough for the raw data to reach a plateau after the rate measurement and 'pump off' period.

For example:
Pump on for 1 min, pump off for 30 sec, reading taken from 1:08 to 1:28, pump starts again at 1:30. There is a slight delay between the pump stopping and the reading being taken to allow for any back-pressure which may affect the reading.

9 Once tubing is primed with running medium, start loading on the sensor chambers. Place fresh running medium in each chamber. Inspect the black cleaning chambers to ensure that no 'O' rings have become detached from the plungers. If they have, use blunt plastic forceps to replace them. Fill these cups with 70% ethanol and place in a sealed box to avoid evaporation–the silicon sensor should not be allowed to dry out.

10 Check that the reference electrodes contain sufficient solution.

11 Run the program without calibrating the machine (alt key & run (front panel)). This will allow you to watch the raw data trace and see if any air bubbles are present and allow you to adjust the machine accordingly.

12 If all traces look good, then attempt to calibrate the machine.

13 Ignore attempts to save data. If the idle speed is set to 0.5% then use the front panel to increase the pump speeds to 50% before starting calibration.

14 If all chambers calibrate immediately, cells can then be loaded onto the machine.

15 If any of the chambers fail to calibrate, error messages will indicate the type of problem.

    No inflection point . . . . suggests bad electrical connection

    Photoresponse is too low . . . . suggests air bubbles present

16 If machine fails to calibrate then lift gantry off affected chamber and remove sensor cup. If air bubbles are present in running medium use a plastic pipette to disperse bubbles. DO NOT TOUCH THE BLUE CIRCLE IN THE CENTRE OF THE CHAMBER: this is the silicon chip and is susceptible to damage.

17 Using an ethanol wipe (e.g. Kimwipe sprayed with 70% ethanol) to clean the silver spot on the sensor cup, the rod and contact pin on the gantry, and sapphire glass window (be careful when cleaning the contact pin as this is delicate). Avoid touching the electrical contact points as oil from skin can interfere with conductance.

18 Reassemble the chamber and disconnect the outward tubing (use a piece of tissue paper to prevent running medium running back onto the plunger), raise the tubing to allow an air bubble to form and then reconnect. This air bubble should be large enough to carry any bubbles away from the reference electrode. Wiggle the sensor cup in its position to ensure electrical contact.

Protocol 3 continued

19 Allow a couple of minutes to elapse before attempting to re-calibrate, check to see if any air bubbles are present in the reference electrodes.

20 If calibration continues to fail then the hole at the bottom of the white chamber can be wiped, but care must be taken as this is a delicate area.

21 If you let the raw data run whilst you are trying to calibrate the machine you will be able to see from the trace if the data have stabilized and which chambers are still affected by air bubbles.

22 Cells require a spacer and capsule insert to be fully assembled. Care must be taken that no air bubbles become trapped in the cell capsule. If the components are placed quickly into the capsule cup this reduces the chances of air bubbles getting trapped. Dry the forceps between each cell chamber and push the spacers to the bottom before adding the inserts. Running medium (0.4 ml) is placed in the capsule insert before assembly into the sensor cup (for non-adherent cells use the agarose entrapment method).

23 Allow a few minutes for the cells to stabilize before calibration. Be careful when opening up the sensor cups as capsule inserts may become attached to the plungers.

24 If you are running a protocol on the microphysiometer close down the Cytosoft file and start the protocol.

25 Agarose cell entrapment (for non-adherent cell cultures) kits and details are available from Molecular Devices (R8023).

26 Closing down (approx. 1 h before end of experiment): follow Molecular Devices procedure for sterilizing and cleaning.

2.3 Experimental data

A typical experimental trace is shown in *Figure 3*. This shows the principle of the cycling of pump on and pump off and the real time measurement of receptor function. During the period when the pump is on, acid metabolites are removed and a stable baseline is seen. When the pump is switched off, there is a build-up of acid metabolites in the chamber which results in a reduction in the photo-current. This rate of acidic metabolite build-up is increased when the cell is activated, in this case by the presence of 10 μM carbachol. It is the rate of acidification, i.e. slope of the change in photocurrent, which is reported. It can also be seen from *Figure 3* that the acidification response to carbachol at human M1 receptors was extremely rapid; indeed the acidification rate measurement was taken during the initial rapid phase (first 13 sec) and with a reduced car-bachol exposure time, otherwise a reduced functional potency and efficacy to carbachol was seen. This suggests that receptor desensitization is occurring, even over these short time periods and that the microphysiometer can be used to follow desensitization.

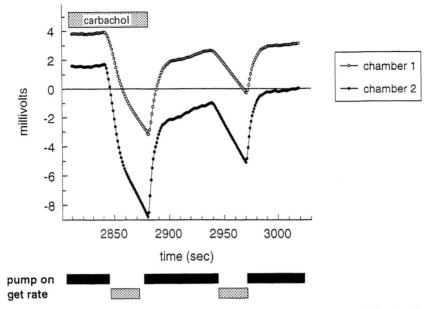

Figure 3 Experimental trace from human muscarinic M1 receptors expressed in CHO cells using the microphysiometer. Shown are typical traces from two individual chambers where carbachol (10µM) was introduced to the chambers 30 sec prior to pump off and was present throughout the 30 sec pump off cycle. Time (sec) is from start of the experiment. Taken from (5).

3 Fluorometric imaging plate reader (FLIPR)

3.1 Principles of the FLIPR

Recently, the fluorometric imaging plate reader (FLIPR: Molecular Devices) (*Figure* 4) has become available for measurement of rapid kinetic changes in fluorescence in live cells in a 96-well format (9). The FLIPR allows simultaneous addition (via an integral 96-well pipettor) and fluorescence measurements, via an argon ion laser and charge-coupled device camera, in all 96 wells. The most common use of the FLIPR is with the Ca^{2+}-sensitive fluorescent dyes such as Fluo-3 AM (Molecular Probes). This method has been used extensively to study G_q-coupled receptors (10, 11) and ion channels (12). Though practically more challenging, the FLIPR can also be used to monitor changes in membrane potential and intracellular pH (9). Although the use of changes in intracellular Ca^{2+} concentrations as a reporter limits the FLIPR's utility in signal transduction studies, a number of different second messengers are capable of modulating Ca^{2+} responses, including cyclic nucleotides, inositol (1,4,5) trisphosphate (IP_3), and cyclic ADP ribose (see ref. 13). Further, although receptor-mediated generation of IP_3 mediated by phospholipase C is probably the most common pathway for release of intracellular Ca^{2+}, it is not the only one. For example, in RBL cells, transfected with CCR2b receptors, activation of the receptor increases Ca^{2+} mobilization as measured on the FLIPR, via a pertussis toxin-sensitive mechanism (14). However, the response is blocked by pre-treatment with the phospho-

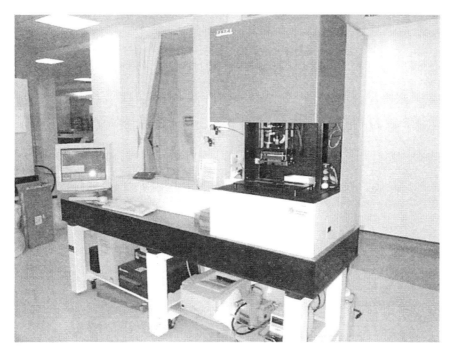

Figure 4 The fluorometric imaging plate reader (FLIPR).

lipase C inhibitor U-73122 (unpublished observations). This suggests that, in these cells, G_i stimulation results in liberation of β,γ-subunits which then activate phospholipase C isoforms. One important extension of the FLIPR has been the use of chimeric G proteins allowing coupling of non-G_q linked G protein-coupled receptors to intracellular Ca^{2+} mobilization, even though they do not do so normally (15, 16).

3.2 Practical aspects

As with the microphysiometer, the preparation of high quality cells is critical for the generation of robust responses and the method used for cell detachment is determined empirically. We have found that it is best to use scraping or trypsin for CHO cells and trypsin for HEK 293 and 132 1N1 cells. However, care has to be taken with exposure times as trypsin can interfere with some assays. Details are similar to *Protocol 1*, steps 1–8 with the following amendment to step 9: cells will need to be subcultured every three or four days, and should be regularly checked until the operator is familiar with the cell line growth. Recommended plating densities:

96-well black-walled, clear-based plates (Costar): 0.1 ml volume giving 20 K/well.

Experimental details for a typical run on the FLIPR are given in *Protocol 4*.

Protocol 4

Typical experimental details for FLIPR studies using Chinese hamster ovary (CHO) cells

Equipment and reagents

- Black plates with a transparent base for FLIPR system (Costar: Life Technologies)
- Black disposable pipette tips (Molecular Devices)
- Denley Cellwash (Life Sciences International)
- Tyrode's buffer: 145 mM NaCl, 2.5 mM KCl, 10 mM Hepes, 10 mM glucose, 1.2 mM MgCl$_2$, 1.5 mM CaCl$_2$, 2.5 mM probenecid, pH 7.4 with 1 M HCl or NaOH

- Probenecid (Sigma), dissolved in 1 M NaOH (5 ml)
- Fluo-3 acetoxymethyl (AM, TefLabs): 1 mM stocks are prepared by dissolving in anhydrous DMSO
- Pluronic F-127 (Molecular Probes) is dissolved as a 20% solution (w/v) in DMSO with overnight stirring
- Fluo-3 AM is then mixed with pluronic acid 1:1, aliquoted (usually 1 ml) as appropriate, and stored at −20 °C until required

A. Cells

1   The cells are prepared from subconfluent/confluent flasks. Passage number between 6 and 20 for best results. Cells are plated out 24 h prior to use on the FLIPR.

2   CHO cells are seeded into black-walled clear-based 96-well plates (e.g. Costar) at a density of 20 000 cells per well in MEM Alpha medium (0.1 ml), supplemented with 10% FCS and 400 μg/ml G418 (antibiotic for selection pressure), and then incubated under 95%/5% O$_2$/CO$_2$ at 37 °C overnight.

B. Method for adherent cells

1   Make up loading medium. This is the culture medium supplemented with 4 μM Fluo-3 AM, 0.04% Pluronic F-127, and 2.5 mM probenecid.

2   Add 50 μl of loading medium per well and incubate the cells at either 37 °C for 60–90 min, or at room temperature for 90–120 min.

3   Aspirate loading medium and wash four times with 200 μl of Tyrode's buffer. This step may be performed manually or using a cell washer (such as the Denley Cellwash). A final volume of 125 μl of buffer should be left in the wells at the end of the final wash.

4   Antagonists made up in Tyrode's buffer or DMSO at 6 × desired final concentration, or buffer alone (control), are then added (25 μl) to the plates, and the cells returned to the incubator for 30 min.

5   Each plate is then transferred in turn to the FLIPR for assay. The cell plate is placed in the central position in the FLIPR drawer.

6   A clear 96-well plate containing the appropriate agonist(s) made up in Tyrode's buffer at 4 × final desired concentration is placed in the right-hand compartment of the drawer and the drawer closed.

7 To establish that cells are loaded with dye, a signal test is carried out, in which the CCD camera captures an image of the plate. The FLIPR automatically converts this image into a numerical fluorescence reading for each well, which is displayed on the computer screen. This allows the operator to establish that cells have loaded with dye and that the loading is uniform across the plate. For most cell types an average signal test of 8000–12 000 fluorescent units/well is optimal (a standard deviation of 10% is acceptable). The CCD camera on the FLIPR saturates at ~65 000 fluorescent units.

8 If required, the sensitivity of the FLIPR can be adjusted by altering the laser output, the camera aperture, or the image exposure length. The camera aperture is usually set to F2 and the camera exposure to 0.4 sec, with the laser output being varied between 300 and 750 mW.

9 The experimental run is then initiated. The integrated robot in the FLIPR will transfer agonist (50 μl) from the agonist plate and add it to the cell plate at a set time. The CCD camera captures images every second both before and after the agonist addition. The timing of the addition, the volume dispensed, the rate and number of images captured can all be varied. We typically obtain a baseline reading for the first 20 sec whereupon the agonist is added. Readings are taken every sec for the next 40 sec with subsequent readings taken at 5 sec intervals.

C. Data acquisition

1 Data (FID and SQU files) are automatically archived to a PC. FID files are subsequently transferred via networked connections to be archived on CD-R.

D. Data analysis

1 For each well, fluorescent readings are extracted from the SQU files and tabulated versus time within an ASCII text file.

2 From each well, peak fluorescence is determined over the whole assay period. From this figure the mean of readings 1–19 inclusive, are subtracted. This figure is plotted versus compound concentration and iteratively curve-fitted using a four parameter logistic fit (17) to generate a concentration effect value.

3 Alternatively, the peak height minus basal can be determined directly using the FLIPR software and then exported for incorporation into data analysis programmes such as Prism (GraphPad software) or Grafit (Erithacus Software).

4 Agonist potencies are typically described with EC_{50} values (concentration yielding half-maximal effect) as well as intrinsic activity (maximal effect as a proportion of the maximal effect of a reference full agonist).

5 Antagonist potencies are typically described with IC_{50} values (concentration yielding 50% inhibition of reference agonist stimulation) and/or K_b values calculated thus:

$$K_b = IC_{50}/(1 + ([agonist]/EC_{50}))$$

where: IC_{50} is as above in molar terms, [agonist] is agonist concentration used in molar terms, EC_{50} is the concentration, in molar terms, of agonist causing a 50% of maximal response) in the stimulated cell.

Protocol 4 continued

E. Safety issues

1 The use of chemical and biological substances described in the methodology necessitates COSHH assessment.

2 Use of human derived genetic material and cell lines. It is required that any part of the procedure that results in a significant production of aerosols must be carried out in a cabinet designed to protect the user.

The use of the FLIPR requires skilled operators and familiarization, therefore some tips and advice are given here.

The laser on the FLIPR has three main excitatory wavelengths, 457 nm, 488 nm, and 510 nm. In general, the optimal dyes for studying Ca^{2+} in the FLIPR are Calcium Green-1 AM and Fluo-3 AM. The best dye to use with a given cell line should be determined experimentally, although Fluo-3 generally produces a larger fluorescent signal than Calcium Green-1 with most cell types tested. Both dyes excite with the 488 nm excitation line of the argon laser and emit in the 500–560 nm range. Fura-2, which requires excitation at 340 nm, is not suitable for use in the FLIPR. Fluo-4 AM can also be used which gives a larger fluorescent signal (see Molecular Probes Handbook), but which is more expensive.

Weak or non-adherent cell lines can also be used in the FLIPR by using poly-D-lysine coating of the plates (pre-coated plates available from Becton Dickinson) and more gentle wash and agonist addition protocols (see below). For most cell types, dyes can be loaded in Tyrode's buffer. Regardless, fresh loading medium-buffer should be prepared each day and stored in the dark until use. This is particularly important if a loading medium containing serum is employed, as esterase, present in serum, will hydrolyse ester groups from Fluo-3 AM and result in impaired loading. The loading medium should be pre-warmed prior to addition to cells, to avoid shocking them. Probenecid should be prepared on the day of use. In some cells, e.g. HEK 293 cells expressing the rat vanilloid receptor, probenecid can interfere with the Ca^{2+} signal and/or reduce cell viability. In these cells the anion transport inhibitor sulfinpyrazone (Molecular Probes, Sigma; 250 µM final concentration) can be used instead, although this is generally less effective in preventing dye efflux than is probenecid. Alternatively, the cells can be loaded in the absence of probenecid if incubated at room temperature in the dark for 2 h. Optimization of the wash protocol for the Denley Cellwash (Life Sciences International) is critical for all cell types. Aspirate/dispense heights, dispense speeds, vacuum level, and number of washes can all be adjusted to ensure that the cell monolayer is not disrupted. This is particularly important for weak/non adherent cell lines. To ensure consistent mixing of reagents when added to the test plate, we add relatively large volumes of agonist/antagonist.

The setting of the FLIPR pipettor's height and speed also requires optimization. The dispensing height of the pipettor should be set to ensure that the tip of the pipette is below the meniscus level of fluid in the well following dispensation, to ensure that no droplets remain on the tips. However, disruption of the cell monolayer occurs if the pipette tip is too close to the bottom of the well. The optimal dispense height should be determined empirically. It is important to note that different makes of plates vary in depth even though they all conform to the Society for Biomolecular Screening footprint. For strongly adherent cell lines, the pipettor dispense speed of the instrument may be set to 60 μl/sec to ensure good mixing. However, if cells are weakly/non-adherent then the dispense speed must be reduced (e.g. 30 μl/sec) to avoid cell disruption. The optimal dispense speed is one that does not remove cells from the base of the well but is sufficiently rapid to ensure adequate mixing. The plate viewer is useful in visualizing the monolayer to identify whether there have been problems with cell detachment following agonist addition.

When setting up any FLIPR assay, certain conditions need to be optimized. First, the seeding density should be optimized to give both suitable basal fluorescent units (8000–12 000) and a reasonable agonist-induced response (usually 8000–25 000 over basal). Responses greater than 40 000 fluorescent units should be avoided due to the risk of camera saturation. For most cell types the optimal density lies in the range 10 000–40 000 cells/well. Secondly, the loading conditions should be optimized. This usually involves altering the incubation time between 30–60 min at 37°C and 60–120 min at ambient temperature. The composition of the loading medium may also be varied, such as by inclusion of probenccid or, alternatively, sulfinpyrazone. Thirdly, the cell washing protocol needs to be optimized to avoid detaching the cells whilst removing the extracellular dye. We use the Packard-Carl Creative 96/384 washer which allows control of dispense speeds and heights. Finally, the running protocol on the FLIPR should be optimized. This mainly involves the pipetting characteristics such as pipette height, speed, and mixing, but also includes image capture and washing of tips. We find the tip washing to be effective for simple organic molecules such as 5-hydroxytryptamine, but it is usually not sufficient to avoid carry-over when using lipids or peptides which stick to the plastic.

Although most FLIPR assays are Ca^{2+}-based the FLIPR can be used for any assay where a suitable dye is available, e.g. membrane potential (Dibac(3,4)) or intracellular pH (BCECF-AM). Guides on the use of these are available from Molecular Devices and the dyes can be obtained from Molecular Probes. Molecular Devices have recently released a 'No Wash' calcium assay which can be used for many cell lines and obviates the need to wash the cells to remove extracellular dye. This is relatively expensive, but does offer advantages for poorly adherent cells and suspension cells. Molecular Devices have also recently released a fast membrane potential dye which should improve this type of assay on the FLIPR.

For adaptation of assays from 96-well format to 384 format, as used in the FLIPR 384, the above points regarding assay optimization are crucial, especially those referring to cell washing, pipetting heights, and speeds. For 384 assays, we often use Fluo-4 AM (Molecular Probes) as the dye of choice, due to its larger fluorescent output on binding Ca^{2+}. Typically cell numbers are 5000–10 000 cells per well in 25–50 μl medium, and loading medium (15 μl), is added using a Multidrop dispenser (LabSystems, Life Sciences International). Cells are washed using either an EMBLA (Skatron, Molecular Devices) or Packard 96/384 washer, leaving a residual volume of 40 μl. Antagonists, 10 μl, are added using a Quadra (Tom Tec). Agonists, 10 μl, are added via the FLIPR at a height of 55 μl and a dispense speed of 10 μl/sec. Sometimes faster dispense speeds or the mix option may be necessary to achieve sufficiently rapid mixing. Alternatively the FLIPR can be programmed to make both additions using the integral tip washer to remove carry-over between additions. As discussed previously, whilst the tip washer is effective when using readily soluble compounds, such as carbachol, it is not effective in removing all traces of peptide or lipid ligands.

3.3 Experimental data

Figure 5 shows typical traces for the three most common forms of temporal response profiles found in FLIPR. *Figure 5a* depicts the response to 3 μM UTP and its needle-like profile is typical of responses where the Ca^{2+} is mobilized from intracellular stores but there is no subsequent extracellular Ca^{2+} influx. *Figure 5b* depicts an orexin-1 receptor-mediated response; its broader biphasic profile is typical of responses where Ca^{2+} mobilization is followed by secondary Ca^{2+} influx through store- or receptor-operated Ca^{2+} channels. *Figure 5c* shows the capsaicin-induced response in cells expressing the rat vanilloid-1 receptor; its relatively slow onset followed by a maintained plateau phase is typical of Ca^{2+} influx through voltage-sensitive or ligand-gated ion channels. In all cases the increase in fluorescence, caused by the binding of the dye Fluo-3 AM to Ca^{2+} in the cell, indicates increased intracellular Ca^{2+} concentrations. Furthermore, as the kinetics of this binding are so fast, this enables the measurement of the response in real time.

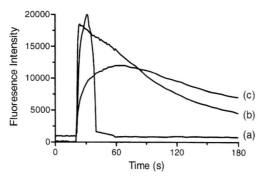

Figure 5 Temporal profiles of agonist-induced Ca^{2+} responses in the FLIPR. $[Ca^{2+}]_i$ (as fluorescence intensity units) was measured using Fluo-3/AM: (a) depicts the response to UTP (3 μm) and (b) the response to prexiu-A (100 nM) in CHO cells stably expressing crexin-1 receptors and (c) the response to capsaicin (100 nM) in HEK293 cells stably expressing rat vanilloid-1 receptors. Graphs show representative traces, typical of at least n = 20.

4 Conclusions

Both the microphysiometer and the FLIPR can be used to measure real time receptor pharmacology. Although the FLIPR has a much greater throughput, it is limited by the fluorescence reporters it can use. The microphysiometer therefore has the greater versatility as it is independent of the signal transduction pathway and has greater flexibility in signal transduction elucidation. If the receptor response can be measured on the FLIPR, i.e. it gives rise to changes in intracellular Ca^{2+}, then there is a huge ability to perform pharmacological studies, particularly with the advent of the FLIPR 384.

Acknowledgements

The authors would like to acknowledge the contribution of Izzy Boyfield, Karen Murkitt, Tracey Gager, Martyn Coldwell, Stephen Brough, and Jeff Jerman to work used in this article.

References

1. Owicki, J. C., Bousee, L. J., Hafeman, D. G., Kirk, G. L., Olson, J. D., Wada, G., *et al.* (1994). *Annu. Rev. Biophys. Biomol. Struct.*, **23**, 87.
2. McDonnell, H. M., Owicki, J. C., Parce, J. W., Miller, D. L., Baxter, G. T., Wada, H. G., *et al.* (1992). *Science*, **257**, 1906.
3. Smart, D. and Wood, M. D. (2000). *Biochem. Cell Biol.*, **78**, 281–288.
4. Owicki, J. C., Parce, J. W., Kercso, K. M., Sigal, G. B., Muir, V. C., Venter, J. C., *et al.* (1990). *Proc. Natl. Acad. Sci. USA*, **187**, 4007.
5. Wood, M. D., Murkitt, K. L., Ho, M., Watson, J. M., Brown, F., Hunter, A. J., *et al.* (1999). *Br. J. Pharmacol.*, **126**, 1620.
6. Smart, D., Coppell, A., Rossant, C., Hall, M., and McKnight, A. T. (1999). *Eur. J. Pharmacol.*, **379**, 229.
7. Redish, D. M., Raley-Susman, K. M., and Sapolsky, R. M. (1992). *Horm. Metab. Res.*, **25**, 264.
8. Tilford, N. S., Bowen, W. P., and Baxter, G. S. (1995). *Br. J. Pharmacol.*, **115**, 160P
9. Shroeder, K. S. and Neagle, B. D. (1996). *J. Biomol. Screen.*, **1**, 75.
10. Medhurst, A. D., Hirst, W. D., Jerman, J. C., Meakin, J., Roberts, J. C., Testa, T., *et al.* (1999). *Br. J. Pharmacol.*, **128**, 627.
11. Smart, D., Jerman, J. C., Brough, S. J., Rushton, S. L., Murdock, P. R., Jewitt, F., *et al.* (1999). *Br. J. Pharmacol.*, **128**, 1.
12. Smart, D., Gunthorpe, M. J., Jerman, J. C., Nasir, S., Gray, J., Muir, A., *et al.* (2000). *Br. J. Pharmacol.*, **129**, 227.
13. Taylor, C. W. and Broad, L. M. (1998). *TIPS*, **19**, 370.
14. Murkitt, K., Nasir, S., and Wood, M. D. (1999). *Br. J. Pharmacol.*, **128**, 148P.
15. Coward, P., Chan, S. D. H., Wada, H. G., Humphries, G. M., and Conklin, B. R. (1999). *Anal. Biochem.*, **270**, 242.
16. Wood, M. D., Chaubey, M., Atkinson, P., and Thomas, D. R. (2000). *Eur. J. Pharmacol.*, **396**, 1.
17. Bowen, W. P. and Jerman, J. (1995). *TIPS*, **16**, 413.

Chapter 9

Regulation of receptors and receptor-coupled ion channels by protein phosphorylation

Bernard J. McDonald and Richard L. Huganir

Howard Hughes Medical Institute, Department of Neuroscience, Johns Hopkins University School of Medicine, 725 North Wolfe Street, PCTB 905, Baltimore, MD 21205-2185, USA.

1 Introduction

Phosphorylation of serine, threonine, or tyrosine residues is accepted to be a key mechanism for the regulation of protein structure and function in the cellular environment. This highly dynamic process is catalysed by a diverse group of enzymes termed protein kinases (1) and is reversed by another group called phosphatases (2). Phosphorylation status has been shown to regulate the function, distribution, and interactions of numerous classes of proteins including those involved in transmembrane signalling, cytoskeletal dynamics, cell division, and transcriptional regulation (3–6).

The importance of protein phosphorylation in the transduction of extracellular stimuli and the generation of appropriate cellular responses is well established. Such responses depend critically on the correct phosphorylation of a multitude of proteins in response to signals arriving at the plasma membrane; for example hormones, growth factors, neurotransmitters, and alterations in membrane potential. Given the diversity of such stimuli and the complexity of the responses they elicit throughout the cell, it is crucial that the initial response at the plasma membrane is tightly regulated. Direct phosphorylation of certain proteins at the plasma membrane is a critical component in the regulation of cellular responsiveness. Cell surface molecules which are regulated in this manner include the receptor tyrosine kinases (RTKs), which mediate responses to growth factors; for example the epidermal growth factor (EGF) receptor (3), the G protein-coupled receptors (GPCRs), which stimulate metabotropic responses *via* G proteins; for example the β-adrenergic receptor (7), voltage-gated ion channels which respond to changes in membrane potential (8), and ligand-gated ion channels which open in response to neurotransmitter binding (9). Given the myriad signalling systems regulated by phosphorylation and our increased understanding of their importance, a number of methods have been developed to characterize phosphorylation of receptors and to examine the functional consequences of these events. This chapter will focus on the biochemical analysis of ligand-gated ion channels of the nervous system, however the techniques described are applicable to a much wider variety of proteins.

Ligand-gated ion channels mediate the vast majority of fast synaptic transmission in the nervous system by coupling extracellular ligand binding to opening of integral ion channels. Cells of the nervous system express ionotropic receptors responsive to neurotransmitters including glutamate, γ-aminobutyric acid (GABA), and glycine. Molecular cloning studies have shown these receptors to comprise a number of distinct polypeptide subunits, many of which contain consensus sequences suggesting that they serve as substrates for protein kinases. Biochemical and electrophysiological examination has supported these observations by showing multiple protein phosphorylation sites within these proteins, as well as demonstrating regulation of receptor function as a consequence of such post-translational modification. Given the central role of ionotropic receptors in synaptic function, it has long been suspected that protein phosphorylation could act as a key regulatory mechanism in the control of synaptic plasticity. As evidence for this mounts, protein phosphorylation will remain at the forefront of research on receptor function and synaptic plasticity. With this in mind we review strategies and methods used in the biochemical characterization and identification of phosphorylation sites.

2 Biochemical determination of receptor phosphorylation

2.1 Examination of endogenous full-length receptor proteins

Ideally, initial characterization of receptor phosphorylation should be carried out in a cell type where the receptor is endogenously expressed. The first step is to determine if the protein is indeed a kinase substrate in the endogenous environment. This analysis requires a number of suitable conditions. The cell type to be analysed should be easy to isolate in relatively pure preparations and it should express the protein of interest at a suitably high level to allow biochemical analysis. The investigator will require a reliable method of isolating the protein of interest from the experimental system; this usually takes the form of a specific high affinity antibody but other methods such as immobilized ligands may be used.

The investigator should characterize every aspect of the system prior to beginning the analysis of phosphorylated proteins. This may be carried out using non-radioactive samples and Western blot detection of purified proteins. A number of purification conditions should be tested and the most stringent giving acceptable protein yields should be used; the experimenter must decide on the compromise between protein yield and protein purity. These preliminary studies are advised for a number of reasons:

(a) To gain familiarity with the technique and prevent laboratory contamination with ^{32}P.

(b) To ensure reagents are functional and avoid unnecessary experiments.

(c) To optimize protein isolation and thereby minimize contamination with non-specific proteins and enhance signal-to-noise ratio within the system.

Once the system has been satisfactorily characterized, protein phosphorylation may be examined directly using *Protocol 1*, which has been used extensively to examine native ion channels in cultured neuronal cells as well as recombinant channels in heterologous expression systems. Solubilization and immunoprecipitation conditions are suggestions and must be optimized for each individual system.

Protocol 1

Examination of protein phosphorylation in cells pre-labelled with ^{32}P

Equipment and reagents

- Screw-top 1.5 ml microcentrifuge tubes with gasket seals (Fisher Scientific)
- Cellophane (Bio-Rad Laboratories Ltd.)
- Adequate safety equipment for personnel shielding, source shielding, and radioactive waste disposal (consult local radiation safety officer)
- Cells expressing protein of interest on 60 mm culture dishes (Falcon)
- Receptor-specific antibody
- Minimal essential medium (MEM) without sodium phosphate (Life Technologies Inc.)
- [^{32}P]orthophosphoric acid (10 mCi/ml in H$_2$O) (NEN™)
- Kinase activators and/or inhibitors (see *Table 1*) (Calbiochem)
- Immunoprecipitation buffer: 10 mM sodium phosphate pH 7, 100 mM sodium chloride, 10 mM sodium pyrophosphate, 50 mM sodium fluoride, 1 mM sodium orthovanadate, 1 mM sodium molybdate, 5 mM EDTA, 5 mM EGTA, 1 μM okadaic acid, 1 μM microcystin-LR, 100 μM PMSF, 10 μg/ml leupeptin, 10 μg/ml antipain, 10 μg/ml pepstatin (all Sigma Chemical Co.)
- Protein A Sepharose beads (Amersham Pharmacia Biotech)
- Non-immune IgG from species in which receptor antibody was raised (Pierce Chemical Co.)
- Denaturing SDS–PAGE sample buffer: 2% (w/v) SDS, 100 mM β-mercaptoethanol (BME), 50 mM Tris–HCl pH 6.8, 0.1% (w/v) bromophenol blue, 10% (w/v) glycerol

Method

1 Aspirate medium from each dish of cells to be labelled.

2 Wash cells twice with 5 ml pre-warmed MEM without sodium phosphate.

3 Incubate cells with 1.5 ml [^{32}P]orthophosphate (1–2 mCi/ml in MEM without sodium phosphate) at 37 °C in 5% CO$_2$/95% O$_2$ for 3–4 h.[a]

4 Stimulate cells to activate signal transduction pathway required (see *Table 1*).

5 Place dishes on ice, remove radioactive ('hot') media to waste, and wash cells twice with ice-cold PBS-A.[b]

6 Add 500 μl ice-cold immunoprecipitation buffer (IPB) containing 2% (w/v) Triton X-100 to each dish of cells on ice.

7 Scrape cells and transfer lysates individually into labelled 1.5 ml screw-top microcentrifuge tubes.

8 Add a further 500 μl ice-cold IPB containing 2% (w/v) Triton X-100 to each dish and remove the remainder of the cell lysates to the appropriate screw-top tubes.

9 Incubate lysates on ice for 10 min.

10 Centrifuge samples at 22 000 g for 10 min at 4°C and discard pellet fractions to radioactive waste.

11 Transfer supernatants to fresh, labelled 1.5 ml screw-top microcentrifuge tubes containing 100 μl 1:1 slurry of protein A Sepharose beads (in ice-cold IPB containing 2% (w/v) Triton X-100, 10% (w/v) BSA, 15 U DNase, and 150 μg RNase) and 10 μg nonimmune IgG.

12 Rotate tubes at 4°C for at least 1 h.

13 Centrifuge tubes at 22 000 g for 1 min at 4°C.

14 Transfer supernatants carefully to fresh, labelled 1.5 ml screw-top microcentrifuge tubes containing 100 μl 1:1 slurry of protein A Sepharose beads (in ice-cold immunoprecipitation buffer containing 2% (w/v) Triton X-100, 10% (w/v) BSA), and 10 μg specific IgG directed against the protein to be examined.

15 Rotate tubes at 4°C for at least 2 h (overnight is acceptable).

16 Centrifuge tubes at 22 000 g for 1 min at 4°C.

17 Discard supernatant to radioactive waste.[b]

18 Wash the Sepharose beads twice with 1 ml ice-cold IPB containing 2% (w/v) Triton X-100, three times with 1 ml ice-cold IPB containing 2% (w/v) Triton X-100 plus 600 mM NaCl, and twice with 1 ml ice-cold IPB alone.[b,c]

19 Elute immunoprecipitated protein from the beads by boiling in denaturing SDS–PAGE sample buffer for 5 min.

20 Separate phosphorylated proteins by SDS–PAGE, fix and stain the gel using Coomassie blue, then dry between two sheets of cellophane.

21 Visualize phosphoprotein bands by autoradiography on X-ray film using intensifying screens at −70°C.

[a] The duration of labelling incubation depends on cell type; steady-state accumulation of ^{32}P into the cellular ATP pool typically takes 3–4 h but this may be determined empirically by the investigator.

[b] Exercise extreme caution when transferring 'hot' medium and wash solutions to radioactive waste.

[c] Monitor the number of counts eluted from the beads and the number of counts remaining after each wash. Washes should be optimized as stressed earlier.

Table 1 Protein kinase activators and inhibitors

Protein kinase	Activators[a]	Inhibitors
PKA	cAMP, forskolin, IBMX	PKA inhibitor peptide, H-89, Rp AMPs
PKG	cGMP, SNAP, nitric oxide, IBMX	
PKC	Ca^{2+}, diacylglycerol, phorbol esters	PKC inhibitor peptide, chelerythrine, staurosporine
CamKII	Ca^{2+}, calmodulin (Cam)	Cam inhibitor peptide, KN62, KN93
v-*src*	Insulin, growth factors	Genistein, tyrphostins, lavendustin A
MAP kinase	Serum, phorbol esters	PD-98059, UO126

[a] IBMX, isobutylmethylxanthine (a phosphodiesterase inhibitor); SNAP, *S*-nitroso-*N*-acetylpenicillamine.

The experiment outlined in *Protocol 1* will indicate whether the protein of interest is phosphorylated within the cell type being studied. The use of appropriate controls should also show if the protein is basally phosphorylated and which signal transduction pathways increase or reduce its phosphorylation.

2.2 Examination of recombinant full-length receptor proteins

Recombinant receptor analysis in heterologous cell systems is a further powerful method in the characterization and identification of protein phosphorylation sites. In the place of cultured cells expressing native receptors, a transfected cell line expressing recombinant receptors is labelled with [^{32}P]orthophosphoric acid and, after kinase stimulation, the recombinant receptors are immunoprecipitated and separated by SDS–PAGE (see *Protocol 1*). The major difference between the two experiments is the potential absence of complete signal transduction pathways in heterologous cells. Therefore these results should be compared to those obtained for native receptors.

The recombinant experimental system offers a number of advantages to the investigator. First, many transfectable cell lines are readily available and offer a more accessible system than primary cultures of cells expressing native receptors. Also, levels of recombinant receptor expression in transfected cells are often significantly higher than those of receptors in the native environment; cell lines which do not express the receptor allow comparison of wild-type and mutated receptors in the same environment. This is of particular importance in the identification of phosphorylation sites (see Section 5.1).

Recombinant expression also allows the introduction of epitope tags and the use of specific antibodies to the epitope tag to purify the receptor (10). This is particularly advantageous where good anti-receptor antibodies are not available, or where cells express endogenous receptors that complicate the biochemical analysis of mutant phenotypes.

2.3 Phosphorylation of recombinant receptor fragments *in vitro*

At this point in the investigation there should be considerable evidence for phosphorylation of the receptor as well as the identity of the protein kinase responsible. The next step ought to demonstrate direct phosphorylation of the receptor by this protein kinase and help to identify further the target residue in the receptor. The most commonly used method involves purification of receptor intracellular domains as fusion proteins in bacteria for use as substrates for *in vitro* phosphorylation assays with purified protein kinases (see *Protocol 2*) (11). This method offers a number of advantages not available in a whole cell experimental system.

In combination with further analysis (see Section 4) it provides compelling evidence of direct receptor phosphorylation by the identified protein kinase. Also, the location of phosphorylation sites can be significantly narrowed down in the circumstance where a receptor has a number of intracellular domains or where a large intracellular domain can be expressed as a number of smaller substrate proteins. This system also allows determination of the kinetic properties of phosphorylation at different sites within the receptor; to do this with native proteins in the cellular environment is difficult and only a limited amount of information can be obtained.

Overall, this method has been remarkably accurate in the prediction of phosphorylation sites within receptor proteins and it remains the method of choice for *in vitro* characterization of protein phosphorylation. The production of recombinant proteins from bacterial cells has been described elsewhere and materials for this purpose are commercially available (11).

Protocol 2

In vitro phosphorylation of receptor fragments

Equipment and reagents

- Screw-top 1.5 ml microcentrifuge tubes with gasket seals (Fisher Scientific)
- Cellophane (Bio-Rad Laboratories Ltd.)
- Adequate safety equipment for personnel shielding, source shielding, and radioactive waste disposal (consult local radiation safety officer)
- Purified receptor domains expressed as fusion proteins in *E. coli* (100 μg/ml)

- Purified protein kinase of known activity (10–50 μg/ml)
- 5 × reaction buffer (see *Table 2*)
- 1 mM ATP (Sigma), containing 500–1000 c.p.m./pmol [^{32}P]-γ-ATP (NEN™)
- 2 × SDS–PAGE sample buffer: 4% (w/v) SDS, 200 mM β-mercaptoethanol (BME), 100 mM Tris–HCl pH 6.8, 0.2% (w/v) bromophenol blue, 20% (w/v) glycerol

Method

1 Place 1 μg purified fusion protein in a microcentrifuge tube on ice.   Total 10 μl

Protocol 2 continued

2 Add 10 µl of 5 × reaction buffer to each tube (see *Table 2*).   Total 20 µl

3 Add H_2O to each tube to a final volume of 39 µl.   Total 39 µl

4 Add 10 µl of 1 mM ATP/[^{32}P]-γ-ATP to each tube.   Total 49 µl

5 Add 1 µl protein kinase (1–5 ng) to each tube.   Total 50 µl

6 Begin reaction immediately by incubating tubes at 30°C.

7 Stop the reactions at the appropriate time by addition of 50 µl of 2 × SDS–PAGE sample buffer to each tube and boil samples for 5 min.

8 Separate phosphorylated proteins by SDS–PAGE then fix and stain the gel using Coomassie blue.[a]

9 Photograph the stained gel while wet as a record to confirm equal protein loading for each sample.

10 Dry the gel between two sheets of cellophane under vacuum and visualize phospho-protein bands by autoradiography on X-ray film using intensifying screens at −70°C.

[a] Unincorporated [^{32}P]-γ-ATP migrates with the dye front in this gel system. Stopping the electrophoresis before elution of the dye will greatly reduce contamination of electrophoresis equipment. Cut the dye front from the wet gel and treat this material as solid radioactive waste.

Table 2 Reaction buffers for *in vitro* phosphorylation using purified protein kinases

Protein kinase	5 × reaction buffer
PKA	200 mM Hepes pH 7, 100 mM $MgCl_2$
PKG	200 mM Tris–HCl pH 7.4, 100 mM $MgCl_2$, 10 mM cGMP
PKC	100 mM Hepes pH 7.5, 50 mM $MgCl_2$, 2.5 mM $CaCl_2$, 50 µg/ml phosphatidylserine, 5 µg/ml diolein
CamKII	250 mM Hepes pH 7, 400 µM EGTA pH 8, 800 µM $CaCl_2$, 50 mM $MgCl_2$, 50 ng/µl calmodulin
MAP kinase	150 mM Tris–HCl pH 8, 100 mM $MgCl_2$, 10 mM $MnCl_2$

3 Biochemical characterization of receptor phosphorylation

3.1 Proteolytic digestion of purified phosphoproteins

Further information can be derived from phosphoproteins by generation of phosphopeptides using specific proteolytic enzymes or chemical treatments (see *Protocol 3*). Chemical or enzymatic cleavage of phosphoproteins purified by SDS–PAGE (see *Protocols 1* and *2*) generates a characteristic mixture of phospho-peptides for each substrate/kinase combination. Phosphopeptide properties de-pend on phosphorylation sites and cleavage sites available for their generation;

chromatographic techniques may then be used to identify the characteristic peptide pattern for any individual receptor protein. Therefore protein cleavage is a crucial step before analysis of phosphopeptides can be undertaken. The most commonly used proteolytic agent is the protease trypsin, see *Protocol 3*.

Protocol 3

Trypsin cleavage of purified phosphoproteins

Equipment and reagents

- Razor blades
- Screw-top 1.5 ml microcentrifuge tubes with gasket seals (Fisher Scientific)
- Vacuum drying apparatus
- Scintillation counter
- 25% (v/v) methanol, 10% (v/v) acetic acid

- 50% (v/v) methanol
- Trypsin solution: 0.3 mg/ml TPCK treated trypsin (Sigma Chemical Co.) in fresh 50 mM ammonium bicarbonate
- 10 × trypsin solution: 3 mg/ml TPCK treated trypsin in fresh 50 mM ammonium bicarbonate

Method

1 Excise the phosphoprotein band of interest from the dried SDS–PAGE gel (from *Protocol 1*) and place in a disposable 50 ml centrifuge tube.

2 Wash 3 × 30 min at room temperature in 40 ml of 25% (v/v) methanol, 10% (v/v) acetic acid.

3 Wash 2 × 30 min at room temperature in 40 ml of 50% (v/v) methanol.

4 Transfer each gel piece to a fresh microcentrifuge tube and dry in a vacuum drying apparatus for approx. 2 h.

5 To each tube add 1 ml trypsin solution and incubate overnight (more than 12 h) at 37°C.

6 Add 100 μl of 10 × trypsin solution to each tube and continue incubation at 37°C for a further 4 h.

7 Transfer the supernatant from each tube into a fresh microcentrifuge tube and begin drying in a vacuum drying apparatus.

8 While the supernatants are drying, add 1 ml H_2O to each gel piece and resume incubation at 37°C until supernatant is dry.

9 Transfer the supernatant from each gel piece into the microcentrifuge tube containing the appropriate dried trypsin supernatant and dry once again in the vacuum drying apparatus.

10 Dry the gel pieces thoroughly in the vacuum drying apparatus and store.

11 Resuspend the dried supernatants in 1 ml H_2O and dry down once more.

12 Repeat step 11 twice.

13 Count the ^{32}P present in each dried supernatant and compare with that of its corresponding gel slice.[a]

[a] Greater than 80% of the total counts should be in the supernatant fraction, indicating that the trypsin digestion has been successful and the sample is suitable for further analysis.

Subsequent phosphopeptide mapping (see *Protocol 4*) and/or phosphoamino acid analysis (see *Protocol 5*) can provide a fingerprint of the phosphorylation events that have taken place and identify which amino acids within the protein have been modified (12).

3.2 Phosphopeptide fingerprint mapping

Proteolytic digestion of a purified phosphoprotein generates a mixture of individual phosphopeptides (see Section 4.1). The exact composition of this mixture of peptides will be determined by the nature of the phosphoprotein itself and, if the peptides are suitably separated, they generate a characteristic 'peptide map' for each substrate/kinase combination (Figure 1). A commonly used separation method for generation of such a peptide map is outlined below.

Protocol 4

Peptide map analysis of trypsin digested phosphoproteins

Equipment and reagents

- Cellulose thin-layer chromatography (TLC) sheets without fluorescent indicator (20 × 20 cm) (Kodak, 13255)
- Hair dryer
- TLC electrophoresis chamber (Fisher Scientific)
- Ascending chromatography tank for TLC (Kodak)
- Phenol red (1 mg/ml) (Sigma Chemical Co.)
- Two sheets of Whatman 3MM filter paper measuring 25 × 25 cm
- Basic fuchsin (1 mg/ml) (Sigma Chemical Co.)
- Electrophoresis buffer (pH 3.4): acetic acid/pyridine/H_2O (19:1:89)
- Chromatography buffer: pyridine/butanol/acetic acid/H_2O (15:10:3:12)

Method

1 Mark a cellulose TLC plate with a cross, the origin, at a point 4 cm from the bottom and 10 cm from each side of the plate.

2 Make two additional pencil marks 4 cm from either edge and 4 cm from the bottom of the TLC plate.

3 Label the plate clearly to identify the sample to be run.

4 Resuspend the dried phosphoprotein digest from *Protocol 2* in 10 μl H_2O and vortex briefly.

5 Centrifuge this sample at 22 000 g for 15 min at room temperature and transfer the supernatant to a fresh microcentrifuge tube.

6 Apply 1 μl of each dye solution (phenol red and basic fuchsin) to the sample origin on the TLC plate.

7 Between dye applications, dry the applied sample using a hair dryer.[a]

Protocol 4 continued

8 Apply phosphoprotein digest sample to the origin in 1 µl increments using the hair dryer to dry the origin between applications.[a]

9 Apply at least 250 c.p.m. but no more than 2000 c.p.m. of sample.

10 Place the TLC plate with the sample facing upward on top of one sheet of Whatman 3MM filter paper.

11 Cut out a circle of diameter 3 cm with its centre 6.5 cm from the bottom and 12.5 cm from either side of the other sheet of filter paper.

12 Wet this second piece of filter paper with electrophoresis buffer (pH 3.4) and place it over the TLC plate so that the sample origin appears in the centre of the excised circle.

13 Carefully add electrophoresis buffer to the upper filter paper using a Pasteur pipette until the TLC plate is entirely wet.[b]

14 Place the wet TLC plate in the electrophoresis tank with the sample origin between the two electrodes and separate the peptides by electrophoresis at 500 V until the dyes reach the pencil marks 4 cm from the edge of the plate.

15 Remove the TLC plate from the tank and stand in a ventilated fume-hood to dry.

16 Place the dried TLC plate in a chromatography tank containing sufficient chromatography buffer to cover the bottom 1 cm of the plate and allow the chromatography to proceed until the solvent reaches a point 1 cm from the top of the TLC plate.[c]

17 Remove the TLC plate from the chromatography tank and stand in a ventilated fume-hood to dry overnight.

18 Wrap the dry TLC plate in Saran wrap and visualize phosphopeptides by autoradiography on X-ray film using intensifying screens at −70°C.

[a] The small application volume and drying step is essential to prevent excessive sample diffusion from the origin.

[b] Carry out this step slowly to prevent diffusion of the sample from the origin.

[c] This takes 4–6 h. Buffer should be replaced before each plate is run and the same volume should be used for each run. The conditions used should be kept as close to constant as possible to minimize variations between plates.

3.3 Phosphoamino acid identification

The phosphopeptides generated from the phosphorylated protein may next be used to identify amino acids which are phosphorylated using a relatively simple separation technique outlined in *Protocol 5*. This step requires complete hydrolysis of phosphopeptides followed by chromotographic separation and identification of individual phosphoamino acids.

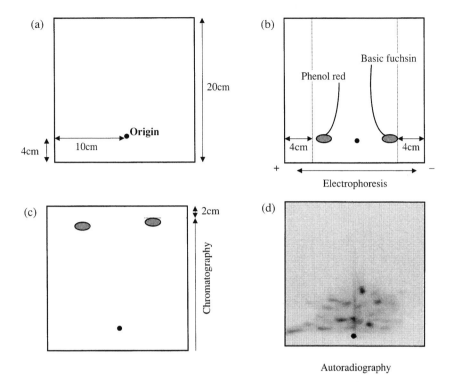

Figure 1 Phosphopeptide map preparation after proteolytic digestion. (a) The origin is prepared as described in the text, 10 cm from each side and 4 cm from the bottom of a crystalline cellulose chromatography plate. (b) After application at the origin, the sample and dyes are separated by electrophoresis until the dyes have migrated approximately 6 cm. (c) After drying, the samples are further separated in the Y dimension by thin-layer chromatography until the solvent front has travelled to within 1–2 cm of the top of the plate. (d) Once dry the plate is exposed by autoradiography onto X-ray film to visualize individual phosphopeptides. Shown here is an example of a phosphopeptide map of the NMDA receptor NR2B subunit protein isolated from ^{32}P-labelled primary cortical neurons.

Protocol 5

Phosphoamino acid analysis of trypsin digested phosphoproteins

Equipment and reagents

- Cellulose thin-layer chromatography (TLC) sheets without fluorescent indicator (20 × 20 cm) (Kodak, 13255)
- Hair dryer
- TLC electrophoresis chamber (Fisher Scientific)
- Compressed N_2 gas
- Sheets of Whatman 3MM filter paper, one of 25 × 25 cm, one of 25 × 17 cm, and one of 25 × 5 cm
- Klimax tubes (16 × 75 mm) with Teflon screw-caps (Fisher Scientific)
- 105 °C oven
- Vacuum drying apparatus

Protocol 5 continued

- Phenol red (1 mg/ml) (Sigma Chemical Co.)
- Fresh phosphoserine, phosphothreonine, and phosphotyrosine standards (10 mg/ml in H_2O) (Sigma Chemical Co.)
- Electrophoresis buffer (pH 3.4): acetic acid/pyridine/H_2O (19:1:89)
- Electrophoresis buffer (pH 1.9): formic acid/acetic acid/H_2O (1:10:89)
- Ninhydrin solution: 1% (w/v) in acetone (Sigma Chemical Co.)
- 6 M HCl (Sigma Chemical Co.)

Method

1 Place at least 500 c.p.m. of trypsin digested phosphoprotein in a glass Klimax tube with 0.5 ml 6 M HCl.

2 Blow N_2 gently over the sample then seal the tube and incubate at 105 °C for 1.5 h to allow hydrolysis of the peptides to individual amino acids.

3 Transfer the hydrolysate to a fresh microcentrifuge tube and dry under vacuum.

4 Resuspend the sample in 1 ml H_2O and dry once more under vacuum.

5 Resuspend the sample in 10 μl H_2O, vortex for 30 sec, centrifuge at 22 000 g for 15 min at room temperature, and transfer the supernatant to a fresh microcentrifuge tube.

6 Make five pencil marks (origins) 4 cm from the bottom of a Kodak cellulose TLC plate. These should be separated by 4 cm intervals, the first and fifth should be 2 cm from the sides of the plate. Make further visible marks at the sides 9 and 18 cm from the bottom of the plate.

7 Apply 1 μl each of phenol red, phosphoserine, phosphothreonine, and phosphotyrosine at each origin, dry each using the hair dryer between applications.[a]

8 Apply ^{32}P samples to the origins in 1 μl volumes, drying between each application. Again, care should be taken to minimize sample diffusion.

9 Place the TLC plate with the sample facing upward on top of one sheet of Whatman 3MM filter paper.

10 Wet the smaller pieces of filter paper with electrophoresis buffer (pH 1.9).

11 Lay the 25 × 17 cm piece of paper over the TLC plate such that its bottom edge is 1.5 cm from and parallel to the row of sample origins.

12 Lay the 25 × 5 cm piece of paper over the TLC plate such that its top edge is 1.5 cm from and parallel to the row of sample origins.

13 Ensure close contact then slowly add electrophoresis buffer to the paper wicks until the entire plate is wet.[b]

14 Place the wet TLC plate in the electrophoresis tank containing electrophoresis buffer (pH 1.9) and separate at 400 V until the phenol red indicator migrates 5 cm and reaches the pencil marks 9 cm from the bottom of the plate.[c]

15 Remove the TLC plate and, while it is still wet, place in an electrophoresis tank containing electrophoresis buffer (pH 3.4).

Protocol 5 continued

16 Separate the samples further by electrophoresis at 400 V until the phenol red reaches the final pencil marks 18 cm from the bottom of the plates.[d]

17 Dry the plate thoroughly in a well ventilated hood and then dip in fresh 1% (w/v) ninhydrin.

18 Heat the plate using the hair dryer in a hood until the phosphoamino acid standards become visible as purple spots.[e]

19 Mark the locations of the visible standards, wrap the plate in Saran wrap, and visualize [32P]phosphoamino acids by autoradiography on X-ray film using intensifying screens at −70°C.

[a] Apply phenol red first. This serves as an indicator of sample diffusion from the origin. Smaller sample areas will result in optimal peptide separation and definition.

[b] Care should be taken to add electrophoresis buffer slowly to minimize disturbance of samples at the origins.

[c] This step should take approximately 1 h, depending on the buffer saturation of the TLC plate. The plate should be oriented such that the row of sample origins are close to and parallel with the cathode (−). The samples migrate and separate toward the anode (+).

[d] Again the plate should be oriented such that the samples migrate toward the anode (+).

[e] Phosphoserine migrates faster than phosphothreonine which migrates faster than phosphotyrosine.

4 Phosphorylation site identification

4.1 Site-directed mutagenesis

Having biochemically characterized the phosphorylation of a receptor by a specific protein kinase it is now necessary to confirm the identity of the substrate residues. The most widely used method is site-directed mutagenesis of candidate amino acids, converting them to residues that cannot be modified by phosphorylation. The most common conversions being: serine to alanine, threonine to alanine, and tyrosine to phenylalanine. This method has been described elsewhere and the materials required are widely commercially available (13).

Protein kinase catalysed phosphoryl group transfer to substrate proteins is the result of highly specific protein–protein interactions. Analysis of protein kinase substrates originally identified the presence of common sequence determinants required for protein phosphorylation. These studies gave rise to the concept of a consensus sequence that determined whether a site could be phosphorylated and which protein kinase could carry out this modification (14). This approach was limited by the number of protein kinases and substrates that were identified by classical biochemical means. More recent studies, using oriented peptide libraries, have extended this analysis to determine consensus sequences for many more kinases and identification of many new protein kinase substrates (15). Much information is now available to aid the investigator in determining

Table 3 Protein kinase phosphorylation consensus sequences[a]

Protein kinase[b]	Phosphorylation consensus sequence[c]
PKA	RRX**S/T** >> RXX**S/T** > RX**S/T**
PKG	R/KR/KX**S/T** >> R/KXX**S/T** > R/KX**S/T**
PKC	R/KX$_{1-3}$**S/T**X$_{1-3}$R/K >> **S/T**X$_{1-3}$R/K > R/KX$_{1-3}$**S/T**
CamKII	RXX**S/T**
v-*src*	E/DEEIY**G**/EEF.
MAP kinase	PXX**S/T**P, P × 5/TP

[a] Adapted from refs 9–11.

[b] PKA, cAMP-dependent protein kinase; PKG, cGMP-dependent protein kinase; PKC, Ca^{2+}/phospholipid-dependent protein kinase; CamKII, Ca^{2+}/calmodulin-dependent protein kinase type II; v-*src*, transforming tyrosine kinase of the Rous sarcoma virus; MAP kinase, mitogen-activated protein kinase.

[c] S, serine; T, threonine; Y, tyrosine; P, proline; K, lysine; R, arginine; D, aspartate; E, glutamate; F, phenyl-alanine; G, glycine; I, isoleucine; X, any.

whether a receptor of interest may act as a protein kinase substrate. Many known protein kinase consensus sequences are shown in *Table 3*. While they are useful as guides for phosphorylation site identification, the investigator must bear in mind that many phosphoproteins do not contain consensus sequences.

When selecting residues to alter by mutagenesis a large amount of information should be known about the phosphorylation profile of the receptor. The biochemical data together with the sequence of the receptor should suggest the best candidate amino acids with which to begin. If HPLC analysis of phospho-peptides has been used to identify phosphorylated amino acids, site-directed mutagenesis should still be used to confirm this identification. Generation of phosphorylation site mutants allows biochemical confirmation of site identity and also provides a specific reagent to examine the physiological effect of phosphorylation at the individual site.

To confirm site identity requires the comparison of biochemical phosphorylation profiles of wild-type and mutant receptors. Evidence to support site identification should include reduced phosphorylation by purified kinases and by activated kinases in heterologous cells. Phosphopeptide mapping of the mutated receptor should indicate the elimination of a site-specific peptide or peptides. In more complex situations, this may take the form of altered peptide migration in which case different combinations of site mutations within the protein should be tested.

4.2 Phosphorylation site-specific antibodies

After identification of phosphorylation sites within a receptor, antibodies that specifically recognize phosphorylation at those sites can be generated (16). Such reagents have already been used to great effect in the study of glutamate receptor regulation (17, 18).

Peptides containing chemically phosphorylated serine, threonine, or tyrosine residues should be synthesized. Generally, 12-mer peptides containing the phosphoamino acid at the sixth position can be used successfully. Coupling of peptides to a carrier protein (e.g. thyroglobulin) prior to immunization can be facilitated by inclusion of an amino terminal lysine residue in the peptide. Methods of phosphopeptide synthesis and immunization have been outlined in greater detail elsewhere (16, 19).

Purification of phosphospecific antibodies is usually carried out as a two-step procedure using standard antibody purification methods. The first step involves removal of non-phosphospecific antibodies by passing serum over immobilized unphosphorylated peptides. The required antibodies should then be purified from the flow-through of this column using the phosphorylated peptide immobilized on a second affinity column as described previously. Great care should be taken in the subsequent characterization of these antibodies; their reactivity should be shown to be phosphospecific in detection of fusion proteins and full-length receptors by Western blotting. Reactivity should also be tested using λ-phosphatase treated samples and using antibodies blocked by the immunizing phospho-peptide.

Once the antibody has been tested and its phosphospecificity confirmed it is an immensely valuable reagent:

(a) It reduces the danger and cost of biochemical examination of protein phosphorylation in a wide variety of experimental systems.

(b) A good phosphospecific antibody will allow detection of phosphoprotein at very low levels where biochemical detection would be problematic; this is particularly useful when looking at less abundant native receptors.

(c) It may also allow examination of phosphospecific alterations in protein distribution and protein–protein interaction that would be difficult to detect using traditional biochemical techniques.

4.3 Candidate protein kinase identification

The presence of a specific consensus sequence indicates the likely identity of the protein kinase responsible for phosphorylation at a given amino acid residue. It must be remembered that this is simply a preliminary indication: for example, the pattern of basic amino acids surrounding a potential substrate site may allow phosphorylation by a number of protein kinases (see *Table 3*). Therefore, more information should be collected and examined prior to embarking on the difficult task of biochemical characterization of receptor phosphorylation. Where possible, specific protein kinase activation should be examined for effects on receptor function, distribution, or protein interaction. Some commonly used methods for activation of protein kinases are indicated in *Table 1*. Such examination should support the role of protein phosphorylation in the physiological regulation of the receptor and should greatly assist in the identification of the protein kinases responsible.

5 Conclusions

It should be clear from the preceding sections that there are a number of experimental approaches available to the investigator when examining the role of protein phosphorylation of cell surface receptors. The biochemical techniques outlined provide a powerful means for the definitive identification of phosphorylation sites and the protein kinases responsible. It must be stressed however that reliable conclusions require the use of many, or all, of the outlined approaches. Further evidence to support data generated by biochemical experiments can be found by analysis of wild-type and mutated receptors under conditions where protein kinase activation is believed to regulate their function. Together, biochemical and functional approaches have given us a clearer picture of receptor function and allow us to begin to understand the interactions of diverse signals arriving at the cell surface.

Acknowledgements

The author's work is supported by the National Institutes of Health and Howard Hughes Medical Institute (R. L. H.) and a Wellcome Trust Postdoctoral Research Fellowship (B. J. M.).

References

1. Edelman, A. M., Blumenthal, D. K., and Williams, L. T. (1987). *Annu. Rev. Biochem.*, **56**, 567.
2. Cohen, P. (1989). *Annu. Rev. Biochem.*, **58**, 453.
3. Fantl, W. J., Johnson, D. E., and Williams, L. T. (1993). *Annu. Rev. Biochem.*, **62**, 453.
4. Hu, S. and Reichardt, L. F. (1999). *Cell*, **22**, 419.
5. Marshall, C. J. (1994). *Curr. Opin. Genet. Dev.*, **4**, 82.
6. Ginty, D. D. (1997). *Neuron*, **18**, 183.
7. Lefkowitz, R. J. (1998). *J. Biol. Chem.*, **273**, 18677.
8. Levitan, I. B. (1994). *Annu. Rev. Physiol.*, **56**, 193.
9. Smart, T. G. (1997). *Curr. Opin. Neurobiol.*, **7**, 358.
10. Evan, G. I., Lewis, G. K., Ramsay, G., and Bishop, J. M. (1985). *Mol. Cell Biol.*, **5**, 3610.
11. McDonald, B. J. and Moss, S. J. (1994). *J. Biol. Chem.*, **269**, 18111. Moss, S. J., Doherty, C. A., and Huganir, R. L. (1992). *J. Biol. Chem.*, **267**, 14470.
12. Boyle, W. J., van der Geer, P., and Hunter, T. (1991). In *Methods in enzymology* (ed. T. Hunter and B. M. Sefton), Vol. 201, p. 110. Academic Press, London.
13. Czernik, A. J., Girault, J. A., Nairn, A. C., Chen, J., Snyder, G., Kebabian, J., *et al.* (1991). In *Methods in enzymology* (ed. T. Hunter and B. M. Sefton), Vol. 201, p. 264. Academic Press, London.
14. Kennelly, P. J. and Krebs, E. G. (1991). *J. Biol. Chem.*, **266**, 15555.
15. Songyang, Z., Blechner, S., Hoagland, N., Hoekstra, M. F., Piwnica-Worms, H., and Cantley, L. C. (1994). *Curr. Biol.*, **4**, 973.
16. Sambrook, I., Fritsch, E. F., and Maniatis, T. (ed.) (1989). *Molecular cloning: a laboratory manual* (2nd edn). Cold Spring Harbor Laboratory Press, NY.
17. Tingley, W. G., Ehlers, M. D., Kameyama, K., Doherty, C., Ptak, J. B., Riley, C. T., *et al.* (1997). *J. Biol. Chem.*, **272**, 5157.
18. Mammen, A. L., Kameyama, K., Roche, K. W., and Huganir, R. L. (1997). *J. Biol. Chem.*, **272**, 32528.
19. Harlow, E. and Lane, D. (ed.) (1999). *Using antibodies: a laboratory manual*. Cold Spring Harbor Laboratory Press, NY.

Chapter 10

Using confocal imaging to measure changes in intracellular ions

Ghassan Bkaily, Danielle Jacques, Ghada Hassan, and Sanaa Choufani

Department of Anatomy and Cell Biology, Faculty of Medicine, University of Sherbrooke, Sherbrooke, Quebec, Canada J1H 5N4.

Pedro D'Orléans-Juste

Department of Pharmacology, Faculty of Medicine, University of Sherbrooke, Sherbrooke, Quebec, Canada J1H 5N4.

1 Introduction

Until the development of three-dimensional (3D) imaging, many questions were raised concerning ion traffic and localization, but these were never fully answered because of the absence of a technique enabling the visualization of dye distribution at the 3D level. The recent development and use of confocal microscopy, coupled to high performance hardware and software systems, has provided scientists with the capability of overcoming some of the limitations of standard microscopic imaging measurements. Until recently, access to confocal microscopy was fairly limited. Today, the majority of laboratories dealing with molecular and cell biology, pharmacology, biophysics, biochemistry, etc., are equipped with this powerful scientific tool. This increase in the need for confocal microscopy has been highly supported by a growing industry for developing new fluorescent probes. However, in order to optimize this technique, scientists need to be familiar with the basic approaches and limitations of confocal microscopy. In this chapter, we will discuss sample preparation, labelling of structures and functional probes, the settings of parameters, development of specific ligand probes, as well as protocols for measurements and limitations of the technique. For further information, the reader should refer to recent key references (1–4).

2 Preparation of tissue

2.1 Choice of cell types for confocal microscopy

To date, several cell types have been used routinely for confocal microscopy, including cells from the heart, vascular smooth muscle, vascular endothelium, nerve, bone, liver, *Xenopus* oocytes, sea urchin eggs, fibroblasts, T lymphocytes, and basophilic leukaemia cells. Most of the work has been performed using freshly isolated and/or cultured single cells either in primary culture or from cell lines. There is no doubt that the choice of the cell type, as well as the thin tissue section, is crucial for obtaining satisfactory results. For example, in order to obtain sufficient details from 3D reconstructions, the following criteria should be taken into account:

(a) Isolated cells should be plated on glass coverslips (plastic should be avoided).

(b) Cells should be attached in order to avoid displacement during serial sectioning.

(c) Serial sectioning should be taken at rest or when the steady-state effect of a compound is reached.

(d) In order to eliminate electrical and chemical coupling between cells, they should be grown at low density.

(e) Cells should be relatively thick and not flattened in order to allow a minimum of 75–150 serial sections (according to the cell type). This could be done by using the cells as soon as they are attached and/or by lowering serum concentration.

(f) When cell lines or proliferative cells are used, it is recommended to replace the culture medium overnight with a medium free of serum and growth factors before starting the experiments. This will allow the cells to settle in a latent phase of mitosis and will increase the accessibility of receptors and channels for agonist and antagonist actions, by washing out the effects of compounds present in the culture medium.

(g) Using specific fluorescent markers, the origin, purity, and phenotype of the cells should be checked continually.

Since ions and other free functional components of the cells are not homogeneously distributed, and since the increase or decrease of the probed free component seems to be faster than computer sampling and acquisition, it is difficult to assess these fast occurring phenomena at the 3D level. Hence, large rapid scans of the area under interest should be done.

2.2 Co-culture of cells

Reconstitution of the interaction between two cell types is an excellent preparation for confocal microscopic studies. For example, single vascular endothelial and smooth muscle cells can be used in co-culture conditions as described under *Protocol 1*.

Once the co-culture is achieved, use *Protocol 2* to load with Fluo-3 (or other fluorescent probes) for 3D Ca^{2+} measurement using confocal microscopy. The simple technique of *Protocol 1*, method A or B, enables a better study of the development of cell-to-cell contact between VECs and VSMCs as a function of time and can be used within an hour or more. Although this technique has a disadvantage, in that cell contacts are generated at random, it may more adequately represent a non-homogeneous organizational contact between two different cell types such as that observed in certain pathological conditions, like atherosclerosis, as well as healing processes following surgery or vascular by-pass. Also, interactions between cells of the immune system and the endothelial system or any other cell type can be studied in a similar fashion with the exception that the immune cells in suspension should be loaded with the dye, collected by centrifugation, and then added to the attached pre-loaded mono-layered cells (VECs or VSMCs).

Protocol 1

Co-culture of vascular endothelial cells (VECs) and vascular smooth mucscle cells (VSMCs)[a]

Equipment and reagents

- Culture medium for VECs and VSMCs: Dubelcco's minimum essential medium (DMEM), supplemented with 10% fetal bovine serum (Gibco) containing penicillin
- Sterile glass coverslip and culture dish

Method A

1 Culture first the VECs for 24 h in the middle of the coverslip using a custom-made watertight compartment 9 mm in diameter.

2 Remove the compartment leaving a confined area of confinent VECs on the coverslip.

3 Add the VSMCs in an area adjacent to the VECs in order to establish, within 12 h, side to side and/or overlapping contacts.

Method B

1 Put a drop of culture medium containing high density VECs in the middle of the coverslip and allow the cells to attach overnight.

2 Wash gently the VECs with fresh culture medium.

3 Add to the VECs a culture medium containing high density of VSMCs for 1 h.

[a] For isolation and culturing of VECs and VSMCs, see ref. 1.

3 Choice and properties of intracellular markers

3.1 Ca^{2+} probes

Freshly cultured single cells and cells in primary cultures are excellent preparations for studying cytosolic and nuclear free Ca^{2+} using 3D confocal microscopy. Several visible wavelength Ca^{2+} dyes were initially tested in order to select the ion indicator best suited to laser-based confocal microscopy. Of the various Ca^{2+} dyes available, we tested Fluo-3, Fura Red, Calcium Green-1, and Calcium Orange (Molecular Probes). Calcium Orange was the least successful with respect to loading and Ca^{2+} fluorescence: the dye was difficult to load in all cell preparations tested, regardless of temperature, time of incubation, or dye concentration. Moreover, addition of Pluronic F-127C, a dispersing agent used to aid solubilization, did not improve loading of the dye. Calcium Green-1 loaded reasonably well and displayed good baseline fluorescence. However, in most preparations tested, little or no Ca^{2+} response was observed in the presence of a number of pharmacological agents known to increase intracellular Ca^{2+}. Fura Red is a long wavelength analogue of fura-2 with a high emission spectrum. Loading was performed at room temperature for 45 min at a concentration of 20 µM. At baseline Ca^{2+} levels, labelling was relatively uniform in the cytoplasm and nucleus in heart cells but appeared slightly higher in the nucleus of vascular endothelial and smooth muscle cells. Probe fluorescence appeared relatively stable and showed little photobleaching during our experiments. However, one characteristic of Fura Red is that its fluorescence emission decreases upon Ca^{2+} binding in contrast to other indicators where emission intensity increases. This property of Fura Red allows its use in combination with Fluo-3 for dual emission ratiometric measurements, thus enabling expression of fluorescence intensity in terms of Ca^{2+} concentration. However, there are some drawbacks to this dye:

(a) Being much weaker than other visible wavelength Ca^{2+} indicators, it must be loaded at higher concentrations.

(b) Intracellular distribution of both indicators must be identical for ratiometric measurements to be valid.

(c) In some of our experiments involving dual-labelling with Fura Red and Fluo-3, both fluorescent dyes increased in response to depolarization of the cell membrane or pharmacological stimulation making Fura Red unreliable, at least with some cell preparations.

Among the Ca^{2+} dyes tested, Fluo-3 provided the best overall loading and fluorescence features in all our cell preparations. Fluo-3 was easily loaded, homo-

Figure 1 (See also Plate 1) (A) Sagittal and cross-sectional view of a 3D reconstructed plasma membrane-perforated ventricular myocyte illustrating intracellular Ca^{2+} distribution in response to increasing concentrations of extranuclear Ca^{2+}. Coverslips were mounted in the bath chamber and after loading with 13.5 µM Fluo-3 AM, cells were bathed in a buffered saline solution containing 100 nM Ca^{2+}. Following rapid perforation of the cell with ionomycin (1–2 min), the cells were washed quickly and stabilized for 5 min in buffer solution without Ca^{2+}. Image acquisitions were performed at the following settings: 488 nm excitation and 510 nm emission wavelengths, laser power at 9 mV, 3% attenuation filter, and PMT set at 700. Z-axis serial sections (0.08 µm) were performed at a step size of 0.9 µm. Nine identical vertical serial series were obtained 2 min after incremental addition of 100 nM, 200, 400, 800 nM, and 1200 nM extranuclear Ca^{2+}. Images are shown in presence of zero and 800 nM Ca^{2+} as well as in presence of Syto-11, as pseudocoloured representations according to an intensity scale from 0 to 255 (bottom). At the end of the addition of the last concentration of Ca^{2+}, mitochondria were stained with Mito Fluor Green (D) followed by Syto-11 staining of the nucleus for 8–10 min and cells were scanned under identical conditions. The reconstructed images depict a gradual

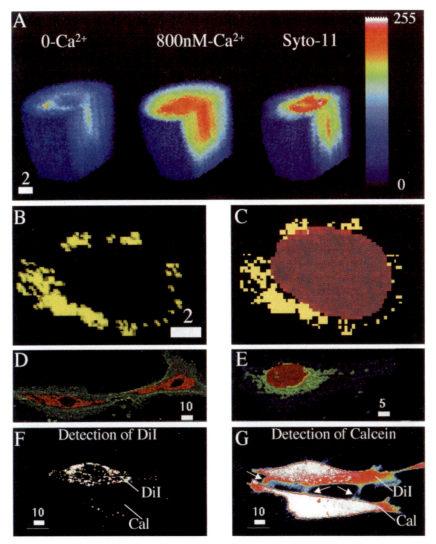

Plate 1. (A) Sagittal and cross-sectional view of a 3D reconstructed plasma membrane-perforated ventricular myocyte illustrating intracellular Ca^{2+} distribution in response to increasing concentrations of extranuclear Ca^{2+}. Coverslips were mounted in the bath chamber and after loading with 13.5 μM Fluo-3 AM, cells were bathed in a buffered saline solution containing 100 nM Ca^{2+}. Following rapid perforation of the cell with ionomycin (1–2 min), the cells were washed quickly and stabilized for 5 min in buffer solution without Ca^{2+}. Image acquisitions were performed at the following settings: 488 nm excitation and 510 nm emission wavelengths, laser power at 9 mV, 3% attenuation filter, and PMT set at 700. Z-axis serial sections (0.08 μm) were performed at a step size of 0.9 μm. Nine identical vertical serial series were obtained 2 min after incremental addition of 100 nM, 200, 400, 800 nM, and 1200 nM extranuclear Ca^{2+}. Images are shown in presence of zero and 800 nM Ca^{2-} as well as in presence of Syto-11, as pseudocoloured representations according to an intensity scale from 0 to 255 (bottom). At the end of the addition of the last concentration of Ca^{2+}, mitochondria were stained with Mito Fluor Green (D) followed by Syto-11 staining of the nucleus for 8–10 min and cells were scanned under identical conditions. The reconstructed images depict a gradual increase of $[Ca^{2+}]i$ within the mitochondria and the nucleus with increasing concentrations of external Ca^{2-}. Scale bar: 2 μm. (B and C) Delimitation of perinuclear space using Mn^{2+} quenching of nucleoplasmic Fluo-3 (C) and nucleoplasmic staining with Syto-11 (B) in single heart cells loaded with Fluo-3 AM. The red colour represents the nucleus and the yellow colour represents the perinuclear space. Scale bars are 2 μm. (D) $DiOC_6$ staining of the ER (green) followed by Syto-11 staining of the nucleus in endocardial endothelial cells from the left ventricle of 21 week-old human fetal heart. Note the filament-like projections extending across the cytoplasm. (E) JC-1 mitochondrial membrane potential staining in fetal human endocardial endothelial cell (light blue/green) followed by Syto-11 staining of the nucleus (red). Tubular and/or round-shaped mitochondria are abundant. The colours had no measurable meaning. The white scale bar is in μm. (F) Detection of the fluorescence emitted by DiI. The cell on top was labelled initially with DiI while the cell below was labelled with Calcein, the fluorescence of which cannot be detected in this image. (G) Detection of Calcein fluorescence. In the cell on top, which was initially labelled with DiI, a Calcein fluorescence was detected. The Calcein was transferred from the cell below to the cell on top. Arrows show the contact between the two cells.

geneously distributed within the cytosol and the nucleus, with low photobleaching at a laser attenuation setting of 3%. More importantly, Ca^{2+} responses to both electrical and pharmacological stimuli were comparable with those using fura-2 in classical Ca^+ imaging studies. Hence, this dye was used in the majority of our Ca^{2+} experiments with the confocal microscope in intact and plasma membrane-perforated cells (*Figure 1A*) of many types (1) and the method of loading is described in *Protocol 2*.

A new fluorescent Ca^{2+} probe, Fluo-4, has been recently introduced commercially by Molecular Probes. This dye is similar to Fluo-3, but requires less time for cell loading and a lower concentration than Fluo-3.

Protocol 2

Loading with Fluo-3 AM

Equipment and reagents

- Single cells cultured on 25 mm diameter sterile glass coverslips to fit a 1 ml bath chamber or on glass-bottom culture dishes (MatTek Corp.)

- Tyrode's salt solution containing 5 mM Hepes, 136 mM NaCl, 2.7 mM KCl, 1 mM $MgCl_2$, 1.9 mM $CaCl_2$, 5.6 mM glucose, buffered to pH 7.4 with Tris base, and supplemented with 0.1% bovine serum albumin (BSA)

- Using an osmometer, the osmolarity of the buffer solution, with and without BSA, should be adjusted with sucrose to 310 mOsm

- Fluo-3 AM probe or other Ca^{2+} probes (Molecular Probes) diluted in Tyrode-BSA from frozen 1 mM stocks in dimethylsulfoxide (DMSO) to a final concentration of 6.5 or 13.6 μM

Method

1   Culture single cells from heart, aortic vascular smooth muscle, endothelium or T lymphocytes, or osteoblasts on coverslips or glass-bottom culture dishes.

2   Wash the cells three times with 2 ml of the Tyrode-BSA solution.

3   Cover the cells on the glass-bottom dish with the Fluo-3 AM (Tyrode-BSA) solution or place the glass coverslip cell-side down on a 50–100 μl drop of the Fluo-3 solution, on a sheet of Parafilm stretched over a glass plate and incubate for 45–60 min in the dark at room temperature.[a]

4   After the loading period, carefully recover the glass-bottom culture dish or the coverslip and wash the cells twice with Tyrode-BSA buffer and twice in Tyrode buffer alone.

5   Leave the loaded cells for an additional 15 min period to ensure complete hydrolysis of acetoxymethyl ester groups.

Protocol 2 continued

6 Optical settings for Fluo-3 AM are shown in *Table 1*.

[a] The inverted coverslip method offers considerable savings with regard to the amount of probe used, especially at high concentrations. More importantly, the use of smaller aliquots of concentrated stock solution needed to prepare the final probe concentration reduces the percentage of DMSO in the incubation medium. We have found that cell blebbing and photobleaching were commonly encountered with final DMSO concentrations greater than 0.6%. Loading should also be performed in a humidified environment when inverting coverslips on less than 100 μl in order to avoid evaporation problems.

Table 1 Probe data table and optical settings for argon-based CLSM

Probe/dye	Application laser line	Ex_{max} beam-splitter	Em_{max} beam-splitter	Argon or barrier[a] filter	Primary	Secondary filter	Detector 1	Detector 2
Fluo-3 AM	Ca^{2+} indicator	464	526	488	510	None	510	None
Fura Red AM	Ca^{2+} indicator	458	600	488	535	None	595	None
Calcium Orange AM	Ca^{2+} indicator	550	567	514	535	None	570	None
Sodium Green	Na^+ indicator	507	532	488	510	None	510	None
Amiloride	Na^+/H^+ pump inhibitor	488	420	488	510	None	510	None
SNARF-1 AM	pH indicator	488–530	Dual 587/636	488/ 514	535	595 DRLP	600	540DF30
$DiOC_6(3)$	Endoplasmic reticulum dye	484	501	488	510	None	510	None
Mito Fluor Green	Mitochondrion probe	490	516	488	510	None	510	None
BODIPY FL C_5-ceramide	Golgi dye	505	511	488	510	None	510	None
JC-1	Mitochondrial membrane potential dye	490	Dual 527/590	488	510	565 DRLP	570	530DF30
Di-8-ANEPPS	Cell membrane	481	605	488	510	None	> 570	None
Syto-11	Live nucleic acid stain	500–510	525–531	488	510	None	510	None
Angiotensin II-FITC	Angiotensin binding sites	494	520	488	510	None	510	None
Ang II-TRITC[b]	Angiotensin binding sites	ND	ND	514	535	None	535	None
PYY-FITC	PYY binding sites	ND	ND	488	510	None	510	None

[a] If secondary beamsplitter is absent, then value indicated applies to barrier filter.

[b] Custom-prepared.

3.2 Na⁺ fluorescent probes

Sodium Green AM is the only argon laser, excitable Na$^+$ indicator commercially available (Molecular Probes). This sodium ion probe exhibits a 41-fold selectivity for Na$^+$ over K$^+$ and, upon binding Na$^+$, increases its fluorescence emission intensity. Loading of Sodium Green into heart and vascular cells is difficult unless coupled with Pluronic F-127 to increase aqueous solubility. The dye, when well-loaded, is relatively stable although slightly more subject to photobleaching than Fluo-3. This can be minimized by setting the laser attenuation filter down to 1%.

Single cells can be loaded with 13.5 μM Sodium Green AM (Molecular Probes) in a similar way to that described above for Fluo-3 (*Protocol 2*) with one modification: that the Na$^+$-sensitive dye should be loaded in the presence of Pluronic F-127 (20%, w/v) (Sigma). When used alone, Sodium Green loads poorly into cells. However, when initially reconstituted in equal volumes of DMSO and Pluronic acid and diluted into the loading buffer (final concentration of 0.1% detergent), the probe offers good fluorescence intensity. Moreover, it was found that the loading temperatures should not exceed 19–29 °C since loading tends to be inconsistent and often compartmentalized. Optical settings for Sodium Green AM are given in *Table 1*.

After loading the ion probe for 30 min at 19 °C, one can see in heart cells that the nuclear Na$^+$ concentration is similar to that in the cytosol. On the other hand, in resting aortic vascular endothelial and smooth muscle cells the concentration of cytosolic Na$^+$ is slightly greater in the immediate cytosolic zone surrounding the nucleus. There is no information concerning a possible role of Na$^+$ in nuclear function or whether the nucleus may play a role in cytosolic Na$^+$ buffering. The mechanism by which Na$^+$ crosses the nuclear membrane, as well as its physiological role in the nucleus, if any, are not known. However, recent work demonstrates that confocal microscopy can be useful in determining the physiological role of Na$^+$ and its modulation at the cytosolic and nuclear levels, as well as in subcellular organelles (5).

3.3 pH fluorescent probes

Several long wavelength pH fluorescent dyes have been developed and are commercially available such as BCECF, SNARF-1, SNARL, DM-NERF, CL-NERF, and HPTS (Molecular Probes). The use of an argon-based confocal microscope limits the use of all these pH indicators. This limitation can be overcome by modifying the laser capabilities into a multiline argon/UV laser or an argon/krypton laser or by using a multiphoton excitation laser source.

Carboxy SNARF-1 is well-suited for all types of laser instrumentation and is efficiently excited by both 488 nm and 514 nm argon laser lines. The emission spectrum of the dye undergoes a pH-dependent shift in wavelength, thus allowing the ratio of fluorescence intensities at two emission wavelengths, typically 580 nm and 640 nm, to be used for accurate determinations of pH. This pH dye has been used extensively in many cell types (6–9). The dye has a pKa of approximately 7.5, making it well-suited to measurement of cytoplasmic pH (7). It is reported to be less susceptible to photobleaching and to exhibit a good pH-dependent shift of its fluorescence emission spectrum (7). Using confocal microscopy, studies on single cells loaded with SNARF-1 suggest that subsarcolemmal and nuclear areas have a pH value near 7.2 but for regions corresponding to the distribution of mitochondria, the pH is 7.8–8.2 (6, 8). This raises the possibility that cytosolic pH is not homogeneous and is a function of the location of organelles within the cytosol. Moreover, measurements of intracellular pH (pH_i) with a H^+-sensitive dye may depend on the cell type used, and isolation and/or culture conditions, as well as the degree of metabolic activity. *Protocol 3* summarizes the method of cell loading with the ratiometric pH indicator, carboxy-SNARF-1 AM.

Protocol 3

Loading with the ratiometric pH indicator, carboxy-SNARF-1 AM

Equipment and reagents

- Cultured cells on glass coverslips or glass-bottom dishes (see *Protocol 2*)
- Tyrode-BSA buffer (see *Protocol 2*)

Method

1 Incubate the cells in the dark for 30 min at ambient temperature with 5 μM of freshly prepared indicator dye in Tyrode-BSA buffer from 1 mM stock solution reconstituted in DMSO.

2 Following the loading period, wash the cells and leave them for 15 min to hydrolyse the AM form as described for the Ca^{2+} dyes in *Protocol 2*.

3 Set the glass-bottom culture dish or the glass coverslip (placed in a 25 mm bath chamber) for visualization with the confocal microscope. Optical settings for SNARF-1 are given in *Table 1*.

The use of SNARF-1 dye in chick embryo heart cells sometimes revealed a heterogeneous, and other times a homogeneous pH_i distribution in the cytosol and the nucleus. In some instances, labelling appeared to be more intense at the point of attachment of the cell to the glass coverslip than in other parts of the cell. In heart and vascular cells, the indicator responds well to changes in extracellular pH from 6.5–8.5. At an external pH of 6.0, no difference could be seen between acidic and basic measurements. Moreover, distribution of pH in heart cells is relatively homogeneous within the cytosol and the nucleus. However, in human vascular aortic smooth muscle cells, the distribution of pH in the cytosol is less homogeneous and appears to be much more basic within, and in the immediate vicinity of, the nucleus than the cytosol pH. This was particularly evident upon lowering the extracellular pH. We should be cautious when expressing pH$_i$ by quantitative values because this may lead to significant errors in comparison to qualitative (fluorescence intensity) or ratiometric measurements (9). Care should also be taken even when expressing pH$_i$ in ratiometric terms because of possible intracellular redistribution of the indicator between the cytosol and lipophilic cell compartments (9).

3.4 Membrane potential dyes: Di-8-ANEPPS and JC-1 probes

Since the late 1960s, measurement of fluctuations in membrane potential using fluorescent indicators, and the search for specific voltage-sensitive dyes with an acceptable signal-to-noise ratio, has sparked the development of several commercially available dyes such as thiazole orange, Di-o-Cn(3), tetramethylrhodamine ethylester, bis-axonal, RH2g2, Di-8-ANEPPS, and the mitochondrial membrane dye, JC-1 (10, 11). The Di-4-ANEPPS and Di-8-ANEPPS indicators (Molecular Probes) belong to the class of fast potentiometric dyes of the styryl type. This family of indicators was found to be extremely optically effective and a less phototoxic transducer of membrane potential (10). Optical imaging of plasma membrane potential can be performed in cells using the fast voltage-dependent potentiometric dye, Di-8-ANEPPS (Molecular Probes) as described in *Protocol 4*.

Protocol 4

Loading with the voltage-dependent potentiometric dye, Di-8-ANEPPS

Equipment and reagents

- Cultured cells on glass coverslips or glass-bottom dishes (see *Protocol 2*)
- Tyrode and Tyrode-BSA solutions (see *Protocol 2*)
- Di-8-ANEPPS (Molecular Probes)
- 13.5 μM final concentration of Di-8-ANEPPS in Tyrode-BSA buffer

Method

1. Wash the glass coverclip (or glass-bottom dish), containing cells, three times with the Tyrode-BSA buffer.

2. Replace the Tyrode-BSA buffer with the Tyrode-BSA containing Di-8-ANEPPS and incubate for 3–5 min maximum at room temperature in the dark.

3. Wash the cells three times with Tyrode-BSA buffer.

4. Replace the Tyrode-BSA buffer with Tyrode solution.

Di-8-ANEPPS is a hydrophobic compound that seems to anchor to the cell membrane and is more stable than other membrane voltage dyes and shows less leakage (10). This dye was reported to offer better time recordings of trans-membrane voltage changes (about 30 min) in cultured neonatal rat heart cells than other fluorescent voltage probes, before its leakage into the cytosol. In experiments with embryonic chick heart single cells, as well as human adult aortic vascular endothelial and vascular smooth muscle cells, Di-8-ANEPPS staining is stable for up to 1 h, with only a diffuse halo of staining observed in a few cells after 30–60 min, but never in the perinuclear region. The quality of Di-8-ANEPPS membrane potential staining could be influenced by several factors including cell type as well as species origin (10). For example, the distribution of the transsarcolemmal membrane potential in heart, VECs, and VSMCs did not appear homogeneous. This irregular distribution of membrane potential in single cells could be due to scattered protein distribution on the sarcolemma membrane, as was suggested for ion channels. This non-homogeneity of labelling can be more apparent when single cells are depolarized with high extracellular $[K^+]$. Sustained depolarization of the human VSMCs with 30 mM $[K]_o$ will rapidly (10 sec) induce a non-homogeneous depolarization of the cell membrane accompanied by cell contracture. This latter contracture remains as long as the cell membrane can be depolarized with 30 mM $[K]_o$.

Recently, J-aggregate formation has also been used to visualize mitochondria in a variety of cells (10, 12). The mitochondrial voltage-sensitive probe, JC-1 (Molecular Probes), a specific energy potential-dependent mitochondrial dye, was reported to display a fairly narrow red peak that was sensitive to a variety of agents that affect mitochondrial membrane potential (10). In preparations such as embryonic chick heart cells, adult human aortic vascular endothelial and smooth muscle cells, mitochondrial membrane potentials are well labelled with the JC-1 indicator. Epifluorescence visualization reveals a decrease in green fluorescence (monomer) compared to orange-red fluorescence (JC-1 aggregate) upon plasma membrane depolarization. The JC-1 probe does not appear to label the nuclear membrane.

Thus, the monomeric mitochondrial voltage-sensitive dye JC-1 can also be used to evaluate intracellular membrane potential responses to depolarization induced by high extracellular KCl concentration. Evaporated stocks of cationic JC-1 (Molecular Probes) can be reconstituted in 100% ethanol and further diluted to a working solution of 10 μg/ml in Hepes–NaCl buffer. Loading should be performed directly into the experimental bath chamber at room temperature in the dark. Good fluorescence is achieved within 10 min of the addition of the probe. Optical settings for Di-8-ANEPPS and JC-1 are given in *Table 1*.

3.5 Quantification of free intracellular ions

Quantification of the intracellular ion concentration using ratiometric fluorescent ion probes is not a problem. However, such a quantification becomes difficult when non-ratiometric dyes are used such as the Ca^{2+} indicator Fluo-3. The procedure for constructing a Ca^{2+} calibration curve for Fluo-3 (or any dye) for both cytosolic and nuclear free Ca^{2+} is described in *Protocol 5*.

Protocol 5

Ca²⁺ calibration curve for a non-ratiometric dye such as Fluo-3[a]

Equipment and reagents

- Cultured cells on glass coverslips or glass-bottom dishes (see *Protocol 2*)
- Fluo-3 acid (Molecular Probes)
- Syto-11 (Molecular Probes) for nuclear staining

- Ca²⁺ calibration buffer kits No. 2 (0–40 μM free Ca²⁺) and No. 3 (1 μM to 1 mM free Ca²⁺) (Molecular Probes)
- 0.1% Triton X-100 solution (Sigma)

Method

1 For each concentration of Ca²⁺ of kits No. 2 and No. 3 add a final concentration of 13.5 μM Fluo-3 acid.

2 Choose the cells under the confocal microscope.

3 Expose the well-attached endured cells to the 0.1% Triton solution for 10 min in order to perforate the sarcolemma and nuclear envelope membranes.

4 Wash the Triton solution with the zero Ca²⁺ buffer of kit No. 2.

5 Set the confocal system for measuring Fluo-3 Ca²⁺ fluorescence.

6 Expose the perforated cells to each different concentration of Ca²⁺ buffer containing 13.5 μM Fluo-3 acid, and record the fluorescence intensity in the whole area under the field of the microscope.

7 When the fluorescence intensity of the last sample (1 mM free Ca²⁺) has been recorded, add Syto-11 to delimit the nuclear regions.

[a] [Ca²⁺] fluorescence intensity curve can be constructed as in *Figure 2*.

Such a Ca²⁺ calibration curve shows that the distribution of Fluo-3 is similar in the medium, the cytosol, and the nucleus (*Figure 2*). This homogeneous distribution and response to variation of the concentration of free Ca²⁺ is not affected by the different viscosities of the cytoplasm and the nucleus.

4 Volume rendering and nuclear Ca²⁺, Na⁺, and pH measurements

Once the scanned images are obtained, they can be transferred onto a workstation similar to a Silicon Graphics Indy 4400 (SGI, CA, USA) equipped with Imagespace analysis and volume workbench software modules (version 3.2, Molecular Dynamics). Using the Imagespace analysis software permits the reconstruction of 3D images of ion-selective dyes. Reconstruction of 3D images of ion-selective dyes are performed on unfiltered serial sections and can be repre-

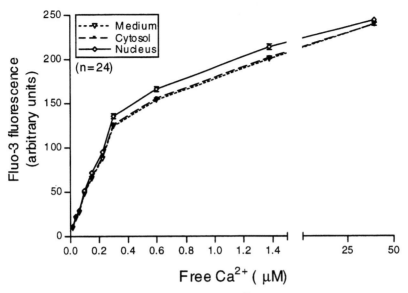

Figure 2 Effect of different concentrations of free Ca^{2+} on Fluo-3 acid fluorescence in water, cytoplasm, and nucleoplasm using laser confocal microscopy.

sented as 3D projections using a number of different volume rendering methods. Each method generates a projection with a different appearance, each having the property of showing or outlining particular features of the scanned specimen or cell. For example, for Ca^{2+}, Na^+, or pH distribution experiments, reconstructed 3D images are represented as 'closest intensity' projections, produced by stacking individual images depicting actual pixel intensities. This model uses the first pixel value above a given threshold value along the line of sight, thus producing an opaque cube or, in this case, an opaque cell in which intensity and distribution of the particular dye can be viewed in any given plane or cross-section of the cell. On the other hand, three-dimensional projections of cell organelles or Ca^{2+} and nucleus co-localization studies use the 'look-through' method, which depicts a specimen as if it were transparent and completely in focus throughout its depth. Pixel intensities are summed along projection rays which lie perpendicular to the plane of the monitor screen. This method is also termed 'extended focus' because the resulting images appear to have been gathered from a transparent specimen by a lens with a large depth of focus. It is useful for studying both surface and interior features of the cell and their relative positions within the cell. One can then associate ion-specific or functional dyes with particular cell organelles and their spatial distribution. The Ca^{2+}, Na^+, or pH (or any functional dye) images can be represented as pseudocoloured representations on an intensity scale of 0–255 with the lowest intensity in black and highest intensity in white. Pseudoscales should be included with all images illustrating pseudocoloured representations of intensity levels.

Measurements of uptake of Ca^{2+}, Na^+, or any free element of the cell within the nucleus can be performed both on 2D images (individual sections) and 3D reconstructs (section series). Following Syto-11 staining, the nuclear area can be distinguished from the rest of the cell by setting a lower intensity threshold filter to confine relevant pixels. A 3D binary image series of the nuclear volume can then be generated for each cell using the same x, y, and z set planes as those used during calcium ion uptake, for example (*Figure 1A*). Then, by applying these binary image patterns of the nucleus to the same cell initially labelled for Ca^{2+}, Na^+, or pH (the binary image serves as a delimitation area), a new 3D projection is created depicting fluorescence intensity levels exclusively within the nucleus. Hence, by 'removing' the nucleus from the surrounding cytoplasm, measurements of mean Ca^{2+}, Na^+, or pH intensity values in the entire nuclear volume can be performed while eliminating a possible contribution of the perinuclear space to the measurements.

Furthermore, in sarcolemma-perforated single cells (sarcolemma perforation with 100 nM ionomycin for 1 min) loaded with Fluo-3 (1), the use of 1–20 mM Mn^{2+} permits the quenching of nucleoplasmic Ca^{2+}-sensitive dye fluorescence, thus leaving intact the fluorescence of the Fluo-3 Ca^{2+} complex within the nuclear envelope (NE) space which will permit the delimitation of the NE (13, 14) and the measurement of the Ca^{2+} level within the NE space. The Mn^{2+} quenching technique coupled to Syto-11 nuclear labelling works well, and *Figure 1B*, and *1C* shows an example.

Rapid line scan imaging can be used to monitor temporal oscillations of a functional probe such as the Ca^{2+} dye, during spontaneous contraction or cytosolic, nuclear, and mitochondrial Ca^{2+} oscillation and/or release. Contracting cells or cytosolic and nuclear Ca^{2+} oscillations can be easily recorded at a rate of 330 msec per scan (3 scans/sec) for up to a total of 1000 frames. Each frame, which consists of 32 lines/scan (512 pixels) and pixel size of 0.42–0.64 μm, allows the visualization of the entire cell throughout the contractile process, with good spatial and temporal resolution. On the other hand, if faster acquisition times are needed, single line scanning (1 line/scan) can be performed at a rate of up to 10 msec per image. However, such a high speed of acquisition allows visualization of only a single line within the cell, usually chosen in the middle region of the nucleus (2). Since free ions such as Ca^{2+} are not homogeneously distributed in the cytosol and the nucleus, fast rate line acquisition does not permit expression of the results in absolute concentration.

5 Organelle, receptor, and channel fluorescent probes

Several commercially available dyes permit the localization and delimitation of several organelles and membranes such as the plasma membrane (Di-8-ANEPPS), mitochondrial membrane (Mito Fluor Green), endoplasmic reticulum (DiOC$_6$(3)), Golgi apparatus (NBDC$_6$-ceramide), and the nucleus (live nucleic acid Syto stains) (Molecular Probes). Once the cultured cell type is available, these probes should be used first in order to get familiar with their distribution and localization. Once this as been done, these structure delimitation probes can be used at the end of each experiment. Other dyes can also be used to delimit organelle structure such as mitochondrial (JC-1) and plasma membrane potential dyes (Molecular Probes). The choice of the structural delimiting probe depends on the excitation and emission of the fluorescent probe used to study the functional element of cells. Double and triple-labelling of functional fluorescent probes can be performed if the two or three probes are excited and emit fluorescence at different wavelengths. Also, certain functional probes, such as membrane potential-sensitive dyes, can be used both as delimiting (structural) and functional probes in conjunction with other functional probes. Furthermore, plasma membrane potential-sensitive probes can be used to study cell coupling as well as voltage distribution and cell shortening (1).

5.1 Endoplasmic reticulum staining

Several short chain carbocyanine dyes (Molecular Probes) such as $DiOC_6(3)$ and $DiOC_5(3)$ and long chain carbocyanines such as $DIIC_{18}(3)$ have been used widely to visualize endoplasmic/sarcoplasmic reticulum (15, 16). These probes pass easily through the plasma membrane, and while some carbocyanine probes have been reported to stain several other intracellular membranes, such as Golgi and mito-chondria, when used at higher concentrations, ER membranes are preferentially labelled and are easily distinguishable by their characteristic morphology. Sarco-plasmic/endoplasmic reticulum (SR, ER) can be visualized using the short chain carbocyanine $DiOC_6(3)$ dye from Molecular Probes (Eugene, OR) as described in *Protocol 6*.

Protocol 6

Dual-labelling of SR (ER) or mitochondria and the nucleus of cells loaded or not with Fluo-3 Ca^{2+} probe[a]

Equipment and reagents

- Cultured cells on glass coverslips or glass-bottom dishes loaded with Fluo-3 AM Ca^{2+} probe (see *Protocol 2*)
- SR (ER) labelling dye: $DiOC_6$ stock solution (Molecular Probes)
- Mitochondria labelling dye: Mito Fluor Green stock solution (Molecular Probes)
- Nuclear staining dye: Syto-11 stock solution (Molecular Probes)

Method

1   After measurements of Ca^{2+} concentration and distribution, add the SR probe $DiOC_6$ or Mito Fluor at a final concentration of 50 nM for 3–5 min.

2   Wash the extracellular probe and then perform serial sections.[b]

> **Protocol 6** continued
>
> 3   Add Syto-11 at a final concentration of 100 nM for 5 min and then perform serial sections.
>
> [a] This stepwise dual-labelling technique enables the localization of both cell organelles using the same excitation wavelengths. *Figures 1D* and *1E* show examples.
>
> [b] Optical settings for $DiOC_6$ and Mito Fluor are given in *Table 1*. Simultaneous labelling with the two dyes cannot be done because of their identical excitation and emission wavelengths.

The $DiOC_5$ and $DiOC_6$ probes were tested in isolated vascular endothelial, smooth muscle cells, embryonic heart myocytes, and human endocardiac endothelium cells. The sarcoplasmic reticulum can be easily recognized by its filament-like morphology. In 3D reconstructions, the SR appears as leaf-like ondulations in heart cells or as tubular structures in VSM and endothelial cells. $DiOC_6$ staining in these cell preparations, when used at a concentration of 10 mM, is more consistent and more intense than that of its equivalent $DiOC_5$. This ER staining can also be used jointly with plasma membrane markers or nuclear stains to evaluate the intracellular distribution of the organelle or its spatial relationship with the nucleus. Unfortunately, these carbocyanine probes, apart from being toxic at high concentrations, cannot be used simultaneously (in double-labelling experiments) with standard long wavelength Ca^{2+} or Na^+ indicators, such as Fluo-3, Calcium Green, or Sodium Green, because of their similar excitatory and emission wavelengths (488 excitation, 500–530 emission range). If one wishes to associate cell function or ionic responses with ER membranes using these particular probes, one approach would be first to complete Ca^{2+} or Na^+ experiments and then to label the organelle with $DiOC_6$ stain to visualize ER localization. Several washes in low $[Ca^{2+}]$ buffer to reduce Ca^{2+} content followed by ER staining with 10–50 mM $DiOC_6$ stain produces a signal sufficiently strong, allowing a reduction in the gain ($\times 2$ to $\times 1$), attenuation filter (3–1%), and/or PMT (two photon multipliers) voltage to levels below Ca^{2+} fluorescence detection. The latter staining, being much stronger than the ionic markers, enables subsequent correlation between structure and function.

5.2 Mitochondria staining

The new Mito Fluor mitochondrial-selective probes developed by scientists at Molecular Probes appear to accumulate preferentially in mitochondria regardless of its membrane potential (Molecular Probes handbook). The method of labelling with Mito Fluor Green is described in *Protocol 6*.

Figure 1E demonstrates an example of mitochondrial labelling in a 21 week-old fetal human endocardiac endothelial cell. As evidenced in this figure, fetal endocardiac endothelial cells are rich in mitochondria surrounding the nucleus. These organelles may contribute substantially to total measured intracellular Ca^{2+} and their contribution to cytosolic Ca^{2+} buffering should be taken into consideration.

5.3 Channel and receptor dyes

Recently, the use of 3D confocal microscopy, a Ca^{2+} channel fluorescent probe, and an angiotensin II (Ang II) fluorescent probe showed that R-type Ca^{2+} channels as well as Ang II receptors are present on sarcolemmal and nuclear membranes of several cell types (1–3).

In our laboratories, we also developed FITC- and Bodipy-conjugated endothelin-1 as well as Ang II fluorescent probes. Using the ET-1 fluorescent probe, ET-1 receptors were found to be localized at the plasma and nuclear membranes in a non-homogeneous cluster-like pattern.

An Ang II fluorescent probe was also developed in our laboratory. This probe was more specific and more effective than the one commercially available. The newly developed PYY fluorescent probe was also found to be more effective than those commercially available. The optical settings for these probes are given in *Table 1*.

Currently, many commercially available receptor labelling dyes do not show a satisfactory labelling of receptors. In order to develop an agonist or an antagonist-coupled fluorescent probe, the following procedures are recommended:

(a) Choose a ligand structure that permits attachment of a fluorescent probe without affecting the known active site of the structure.

(b) Choose a fluorescent probe that is highly stable and does not bind to any cellular structure.

(c) The fluorescent probe alone should be completely washable.

(d) No single probe can be coupled with all different ligand types.

(e) Once the ligand is coupled to the fluorescent probe, the complex must be highly purified and separated from free dye or ligand.

(f) The ligand–dye complex should be tested in isolated cells or preferentially in functional experiments in order to ensure that the ligand is still active and is as specific and powerful as the ligand alone.

(g) One must ensure that the effect of the fluorescent ligand probe is washable and is blocked by a well known antagonist, as well as being displaced by the 'cold' ligand alone but not by the dye alone.

(h) The stability of the complex should be determined in solution as a function of time and storage. It is highly recommended that the ligand–fluorescent probe complex is used as soon as it is purified (i.e. within two days) and to avoid storage if possible. The development of fluorescent ligand probes, although sometimes difficult, is a worthwhile objective.

6 Introducing non-permeable fluorescent macromolecules into cultured mammalian cells by osmotic lysis

The introduction of macromolecules into living cells is a powerful experimental approach. It is based on osmotic lysis of pinocytic vesicles. This commercially available Influx cell loading reagent kit (Molecular Probes) provides a convenient, rapid, and simple procedure for loading water soluble materials into live cells, a technique introduced by Okada and Rechsteiner (17). With the Influx reagent kit, polar compounds can be introduced into many cells simultaneously without significantly altering normal cell function. The Influx reagent kit provides a more gentle cell loading method than the typical cell loading techniques of microinjection, electroporation, hypotonic shock, or scrape loading, which are all physically disruptive to cells.

Park and colleagues (18), using endosomal fluid phase pinocytic markers, such as FITC-dextran, showed that early endosomes are resistant to osmotic lysis, while lysosomes are sensitive. Hypertonic sucrose, presumably due to elevated turgor pressure, inhibits transport of internalized solute from endosomes to lysosomes.

The Influx pinocytic cell loading reagent kit has been designed to load compounds into cells grown on coverslips, in suspension, and in tissue culture flasks. Cell labelling can be accomplished in a single 30 min loading cycle and may be enhanced by repetitive loading. *Protocol 7* describes the cell loading procedures for adherent human venticular heart, vascular smooth muscle, and endothelial cells on glass coverslips.

Protocol 7

Cell loading procedures for adherent cells on glass coverslips using Influx pinocytic cell loading reagent kit[a]

Equipment and reagents

- Adherent cultured cells on glass coverslips (see *Protocol 2*)
- Influx pinocytic cell loading reagent kit (Molecular Probes) for preparation of the hypertonic loading medium
- Cultured cells incubator
- Coverslip mini-rack (Molecular Probes)

- DMEM (Dulbecco's minimum essential medium, Gibco) with and without 5% FBS (fetal bovine serum, Gibco) pre-warmed at 37 °C
- 300 µl FBS (250 µl) containing DMEM (20 µl)
- Tyrode buffer

Method

1 Prepare the hypertonic solution by heating the kit tube at 100 °C in order to liquify the PEG, then homogenize with pre-warmed DMEM (4.7 ml), followed by the addition of the 300 µl FBS solution and keep it in a 37 °C incubator.

2 Add the membrane impermeable dye into 100 µl of the hypertonic solution.

3 Prepare 100 ml of hypotonic solution by diluting the DMEM solution (60 ml DMEM + 40 ml deionized water) and keep it in a 37 °C incubator.

4 Place the glass coverslip cell-side down on a 100 µl drop of the hypertonic

Protocol 7 continued

medium containing the dye as described for Fluo-3 in *Protocol 2* and place it in the incubator for 10 min.

5 Insert the glass coverslip in the mini-rack and place it in the 50 ml beaker containing the hypotonic lysis medium for 2 min at room temperature.

6 Remove the glass coverslip and place it cell-side up in a Petri dish containing DMEM + 5% FBS solution and place it for 10 min in the incubator. Then the cells are ready to be used for confocal microscopic study.

[a] The kit contains polyethylene glycol (PEG) and sucrose crystals.

7 Fluorescent labelling for identification of cell types and communication in co-culture

A technique based on the fact that gap junctions (GJ) allow only the passage of molecules of low molecular weight (< 1000 daltons), is that of double-labelling of donor cells with the probe permeable to GJ, calcein (calcein AM, Molecular Probes) and the recipient cells with the membrane probe, DiI (1,1'-dioactadecyl-3,3,3',3'-tetramethylindocarbocyanine perchlorate, Molecular Probes) which cannot pass through the GJs (19, 20). *Protocol 8* describes the method.

Protocol 8

DiI fluorescent labelling for identification and study of communication between same and different cell types[a]

Equipment and reagents

- One specific type of cells cultured on 25 mm glass coverslips or glass-bottom dishes (donor cells)
- One specific type of cells in a flask (receiver cells)
- Tyrode buffer (or PBS)
- Sarcolemma membrane probe: DiI (Molecular Probes) for labelling of recipient cells

- Gap junction Ca^{2+} probe: calcein AM (Molecular Probes) for labelling of donor cells
- DMEM (Dulbecco's minimum essential medium, Gibco) with and without 0.1% trypsin (Sigma)
- Culture medium: DMEM supplemented with 10% FBS (Gibco)

Method

1 Incubate the donor cells cultured on glass coverslips or glass-bottom dishes with 0.2 μM calcein AM for 15 min at room temperature.

2 Wash the coverslip or glass-bottom dish with Tyrode buffer three times to ensure the complete removal of the extracellular calcein.

Protocol 8 continued

3 Detach the recipient cells initially cultured in flasks, by trypsinization (0.1% trypsin solution).

4 After centrifugation of the recipient cells at 4 °C and 1000 g for 10 min, suspend the recipient cells in 10 ml of DMEM + 10% FBS.

5 Shake the recipient cells in suspension at 37 °C for 2 h, for them to regain their normal state after the trypsinization.

6 Incubate the recipient cells with 2.5 μM DiI, diluted in 1 ml of culture medium, for 15 min at room temperature.

7 Centrifuge the recipient cells at 1000 g for 10 min and wash the cells twice with the culture medium, to ensure the total elimination of the DiI from the suspension medium.

8 Afterwards, suspend the recipient cells in culture medium and add them on top of the labelled donor cells on coverslips, and incubate in the dark at 37 °C for a period of 4–5 h to allow the development of contacts between the donor and recipient cells.

[a] The transfer via the contacts is defined by the passage of the calcein (cal) from donor cells (labelled initially only with calcein) to the recipient cells (labelled initially only with DiI).

For DiI and calcein labelling studies, the argon laser should be set at 20 mV, and at all the excitation wavelengths, and directed to the sample via a dichroique filter of 535, and attenuated by a neutral density filter of 30% to reduce photobleaching. Two secondary filters should be used: a 600 EFLP filter allowing the passage of the fluorescence emitted by the DiI and not that emitted by the calcein (as *Figure 1* shows), and another filter of 540 DF 30 allowing the passage of the fluorescence emitted by the calcein while eliminating the residual fluorescence emitted by the DiI (as *Figure 1G* shows).

Two photon multipliers (PMT) should be used to detect the fluorescence intensity of the two probes: PMT 1 for the fluorescence emitted by the DiI (*Figure 1F*), and PMT 2 for that emitted by the calcein (*Figure 1G*). The cells are scanned by sections and two images representing the fluorescence of each probe are produced.

The laser intensity, as well as the filters used, should be kept constant during the experimental procedures, while the configuration of the PMTs are varied in a way that ensures an adequate fluorescence for both probes. *Figures 1F* and *1G* shows typical experiments that demonstrate more than one human vascular smooth muscle cell in culture developing a contact which enables the passage of small molecule such as calcein.

8 Settings of the confocal microscope

There is no doubt that one of the major elements in confocal microscopy is the setting of optimal parameter conditions. This should be determined before starting any serious experiment and should be maintained for each specific fluorescent probe. This will allow a realistic comparison between results obtained using the same fluorescent probe. *Table 1* shows typical probe data and optical settings for a Molecular Dynamics Multi Probe 2001 confocal argon laser scanning (CSLM) system equipped with a Nikon Diaphot epifluorescence inverted microscope and a ×60 (1.4 NA) Nikon Oil Plan achromat objective. These optical settings are also valid with other confocal microscopes. When the argon laser is used, the 488 nm (9.0–12.0 mV) or 514 nm laser line (15–20 mV) is directed to the sample via the 510 nm or 535 nm primary dichroic filter and attenuated with 1–3% neutral density filter to reduce photobleaching of the fluorescent probe. Pinhole size could be set at 50 nm for structural determination and 100 μm for ion-specific dyes. In most cases, the image size is set at 512 × 512 pixels with pixel size of 0.08 μm for small ovoid-shaped cells and 0.034 μm for elongated cells.

In order to optimize and validate fluorescence intensity measurements, laser line intensity, photometric gain, photomultiplier tube (PMT) settings, and filter attenuation should be determined and kept rigorously constant throughout the duration of the experimental procedures.

9 Conclusion

Confocal microscopy imaging studies in single cells and tissue sections confirm the importance of this non-invasive technique in the study of cell structure and function as well as the modulation of working living cells by various constituents of cell membranes, organelles, and cytosol.

The use of non-ratiometric dyes does not allow the adequate expression of intracellular Ca^{2+} concentrations in absolute values and so care must also be taken when expressing results as either intensity, or even ratio values. One approach to addressing these difficulties is to use different probes for the same ion in order to ensure that the results do not contain an artificial component due to loading artefacts. A complementary approach is by selecting cells which exhibit fluorescence intensities within fixed values and by determining the R_{min} and R_{max} values of these probes. In addition, in instances where chromophores are conjugated to a particular hormone or drug, steps including washouts, use of cold or unlabelled hormones/drugs, as well as testing of the free dye should be performed. Our results also reveal that confocal microscopy can be used successfully in co-culture studies.

Photobleaching of the fluorophore can become a major obstacle in certain instances where the sample must be exposed to a higher intensity of excitation light because of weak signal emission. In the case of mounted material, agents such as propyl gallate (21), Vectashield (Vector Laboratories, CA, USA), and Slow-Fade (Molecular Probes) have been used to reduce light-induced photodamage. In our laboratory, 0.1% phenylenediamine (Sigma) in glycerol/PBS (9:1) and Vectashield are routinely used for mounted slide work. Protecting live material from photodynamic damage, on the other hand, is more difficult. Additives, such as ascorbic acid as well as high doses of vitamin E, have been described as having some antifading properties (22).

Differences in results that may be found between one laboratory and another, may be due to:

(a) Cell type origin and culturing.

(b) Experimental conditions and confocal setting.

(c) Use of 3D measurements and reconstruction.

(d) Method used to determine nuclear volume and localization.

Thus, confocal microscopy is extremely powerful in studying contraction, voltage, and chemical coupling between various cell types. One of the limitations however, is the inability to record serial sections during quick cell responses such as those occurring during cell contraction. Rapid scans can be performed only on single vertical planes, thus limiting access to possible spatial differences in cellular response patterns. On the other hand, results using fluorescent ion and organelle probes illustrate that three-dimensional studies of cell response and function are not only feasible, but also important in evaluating cell responses to external stimuli.

Moreover, confocal microscopy is a highly useful tool in the assessment of cell pathology, whether in single cells or tissue sections. Site-selection probes such as receptor, protein, and second messenger probes, organelle probes and nuclear stains, provide important indicators for the determination of structure and location of cell components and also for the study of subcellular distribution and movement of various ions and molecules in working living cells.

For the time being, fluorescence confocal microscopy does not enable the localization or quantification of bound ionic elements of the cell. Neverthless we hope that new technologies will soon enable us to determine not only free ions but also bound and/or compartmentalized ones.

Finally, with the help of 3D imaging and volume rendering capabilities, confocal microscopy constitutes a powerful state-of-the-art technique in the continuing investigation of cell structure and function in normal and pathological conditions.

Acknowledgements

Supported by CIHR grants to Drs Ghassan Bkaily, Pedro D'Orléans-Juste, and Danielle Jacques. Drs Pedro D'Orléans-Juste and Danielle Jacques are Scholars of the 'Fonds de Recherche en Santé du Québec'. The authors thank Ms Susann Topping for her secreterial assistance and Ghassan Bassam Bkaily Jr. for his technical assistance.

References

1. Bkaily, G., Pothier, P., D'Orléans-Juste, P., Simaan, M., Jacques, D., Jaalouk, D., *et al.* (1997). *Mol. Cell. Biochem.*, **172**, 171.

2. Bkaily, G., D'Orléans-Juste, P., Pothier, P., Calixto, J. B., and Yunes, R. (1997). *Drug Dev. Res.*, **42**, 211.
3. Paddock, S. W. (1996). *Proc. Soc. Exp. Med.*, **213**, 24.
4. Rizzuto, R., Carrington, W., and Tuft, R. A. (1998). *Trends Cell Biol.*, **8**, 288.
5. Bkaily, G., Jaalouk, D., Haddad, G., Gros-Louis, N., Simaan, M., Naik, R., *et al.* (1997). *Mol. Cell. Biochem.*, **170**, 1.
6. Chacon, E., Reece, J. M., Nieminen, A. L., Zahrebelski, G., Herman, B., and Lemasters, J. J. (1994). *Biophys. J.*, **66**, 942.
7. Dunn, K. W., Mayor, S., Myers, J. N., and Maxfield, F. R. (1994). *FASEB J.*, **8**, 573.
8. Lemasters, J. J., Chacon, E., Zahrebelski, G., Reece, J. M., and Nieminen, A. L. (1993). In *Optical microscopy: emerging methods and applications* (ed. B. Herman and J. J. Lemasters), p. 339. Academic Press, New York.
9. Opitz, N., Merten, E., and Acker, H. (1994). *Pflugers Arch.*, **427**, 332.
10. Rohr, S. and Salzberg, B. M. (1994). *Biophys. J.*, **67**, 1301.
11. Smith, J. C. (1990). *Biochim. Biophys. Acta*, **1016**, 1.
12. Smiley, S. T., Reers, M., Mottola-Hartshorn, C., Lin, M., Chen, A., Smith, T. W., *et al.* (1991). *Proc. Natl. Acad. Sci. USA*, **88**, 3671.
13. Perez-Terzic, C., Stehno-Bittle, L., and Clapham, D. E. (1997). *Cell Calcium*, **21**, 275.
14. Stehno-Bittel, L., Perez-Terzic, C., and Clapham, D. E. (1995). *Science*, **270**, 1835.
15. Terasaki, M., Song, J., Wong, J. R., Weiss, J. J., and Chen, L. B. (1984). *Cell*, **38**, 101.
16. Wadkins, R. M. and Houghton, P. J. (1995). *Biochemistry*, **34**, 3858.
17. Okada, C. Y. and Rechsteiner, M. (1982). *Cell*, **29**, 33.
18. Park, R. D., Sullivan, P. C., and Storrie, B. (1988). *J. Cell. Physiol.*, **135**, 443.
19. Koval, M., Geist, S. T., Westphale, E. M., Kemendy, A. E., CiVitelli, R., Beyer, E. C., *et al.* (1995). *J. Cell Biol.*, **130**, 987.
20. Kiany, D. T., Kollander, R., Lin, H. H., LaVilla, S., and Atkinson, M. M. (1994). *In Vitro Cell. Dev. Biol. Anim.*, **30A**, 796.
21. Chang, H. (1994). *J. Immunol. Methods*, **176**, 235.
22. Mikhailov, A. V. and Gundersen, G. G. (1995). *Cell Motil. Cytoskel.*, **32**, 173.

Chapter 11

Monitoring nuclear receptor function

B. Scott Nunez

Department of Marine Science, The University of Texas Marine Science Institute, 750 Channel View Drive, Port Aransas, Texas 78373-5015, USA.

Wayne V. Vedeckis

Department of Biochemistry and Molecular Biology, Louisiana State University Health Sciences Center, 1901 Perdido Street, New Orleans, Louisiana 70112, USA.

1 Introduction

The glucocorticoid receptor (GR), one of the first nuclear receptors to be characterized and to have its gene cloned, serves to illustrate the complexity of nuclear receptor function. The expression of the GR gene is regulated by at least three different promoters and the resulting mRNAs can be differentially spliced into several transcriptional variants (1–3). Following the translation of these transcripts into protein, the activity and/or subcellular distribution of the GR protein may be modified by post-translational modifications such as phosphorylation (4). In the absence of ligand, the GR is found primarily in the cytoplasm associated with protein chaperones, such as heat shock protein 90 (HSP 90), which act to maintain the receptor in an inactive state (5). Other proteins, called co-activators and co-repressors, bind to and modulate the transcriptional activity of nuclear receptors (6). The GR is activated by binding glucocorticoid (GC): it is then translocated into the nucleus where it can alter gene transcription through two avenues. The classical mode of action is for activated GR to alter transcription directly by binding a glucocorticoid responsive element (GRE), a negative glucocorticoid responsive element (NGRE), or a composite GRE. The GR can also alter transcription directly without binding DNA by interacting with other transcription factors that bind DNA. The second avenue of GR action is via transcriptional interference, whereby the activity of a transcription factor is diminished by an interaction with the GR (7).

The regulation of nuclear receptors is obviously a complex matter. However, recent technological and methodological advances have dramatically increased our understanding of this field. This chapter will review some of the methods used to illuminate the actions of GR and other nuclear receptors and will also touch on new technological developments that may refine our understanding of these proteins in the future.

2 Cell culture assays of transfected gene expression constructs

2.1 Creation and isolation of eukaryotic gene expression constructs

The introduction, or *transfection*, of exogenous complementary DNA (cDNA) into cultured cells is a powerful way to characterize nuclear receptors. Following the isolation of a cDNA clone, the initial step in such a characterization is the creation of an expression construct. A typical eukaryotic expression plasmid contains a gene for antibiotic resistance, a eukaryotic promoter region, a multiple cloning site (MCS), and a polyadenylation signal (*Figure 1a*). A cDNA encoding the gene of interest is ligated downstream of the promoter sequence using restriction sites in the MCS. Bacteria are transformed with ligated DNA and plated onto agar plates that contain an appropriate antibiotic (8). Single colonies that harbour the cDNA/plasmid construct are identified and used to propagate the plasmid DNA. The creation of an expression construct represents a substantial investment of time and money. Therefore, a freezer stock of transformed bacteria should be made once a construct is created. If purified constructs are obtained commercially or as a gift from another investigator, bacteria can be rapidly transformed as described in *Protocol 1*. It should be confirmed that a colony harbours the appropriate construct before creating a freezer stock.

The purification of plasmid DNA from bacterial cultures has become quite

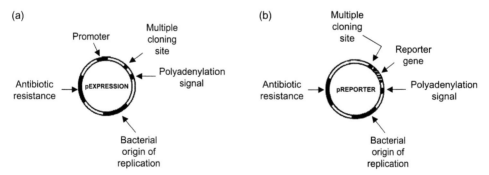

Figure 1 Characteristics of expression and reporter plasmids. A plasmid DNA vector contains several features. A bacterial origin of replication and a gene conferring antibiotic resistance is necessary for propagation and selection of the plasmid in bacteria. Most plasmid vectors have a multiple cloning site that contains unique recognition sequences for several restriction enzymes, simplifying the insertion of exogenous DNA into the vector. (a) A viral promoter such as the CMV promoter typically drives the expression of cDNAs inserted into expression plasmids. (b) In reporter constructs, a putative promoter sequence is inserted into the multiple cloning site upstream of a reporter gene such as luciferase. A polyadenylation signal sequence is included in both types of vectors to enhance the translation of plasmid-encoded transcripts.

standardized, and there are many plasmid preparation kits on the market. In many cases, plasmid DNA is further purified with caesium chloride centrifugation (8). However, many commercial kits now isolate DNA essentially free of bacterial contaminants that can cause cellular toxicity or lower transfection efficiency. Once purified, DNA is transfected into cultured cells (methods of transfection are discussed in Chapter 12 of this volume). For transient transfections, 24–48 h post-transfection is generally a sufficient period before cells are assayed for the presence of the expressed protein. Stable cell lines can be created using plasmids that confer antibiotic resistance to eukaryotic cells, but this process is much more involved (Chapter 12). The presence of an expressed nuclear receptor in transiently or stably transfected cells can be detected by Western blot analysis, biological activity, or binding assays.

Protocol 1

Freeze-shock transformation of bacteria[a]

Equipment and reagents

- Agar plates containing appropriate antibiotic
- Liquid nitrogen
- Bacterial cells competent for transformation
- Liquid bacterial transformation medium: 'SOC' pH 7.0, 20 g/litre Bacto tryptone, 5 g/litre Bacto yeast extract, 0.5 g/litre sodium chloride, 10 mM magnesium chloride,[b] 20 mM glucose[b]

Method

1 Thaw competent cells on ice and gently transfer 50 μl to a pre-chilled microcentrifuge tube. Add 0.25–1 μg of plasmid DNA and mix gently.

2 Immerse the microcentrifuge tube in liquid nitrogen for 10 sec. Be careful when removing the tube from the liquid nitrogen, because nitrogen trapped in the tube will expand quickly and may cause the cap to fly off unexpectedly. Carefully remove the cap of the microcentrifuge tube to release the pressure and then replace the cap. Allow the bacteria to thaw at room temperature.

3 Pipette 400 μl of SOC into the tube and incubate at 37 °C for 10–30 min.

4 Transfer 50–200 μl of the transformed bacteria onto an agar plate. The whole transformation can also be plated by centrifuging the tube at 800 g for 5 min, removing 250 μl of SOC, and resuspending the cells in the remainder.

5 Allow the medium to soak into the agar and place the plate upside down in a 37 °C incubator. Growth should be evident 14–16 h later.

[a] This is a rapid method for transforming bacteria with purified plasmid DNA. It is less efficient than heat shock and should not be used to transform bacteria with products of ligation reactions.

[b] Add from sterile stock concentrates immediately before use.

2.2 Use of Tet-'Off' and Tet-'On' systems

Several regulated gene expression systems have been developed over the years, including systems based upon heavy metal or steroid hormone regulation. Most of these systems have serious drawbacks that limit their utility. In many cell types, specificity of regulation may not be sufficient, and many endogenous genes can be affected by the treatment. The use of native promoter elements to drive gene expression may lead to unintended interactions with other endogenous transcription factors, increasing the background or leading to non-specific regulation. In addition, many of these systems depend upon the repression, rather than the activation, of a transfected gene and the complete repression of a gene usually requires a high level of repressor. Gossen and Bujard (9) developed an elegant method based on prokaryotic tetracycline-resistance for the tight regulation of genes transfected into mammalian cells. The Tet repressor (TetR) negatively regulates the genes conferring tetracycline resistance to *E. coli*. In the absence of tetracycline, the TetR protein binds to tet operator sequences and represses transcription (10). The TetR protein was modified by fusing the N-terminus of the TetR protein with the C-terminus of the herpes simplex virus VP16 protein (11). The C-terminus of the VP16 protein acts as a transcriptional activation domain and its fusion onto the TetR protein forms a transcriptional activator (TetR/VP16) that increases transcription from the TetR responsive element (TRE) in the absence of tetracycline.

The initial step in creating a Tet-responsive system is the production of a stable line of cultured cells expressing the TetR/VP16 protein. Once isolated and characterized, these cells are then stably transfected with another construct containing the TRE and a target gene. The use of the TRE promoter in the expression construct allows for up-regulation of expression by the removal of tetracycline (Tet-Off). Specific mutations introduced into the TetR/VP16 chimera protein produced a protein that activates transcription from the TRE in the presence of tetracycline (Tet-On) (10). The investigator has the option of creating a system in which either the presence or absence of tetracycline can be used to regulate expression.

Because tetracycline-regulated systems involve prokaryotic genes and promoters, they are advantageous when tight control of a single gene is necessary. However, the advantages of this system must be weighed against several considerations. Creation of a tetracycline-responsive cell line is very time- and labour-intensive, involving first the creation of a stable line expressing the TetR/VP16 fusion protein. Once this line has been established and characterized, it can be transfected with the TRE-containing construct. The selection and characterization of a doubly transfected stable cell line may take months, and cells must always be maintained in medium containing at least two antibiotics. Doxycycline is sometimes used as an alternative to tetracycline, because it has a longer half-life in solution, but it is more expensive than tetracycline. The Tet-On system is 100-fold more sensitive to doxycycline than tetracycline, so it is preferred in this system, despite its cost. Another consideration when developing a tetracycline-regulated expression system is the source of serum for tissue culture medium. Because of the common use of high doses of tetracycline in the cattle industry, levels of tetracycline found in bovine serum can elevate background expression in the Tet-On system. This is especially critical when examining clones in which the dynamic range of expression is large, where small changes in the concentration of tetracycline can result in large changes in protein expression (up to 1000-fold). Dialysis of serum stocks can reduce background levels of tetracycline, and serum that has been tested for use in Tet-systems is available (Clontech).

2.3 Use of reporter constructs in promoter characterization

The expression of exogenous DNA in cultured cells is possible because the endogenous transcriptional machinery binds to, and transcribes from, promoter elements in the plasmid DNA. With slight modifications to the expression construct, this phenomenon can be used to characterize putative promoters. Instead of inserting a cDNA into the MCS, a putative promoter sequence is inserted upstream of a gene that encodes a reporter protein (*Figure 1b*). Such proteins include chloramphenicol acetyltransferase (CAT), β-galactosidase (β-gal), firefly luciferase (luc), secreted alkaline phosphatase (SEAP), and green fluorescent protein (GFP). The assays for these reporter proteins are well characterized and are typically simple and cost-effective (see ref. 12 for a comprehensive review of reporter advantages and disadvantages). The reporter construct is transfected into cultured cells, which are collected 24–48 h later. In the case of CAT, β-gal, and luc, the activity of the reporter protein is measured in the cell lysate. Harvesting the cells is not necessary when using SEAP or GFP as a reporter. Transfected cells secrete SEAP into the growth medium; activity can be determined in the medium without killing the cells. GFP requires no reagents for visualization and can be quantified using a fluorometer. Several different variants that fluoresce at different wavelengths are available (Clontech).

Results of reporter assays must be normalized for well-to-well differences in transfection efficiency. The easiest way to accomplish this is to co-transfect a reporter construct, driven by a strong constitutive promoter, along with the experimental reporter constructs. A β-gal construct driven by a cytomegalovirus or other viral promoter is often used as a transfection control. A dual-luciferase system is now available (Promega) that utilizes one isoform of luciferase as the experimental reporter and a second isoform as a transfection control. The reporter luciferase is assayed in a reaction buffer in which the second luciferase is not active. This reaction is terminated with the addition of a reaction buffer in which only the second isoform is active. This allows reporter and transfection control activities to be determined from the same well, simplifying the process and eliminating a source of variability.

3 *In vitro* analysis of protein–DNA interaction

3.1 Isolation of nuclear extract

The concentration of GR protein within a cell is a primary determinant of the cellular response to GC (13). Transcriptional regulation of the GR gene is quite complex, with at least three different promoters driving both constitutive and inducible expression (1–3). Characterization of these promoter regions is critical to our understanding of GR action. Additionally, many effects of the GR are elicited through binding of the activated receptor to specific DNA sequences and the subsequent transcriptional modulation of GR responsive genes (7, 14). Therefore, techniques used to identify protein–DNA interactions have been instrumental in deciphering GR function.

The isolation of a nuclear extract is an initial step in many of these techniques. This extract contains general transcription factors, nuclear receptors, and other soluble nuclear proteins. It should be relatively free of contaminating proteins from the plasma membrane, cytoplasm, and nuclear matrix. The classical method of nuclear extraction involves leaching nuclear proteins from intact nuclei with a high salt buffer (15). Kits containing reagents necessary for nuclear and cytoplasmic extraction are commercially available and the process can be carried out in less than 2 h in some cases (Pierce).

Several things must be considered when extracting nuclear proteins for analysis of nuclear receptor function. Some transcription factors bind DNA only when phosphorylated, so the inclusion of phosphatase inhibitors (2.5 mM sodium orthovanadate, 2.5 mM sodium molybdate, 10 mM sodium fluoride, for example) may be necessary. Many transcription factors bind DNA only as oligomers, so it may be necessary to optimize in order to minimize the disruption of such interactions. The subcellular distribution of a protein must also be considered. In the absence of GC, GR is found predominately in the cytoplasm, but the equilibrium shifts to the nucleus in the presence of GC (16). Obviously, if one wishes to extract a nuclear receptor one must be careful to ascertain in which subcellular compartment the receptor resides. In addition, the activity of most nuclear receptors is ligand-dependent and so it may be important to include ligand in assay buffers.

3.2 *In vitro* DNase I footprinting

DNase I footprinting is a theoretically simple, but technically challenging, method used to detect DNA–protein interactions (17). A short length of DNA (250–500 nucleotides) is labelled with the large fragment of *E. coli* DNA polymerase I (the Klenow fragment). It is important that the DNA is labelled to a high specific activity as footprints are most easily detected when DNA and the protein that binds it are present in nearly equal amounts. Labelled DNA is mixed with nuclear extract or purified protein and digested with DNase I, which randomly nicks the DNA. Over-digestion can result in a diminution of signal, making detection of footprints difficult. The optimal amount of DNase I added to each reaction must be determined empirically, in a preliminary experiment, by maintaining the concentration of nuclear extract and labelled DNA constant and varying the amount of DNase I.

After DNase I treatment, DNA fragments are resolved on a denaturing polyacrylamide gel (PAGE) which is dried and exposed to autoradiography film. Following autoradiography, the DNA is seen as a ladder of different sized fragments. Optimally, all reactions should yield similar band intensities upon exposure to film. Transcription factors and other proteins that bind DNA protect the sequences that they bind from DNase I digestion. This interaction is seen as a hole (the DNA footprint) in the DNA ladder (*Figure 2*). The sequence of a footprint can be determined using the method of Maxam and Gilbert (18). The labelled DNA used in DNase I footprinting reactions is treated with chemicals that cleave DNA at specific nucleotides, resulting in a ladder of labelled fragments which represent the DNA sequence. The sequencing and DNase I footprinting reactions are electrophoresed simultaneously and the sequence of a footprint is determined by comparing the footprint to the corresponding region of the sequencing reaction. Computer analysis is then used to compare the protected sequence to known consensus binding elements to narrow down the prospective list of proteins responsible for the footprint. While in the past such comparisons were tedious and time-consuming, several databases are now available on the World Wide Web that offer rapid, easy analysis of sequence. Web Signal Scan Service, a very useful search engine, is found at *http://bimas.dcrt.nih.gov/molbio/signal/* (19). As the number of known DNA binding motifs increases, such tools will become indispensable.

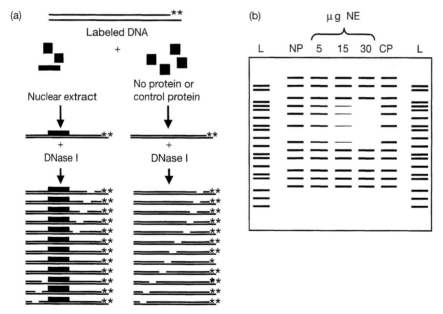

Figure 2 DNase I footprinting. This procedure involves mixing labelled DNA with a nuclear extract (NE) that contains DNA binding proteins. The DNA is partially digested with DNase I (a) and resolved on a sequencing gel (b). Partial digestion results in a ladder of different sized fragments in the absence of protein (NP) or in the presence of control protein that does not bind DNA (CP). If the labelled DNA contains a binding site for a protein present in the NE, that binding site is protected from digestion, resulting in a footprint in the DNA ladder. Adding more NE protects more DNA and yields a better footprint. The sequence of a footprint can be determined by simultaneously running a Maxam and Gilbert sequencing ladder (L).

3.3 Electrophoretic mobility shift assay

Sequences corresponding to DNase I footprints can be further analysed using a technique called electrophoretic mobility shift assay (see *Protocol 2*). This technique, sometimes called EMSA or mobility shift analysis, is less complex than the DNase I footprinting. Double-stranded oligonucleotides are labelled with ^{32}P using polynucleotide kinase or the Klenow fragment. Restriction fragments of promoter DNA can also be used, but longer lengths of DNA may contain several binding sites, complicating interpretation of the results. Labelled DNA is mixed with nuclear extract and the reactions are run on a non-denaturing, low percentage PAGE gel. In the absence of protein, or of protein binding, a labelled oligonucleotide moves rapidly through the gel and is visualized as a band in the lower part of the gel (*Figure 3*). Because oligonucleotide–protein complexes migrate much more slowly than labelled oligonucleotide alone, they are seen (upon autoradiography) as bands shifted to the upper part of the gel. Competition experiments in which unlabelled oligonucleotides are added in excess can help to determine the specificity of the binding.

If an identified sequence displays similarity to a known binding site, the addition

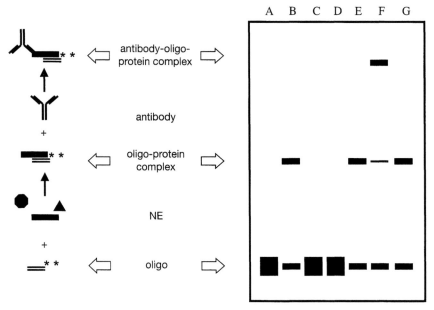

Figure 3 Electrophoretic mobility shift assay. Oligonucleotides are labelled, mixed with nuclear extract (NE), and run on a low percentage, non-denaturing gel. Lane A: in the absence of protein or protein binding, the labelled oligonucleotide migrates quickly through the gel, resulting in a band at the bottom of the gel. Lane B: if protein binds the oligonucleotide, the DNA–protein complex migrates more slowly, resulting in a band further up in the gel. The formation of this complex can be blocked with excess unlabelled oligonucleotide or unlabelled consensus oligonucleotide (lanes C and D, respectively), but not by unlabelled oligonucleotide of unrelated sequence (lane E). Lane F: analysis of the sequence of the labelled oligonucleotide may provide clues to the identity of the protein that binds it, and addition of an antibody to that protein may supershift the complex. Lane G: an antibody to an unrelated protein should not result in a supershift.

of a concentrated antibody to the suspected factor may form an oligonucleotide/protein/antibody complex. The migration of this complex through a PAGE gel is even further retarded and culminates in a supershift further into the upper part of the gel (*Figure 3*). Many antibodies do not supershift, but rather block oligonucleotide–protein complex formation (antibody block). However, a lack of a supershift or antibody block must be interpreted carefully to avoid confusing poor antibody interaction with a truly negative result. Proper controls, such as the use of a labelled consensus oligonucleotide, used in conjunction with either a nuclear extract known to contain the DNA-binding protein or a purified protein, should be used to guard against incorrect interpretations.

Purified preparations of suspected proteins are also used in EMSA to identify binding sites. If the addition of a pure protein results in a gel shift, this is good evidence that the protein binds *in vivo*. Purified protein is not strictly necessary, however; nuclear extract from cells lacking the DNA-binding protein that are transfected with an expression construct containing the sequence for the protein can be used in EMSA. Nuclear extract isolated from cells transfected with empty expression plasmid serves as a control. If the extract from cells over-expressing the protein results in a gel shift, and the extract from control cells does not, this is good evidence that the protein binds the sequence. This approach is helpful when commercial antibodies are not available or when antibodies fail to supershift or block complex formation.

Because the interaction between DNA and protein is sometimes easily disrupted, care should be taken to avoid denaturants such as heat and detergents. Running buffers should be chilled before use to avoid overheating of the gel during electrophoresis. Apparatus that is cleaned with soap should be rinsed well before pouring gels. As when loading all gels, the samples should be loaded down the side and as close as possible to the bottom of the well so as to avoid mixing the sample with the running buffer. Dilution of sample in running buffer can lead to diffuse bands following electrophoresis. Gel composition is also important in maintaining the DNA–protein interaction. Glycerol (2–4%) is sometimes added to help stabilize interactions. In general, the greater the pore size of the gel the better, so 4% gels of an 80:1 mix of acrylamide to bisacrylamide are commonly used. However, these gels have a gelatin-like consistency when polymerized and are extremely sticky. Gels of 40:1 acrylamide to bisacrylamide have a more resilient consistency and can be used in many cases. Care should be taken to avoid stretching the gel when removing it from between the glass plates following electrophoresis. Siliconizing one of the glass plates before pouring the gel can facilitate its removal from between the plates.

The gel shift can be modified to determine the exact nucleotides involved in protein binding. Dimethyl sulfate (DMS) will methylate adenine and guanine residues in DNA, interfering with protein binding. In the methylation interference assay (20), a labelled DNA probe is treated with DMS, such that DNA is methylated once per DNA strand, on average. This probe is then incubated with nuclear extract and electrophoresed as in EMSA. Methylation of guanines and adenines, important for protein binding, will not be shifted. The DNA–protein complexes that form are eluted from the gel and the DNA cleaved with piperidine. The resulting fragments are resolved on a sequencing gel. Nucleotides that are necessary for binding are absent yielding a footprint pattern similar to DNase I footprinting.

Once a footprint has been identified and characterized by DNase I footprinting or EMSA analysis, the protein component of the complex can be purified using classical protein purification methods. The protein can be assayed in column elution fractions using EMSA. The protein can also be purified using affinity chromatography. Biotinylated oligonucleotides are mixed with nuclear extract and then passed over a streptavidin column. The protein is eluted with a high salt buffer. Alternatively, the oligonucleotide can be directly conjugated to a column matrix.

Protocol 2

Electrophoretic mobility shift assay

Reagents

- 1 × TBE: 90 mM Tris–borate, 2 mM EDTA
- N,N,N',N'-tetramethylethylenediamine (TEMED)
- Ammonium persulfate (APS): 20% in water[a]
- Dialysis buffer: 20 mM Hepes pH 7.9 at 4 °C, 20% glycerol, 100 mM KCl, 0.2 mM EDTA, 0.2 mM phenylmethylsulfonyl fluoride,[a] and 0.5 mM dithiothreitol[a]

- 30% acrylamide solution: 80:1 or 40:1 acrylamide/bisacrylamide in water
- Bovine serum albumin (BSA): 4.5 mg/ml
- Polydeoxyinosinic-deoxycytidylic acid (polydI-dC): 1 µg/µl
- 6 × DNA loading buffer: 0.25% bromophenol blue, 0.25% xylene cyanol, 30% glycerol in water

A. Pouring the 4% acrylamide gel

1 Clean the electrophoresis glass plates and rinse them well. Siliconize one side of one glass plate, allow it to dry, and buff away residual silicon. Assemble the glass plates as instructed by the apparatus manufacturer.

2 For 50 ml of 4% gel solution, mix 6.7 ml of 30% acrylamide, 2 ml of glycerol, 0.1 ml of 20% APS, 12.5 ml of 1 × TBE, and 28.7 ml distilled or deionized water in a flask. Degas this solution for 10–15 min. Add 50 µl of TEMED, mix quickly and thoroughly, and pour the gel between the glass plates. Insert the comb into the gel between the plates. Allow the gel to polymerize completely (1 h).

3 Following complete polymerization of the gel, assemble the electrophoretic apparatus. Dilute enough 1 × TBE to 0.25 × TBE to fill the upper and lower buffer reservoirs. Thoroughly wash the sample wells with 0.25 × TBE.

B. Electrophoretic mobility shift assay

1 Mix 1 µl of BSA, 2 µl of polydI-dC, and 5–10 µg of nuclear extract in a microcentrifuge tube. Add dialysis buffer to a total volume of 14 µl. Incubate this solution for 15 min at room temperature or at 4 °C.

2   Add 1 µl of labelled oligonucleotide (20 000 c.p.m.) and incubate for 1 h at room temperature or at 4 °C.

3   Load 2 µl of 6 × DNA loading buffer in the first well of the 4% PAGE gel. Load the samples in the remaining wells. Do not load samples into the well containing the loading buffer, as the presence of the loading buffer may disrupt DNA–protein complexes. Run the gel (3–4 A/cm gel length) until the bromophenol blue band has migrated 3/4 of the length of the gel. Avoid excessive current or run times, as heat can lead to dissociation of the DNA–protein complexes.

4   Dismantle the apparatus and separate the glass plates. Cut a sheet of Whatman 3MM filter paper large enough to completely cover the gel. Place the paper over the gel and press down to affix the gel completely to the paper. Cover the gel with plastic wrap and dry under vacuum at 80 °C. Expose the gel to autoradiography film for 1 h at –80 °C using intensifying screens. Exposure time can be adjusted following this initial exposure.

C. Supershift modifications

1   Prepare nuclear extract mix as in part B, step 1. Add 2 µl of a concentrated antibody (2 µg/µl) to the mix and incubate for 1 h at room temperature or at 4 °C.

2   Add 1 µl of the labelled oligonucleotide and incubate for 15 min.

3   Electrophorese samples on 4% PAGE as in part B.

4   A modification of this protocol is sometimes used, in which the nuclear extract and oligonucleotide are incubated first, followed by incubation with the antibody.

[a] Add immediately before use.

4 Assays of endogenous gene expression

4.1 Northern blot analysis

The classic mode of action for nuclear receptors is by altering the rate of gene transcription, ultimately manifested as a change in mRNA levels. Techniques used to quantify RNA have therefore played a large role in the study of nuclear receptor function. Northern blot analysis is the most frequently used method for examining RNA abundance (8). RNA is size fractionated through an agarose gel and transferred to a nitrocellulose or nylon membrane. The RNA is then immobilized on the membrane by baking under vacuum or irradiating with UV light. After non-specific sites on the membrane are blocked with an excess of low molecular weight nucleic acid, the membrane is probed with a specific antisense nucleic acid probe. This probe is typically labelled with ^{32}P, but probes can also be modified with biotin, digoxigenin, or similar chemical that allows subsequent detection. The membrane is incubated with probe for 2–16 h and then washed in buffers of increasing stringency to eliminate non-specific binding. The blot is then wrapped in plastic wrap and exposed to film. The resulting bands can be

quantified using densitometry. Because the RNA is size fractionated before transfer, this method has the advantage of yielding information about transcript size.
It is therefore possible to simultaneously quantify the expression of different
sized transcripts from the same gene. However, Northern blot analysis is the least
sensitive of the methods discussed in this section. Sensitivity can be increased
with the use of purified mRNA. However, such purification requires the isolation
of large quantities of total RNA. Isolation of mRNA from total RNA also increases
the possibility of contamination and RNA degradation by RNase. In instances
where the amount of starting material is limited, Northern blot analysis may not
be an option.

4.2 RNase protection assay

In the RNase protection assay (RPA), a labelled antisense RNA probe is annealed
with total RNA and then treated with RNase T1, which degrades single-stranded
RNA, but leaves double-stranded RNA intact (21). This method requires the
synthesis of a labelled RNA probe (complementary to the target mRNA) using
purified viral RNA polymerases. Many cloning vectors contain the necessary SP6,
T7, or T3 polymerase promoter sites needed for the *in vitro* synthesis of these
probes. Introducing viral promoter sites into a DNA template using the polymerase chain reaction (PCR) can simplify probe production and avoid possible RNase
contamination (most plasmid preparation procedures include an RNase treatment). Labelled single-stranded RNA probe is mixed with sample RNA and allowed
to anneal 16–24 h. RNase is then added and the solution incubated overnight.
Loading buffer is added and the digested RNA is electrophoresed on a non-
denaturing PAGE gel. Following electrophoresis, the gel is dried and exposed to
film. Several different messages can be analysed simultaneously as long as each
individual probe yields a product of unique size. The intensity of bands representing intact double-stranded RNA is quantified using densitometry. This technique is about tenfold more sensitive than Northern blot analysis.

4.3 Reverse transcription-polymerase chain reaction

Reverse transcription-polymerase chain reaction (RT-PCR) is the most sensitive
method used to detect gene expression. Reverse transcriptase (RT) is a nucleic
acid polymerase that can generate cDNA from RNA templates and is one of the
most versatile enzymes used in molecular biology. Two basic forms of RT are
commercially available, each having unique characteristics (22). Avian myelo-
blastosis virus (AMV) RT is fully active at 42°C and pH 8.3, but retains potent
RNase H activity. The Moloney murine leukemia virus (MMLV) RT has much
weaker RNase H activity, and is thus has a greater advantage in synthesizing
cDNA from long RNA templates. However, the MMLV-RT operates more efficiently
at 37°C and pH 7.6. Temperature resistance is an advantage when secondary
structure of the RNA is a consideration. Several commercially available RT proteins have mutations that reduce RNase H activity and increase the yield of full-
length cDNAs from an RT reaction.

Typically, oligo(dT) is used as a primer for the RT reaction, but oligos of six to ten random nucleotides or a gene-specific primer can also be used. The RT reaction produces a single-stranded DNA that is complementary to the original RNA template. A second gene-specific set of primers is then used in the PCR reaction. The typical PCR cycle includes three steps, a high temperature (94–96 °C) denaturing step, a low temperature annealing step, and a 72 °C extension step. Heating to the denaturing temperature renews the cycle. The temperature of annealing step is primer-dependent, and may require optimization. In some cases, a 1–2 °C change in annealing temperature can greatly alter the efficiency of the reaction.

Because the amount of template increases after each round, PCR amplification occurs at an exponential rate. Theoretically, RT-PCR is sensitive enough to detect a single mRNA transcript from a single cell, although practical limits appear to be in the range of a few hundred copies per cell. However, the utility of RT-PCR in quantifying the relative amounts of mRNA between samples has been of limited use for several reasons. RT-PCR is very sensitive to variations in the quantity and quality of starting template. Differences in reaction mixes (e.g. variation in the concentration of enzyme or primer) can introduce inter-sample variation and confound interpretation. In many cases, sample variability can be minimized by running duplicate reactions of the same sample and by using master reaction mixes. Another major problem of quantitative PCR has been the identification of the proper endogenous standard or control. Housekeeping genes such as actin and glyceraldehyde-3-phosphate dehydrogenase are widely used endogenous standards because these genes are expressed in all cells. However, cell-to-cell differences in the expression of these genes complicate their use as internal RT-PCR standards. This is especially true when quantifying RNA from different tissues or cell types. Because ribosomal RNA (rRNA) is essentially invariant between cell types and is by far the most abundant RNA, it is perhaps the optimal internal standard for quantitative RT-PCR. However, the abundance of rRNA is perhaps the greatest drawback to its use as a standard, because amplification becomes saturated far too quickly to be of much use in the quantification of even high copy mRNAs. The addition of 18S primers modified so that they cannot be extended (competimers), can be used to attenuate the amplification of 18S RNA, thereby enabling it to be used as an internal standard (Ambion).

The synthesis of good primers is critical to successful PCR. Primers can be as short as 13–15 nucleotides and as long as 70–80 nucleotides. In the synthesis of longer primers, a significant percentage of primer will not be full-length. To avoid complications caused by less than full-length primers, long primers should be gel or column purified. While the sequence of a primer will depend on the sequence of the template, a good rule of thumb is that the primer should contain about 50% guanines and cytosines. Ideally, a primer should contain an equal percentage of all four nucleotides. The composition of the primer will dictate the annealing temperature of the primer–template complex. In general, an adenine or a thymine will add 2 °C to the annealing temperature while a guanine or cytosine will add 4 °C. The primer should not include long strings (three or more) of the same nucleotide, because this can lead to non-specific annealing. The use of primer selection software for the design of PCR primers is recommended, as such software can calculate quickly parameters, such as nucleotide composition and annealing temperature, while identifying potential problems such as secondary structure.

Once primers have been designed and synthesized, the PCR reaction should be optimized. Variables in the PCR reaction include Mg^{2+}, nucleotide, primer, and enzyme concentration. The amplification of some templates may require the addition of a stabilizing compound such as glycerol or dimethylsulfoxide. Several commercial optimization kits include PCR buffers containing different concentrations of Mg^{2+} and stabilizing compounds. Once a reaction has been optimized, preliminary experiments should be conducted in order to establish the linear range of the PCR reaction. In the early rounds of PCR, template may be limiting, resulting in very little product; linear amplification then occurs. As template increases further, the reaction components, such as nucleotides and primers, are incorporated into DNA and eventually become limiting. The quantification of starting RNA template depends on measuring PCR product in the linear range of amplification. Using PCR products from cycles outside this range leads to erroneous results.

There are a number of ways that PCR products can be quantified. The most common is the use of ethidium bromide (EtBr). EtBr stained gels are visualized using UV light and band intensities are recorded using photography and images are quantified using densitometry. Many gel documentation systems include a video camera, computer, and associated software so PCR products can be analysed quickly and easily. EtBr, although commonly used, is a potential health hazard. Other stains (SYBR green I from FMC, for example) that are less toxic can also be used with minimal modification.

Real time PCR allows simplified quantification of PCR products (23). EtBr is included in PCR reactions that are cycled in a thermocycler, modified to enable samples to be irradiated with ultraviolet light, and the resulting fluorescence detected. Fluorescence increases as DNA is amplified. Other fluorescent molecules have been developed that yield increases in fluorescence as PCR amplification occurs. Such systems can produce a plot of fluorescence *versus* cycle number that is more representative of the PCR process while eliminating the need to take samples, run gels, and quantify band intensities manually. Because a fluorescent reading is obtained during cycling, the linear range of amplification can be visualized, allowing precise quantitation.

4.4 DNA array technology

While Northern blot analysis, RPA, and RT-PCR allow an investigator to monitor the relative expression of a few specific genes, these techniques cannot identify novel genes that may be regulated by nuclear receptor action. Several techniques have recently been developed to simultaneously assess the regulation of expression of a whole suite of genes. Subtractive hybridization (24) and differential display (25) are techniques commonly used to isolate transcripts that are unique to a cell type or regulated by a specific treatment. As is the case with DNase I footprinting, these techniques are theoretically straightforward but technically demanding. A new technology, the DNA array, uses the growing library of cDNA clones and expressed sequence tags to identify genes that are differentially regulated (26). Individual cDNA clones are spotted onto a discrete region of a membrane, glass slide, or chip, and each region is given a unique identifier. These arrays are then probed with cDNA probes synthesized from mRNA. To study the differential expression of genes in different cell types, labelled cDNA is made from each cell type and used to sequentially probe the array (see *Protocol 3*). To examine the effect of an experimental treatment on gene expression in a cell type or tissue, labelled cDNA is made from RNA from treated and untreated cells of a single type. After hybridization, the array is exposed to a phosphorimaging plate and then analysed in a phosphorimager. Alternatively, the cDNA is synthesized using a variety of fluorescent nucleotides and the fluorescent probes quantified by an array reader with associated computer software. The image that results is a grid of dots, each representing a single, unique gene. The intensity of each dot is proportional to the expression of that gene in the cDNA probe. The array is then stripped and incubated with a different probe. Hybridization is normalized using several housekeeping genes, such as actin and rRNA. By comparing the intensity of dots resulting from hybridization with different probes, it is possible to assess differences in expression. The accurate comparison of thousands of genes cannot be undertaken manually, and so automated analysis of DNA array experiments is typically done by the manufacturer of the DNA array. Arrays dotted with tens of thousands of non-redundant cDNA sequences will soon be available commercially, creating further reliance on computer analysis.

Protocol 3

cDNA array hybridization using membrane filter arrays and radiolabelled probes[a]

Equipment and reagents

- High density cDNA array on a filter membrane
- Hybridization buffer

- $20 \times$ SSPE: 175.3 g/litre NaCl, 27.6 g/litre $NaH_2PO_4.H_2O$, 7.4 g/litre EDTA pH 7.4
- Sodium dodecyl sulfate (SDS): 10% in water

Method

1. Prepare ^{33}P-labelled cDNA probes from mRNA isolated from the cell or tissue types of interest using reverse transcriptase.

2. Pre-hybridize the array in hybridization buffer at 42 °C. There are several hybridization buffers available, some require an incubation time as short as 5 min.

Protocol 3 continued

3   Mix the probe in hybridization buffer to a specific activity of $1\text{-}2 \times 10^6$ c.p.m./ml. Remove the pre-hybridization solution from the membrane and add the probe. Incubate the membrane at 42 °C for 12–16 h.

4   Remove the hybridization buffer containing the probe and briefly wash the membrane twice in $2 \times$ SSPE + 0.1% SDS at room temperature. SSPE should be diluted from $20 \times$ stock with distilled or deionized water.

5   Wash the membrane twice in $1 \times$ SSPE, then once in $0.1 \times$ SSPE at 65 °C.

6   Wrap the membrane in plastic wrap and expose to a phosphorimaging plate for 24–72 h.

7   Analyse the phosphorimaging plate in a phosphorimager and process the resulting signals using quantitative analysis software.

[a] When using fluorescently labelled probes and instrumentation for producing glass slide or chip arrays, follow the instructions provided with the instrument.

4.5 Western blot analysis

In most cases, the ultimate consequence of increased gene transcription is a concomitant increase in protein. Western blot analysis is the most widely used method for determining relative levels of proteins (see *Protocols* 4 and 5). The basis of this procedure is much like Northern blot analysis. Proteins are resolved using SDS–PAGE and are transferred onto a nitrocellulose or nylon membrane. The membrane is then blocked with a protein solution to reduce non-specific binding and probed using an antibody developed against the protein of interest. The protein–antibody interaction is then detected using a secondary antibody that recognizes the class of antibodies to which the primary antibody belongs. The secondary antibody (and in some cases the primary antibody) is typically conjugated to an enzyme such as alkaline phosphatase or horseradish peroxidase whose activity can be visually monitored. In the past, substrates for these enzymes that yield a colour change visible to the naked eye were used in Western blot analysis. However, the sensitivity of these substrates was found lacking and colorimetric techniques have given way to enhanced chemiluminescence. Horseradish peroxidase, in the presence of hydrogen peroxide, oxidizes a compound called luminol into an excited state. The oxidized luminol emits light when it decays to its ground state (27). The light emitted can be increased 1000-fold by chemical enhancers such as phenols (28). After incubation with enhanced chemiluminescence reagents, the membrane is covered with plastic wrap and exposed to film. Resulting bands are then quantified using densitometry.

Protocol 4

SDS–PAGE analysis

Equipment and reagents

- SDS–PAGE mini-gel electrophoresis apparatus
- 10% (w/v) APS in water[a]
- TEMED
- Resolving Tris–HCl: 1.5 M pH 8.8
- Stacking Tris–HCl: 0.5 M pH 6.8
- 10% (w/v) SDS in water

- 30% acrylamide: 37.5:1 acrylamide/bisacrylamide in water
- 4 × sample buffer: 10% glycerol, 2.0% SDS, 5% 2-mercaptoethanol, 0.4% bromophenol blue, 64 mM Tris–HCl pH 6.8
- 5 × running buffer: 15 g/litre Tris base, 72 g/litre glycine, 5 g/litre SDS

Method

1 Assemble the glass plates, spacers, and combs to pour SDS–PAGE mini-gels (7 cm × 8 cm). Before pouring the resolving gel, insert the comb between the glass plates and place a mark 1.5 cm below the bottom of the comb teeth on the outside of the plate, then remove the comb.

2 Mix 2.5 ml resolving Tris–HCl, 3.3 ml acrylamide, 0.1 ml SDS, 0.05 ml APS, and 4 ml distilled or deionized water in a beaker. This volume is sufficient to pour two 10% acrylamide mini-gels of 0.5–1.5 mm thickness. Gels of different percentages can be poured by changing the volumes of 30% acrylamide stock and water while leaving the remaining ingredients constant.

3 Degas this solution for 10–15 min, add 10 μl TEMED to the flask, and mix immediately. Quickly pipette the solution between the glass plates until it is just above the mark from step 1. Carefully overlay the acrylamide with water or n-butanol. Allow the gel to completely polymerize (0.5–1 h). Before pouring the stacking gel, remove the water or n-butanol from the top of the resolving gel and thoroughly rinse with water.

4 Mix 2.5 ml stacking Tris–HCl, 1.3 ml acrylamide, 0.1 ml SDS, 0.05 ml APS, and 6.1 ml water in a beaker and degas the solution for 10–15 min.[b] Add 10 μl TEMED, mix well, and layer the solution over the polymerized resolving gel. Insert the comb and ensure that the teeth of the comb are parallel to the top of the resolving gel. Allow the stacking gel to completely polymerize (0.5–1 h).

5 Mix 5–25 μg of total protein with an appropriate amount of 4 × loading buffer and heat the sample to 95 °C for 5 min. Chill the samples on ice and collect with brief centrifugation. Centrifugation also pellets any precipitates that may have formed.

6 Pour enough 1 × running buffer into the lower buffer reservoir to cover the electrode and to submerge the bottom of the gel when it is placed into the reservoir. Place the upper reservoir containing the gel into the lower reservoir and fill it with 1 × running buffer. Thoroughly rinse the sample wells with running buffer using a transfer pipette.

Protocol 4 continued

7   Load samples and molecular weight standards into the bottom of the wells. Avoid mixing the sample with the surrounding buffer. Attach the electrodes and run the gel at the power settings recommended by the apparatus manufacturer. Dilute 5 × running buffer with distilled or deionized water.

8   If resolved protein is to be transferred to a membrane, cut Whatman 3MM filter paper and membrane (nitrocellulose or nylon) so that it is slightly larger than the gel to be transferred. Equilibrate the Whatman 3MM filter paper and membrane in transfer buffer while the SDS–PAGE gel runs.

9   Run the gel until the lower blue dye (bromophenol blue) reaches the bottom of the gel.

10 Turn off the power to the apparatus and remove the glass plates containing the gel. Separate the glass plates and remove the gel. At this point, the gel can be stained with Coomassie brilliant blue to visualize resolved proteins or equilibrated in transfer buffer for 10–30 min for subsequent Western blot transfer.

[a] Prepare fresh.

[b] A little bromophenol blue can be added to colour the stacking gel and facilitate sample loading.

Protocol 5

Western blot analysis using enhanced chemiluminescence

Equipment and reagents

- Transfer apparatus
- Western blot transfer buffer: 39 mM glycine, 48 mM Tris, 0.04% SDS, 20% methanol
- PBS: 8 g/litre NaCl, 0.2 g/litre KCl, 1.44 g/litre Na_2HPO_4, 0.24 g/litre KH_2PO_4 pH 7.4
- Ponceau S solution: 0.1% in 5% acetic acid
- Blocking solution: 5% non-fat dry milk in PBS
- PBS-T: PSB + 0.1% Tween 20
- Enhanced chemiluminescence (ECL) reagents

Method

1   Assemble transfer sandwich as instructed by the manufacturer of the transfer apparatus. In general, this involves placing the membrane and gel between sheets of filter paper. Be careful not to trap bubbles between any of the layers; these can be removed by rolling a rod over the stack. Transfer the protein from the gel to the membrane at the power settings recommended by the manufacturer for 1–16 h, depending on the apparatus and the transfer method.

2   Turn off the power and dismantle the apparatus. Separate the gel and membrane. The gel can now be stained with Coomassie blue and the membrane with Ponceau S to evaluate transfer efficiency.

3 Destain the membrane with five washes in distilled water for 2 min each wash and place it in blocking solution for 1–3 h at room temperature or overnight at 4 °C.

4 Wash the blocked membrane with PBS for 15 min, and repeat this wash three times. If an antibody–alkaline phosphatase conjugate is to be used for detection, substitute a Tris-buffered saline for PBS, as phosphate buffers inhibit alkaline phosphatase activity.

5 Place the membrane in primary antibody diluted in PBS for 1–2 h at room temperature or overnight at 4 °C. The optimal dilution of primary antibody is determined empirically.

6 Wash the membrane with PBS-T for 15 min, and repeat this wash three times.

7 Incubate the membrane with secondary antibody–horseradish peroxidase conjugate diluted in PBS-T for 1–2 h at room temperature. Wash the membrane with PBS-T for 15 min, and repeat this wash three times.

8 Mix the ECL reagents as instructed by the manufacturer.

9 Remove the membrane from the PBS-T and blot any excess buffer. Cover the entire surface of the membrane with the ECL reagent and incubate for 1–5 min (depending on the reagent) at room temperature.

10 Remove the reagent from the membrane and blot any excess. Wrap the membrane in plastic wrap and expose to autoradiography film for 1 min. Exposure time can be adjusted following this initial exposure.

5 Analysis of protein–protein interaction

5.1 Immunoprecipitation

Antibodies to relevant proteins can be used to identify physiologically significant nuclear receptor–protein interactions (see *Protocol 6*). Cell extract is mixed with a primary antibody to a nuclear receptor and incubated to allow the saturation of antibody binding sites. Protein A or protein G (proteins that bind immunoglobulins), conjugated to an insoluble high molecular weight compound (Sepharose or agarose beads), is then added and the solution centrifuged to collect the resulting complex. Proteins that interact with the nuclear receptor are also pulled down. After thorough washing, the complex is disrupted with the addition of SDS–PAGE loading buffer and heating to 95 °C for 5 min. Samples are centrifuged to pellet the beads and the liberated proteins in the supernatant are resolved using SDS–PAGE electrophoresis. When a specific protein is thought to interact with the nuclear receptor, samples are then transferred to a membrane and Western blot analysis is conducted using an antibody to that protein. Batch preparations of immunoprecipitated proteins can also be used to identify novel protein–protein interactions. For such purification, proteins are metabolically labelled with [^{35}S]methionine before extraction, then used in immunoprecipitation. Following SDS–PAGE, the gel is exposed to autoradiography film and resulting bands eluted from the gel and prepared for protein sequencing.

Protocol 6

Identification of protein–protein interaction by immunoprecipitation

Equipment and reagents

- PBS (see *Protocol 5*)
- Lysis buffer: 50 mM Tris–HCl pH 8.0, 150 mM NaCl, 1.0% NP-40 (or a similar lysis buffer)
- Protein A–Sepharose or protein G–Sepharose beads

- 2 × SDS–PAGE loading buffer: 5% glycerol, 1.0% SDS, 2.5% 2-mercaptoethanol, 0.2% bromophenol blue, 32 mM Tris–HCl pH 6.8

Method

1 Wash adherent cultured cells once with PBS at room temperature. Collect suspension cells with centrifugation and wash once with PBS at room temperature.

2 Add 1 ml of lysis buffer per 100 mm plate or pellet, obtained from 10–20 ml of suspension cells, and incubate 15 min on an orbital shaker at low speed in a 4°C cold room. If a cold room is not available, lysis can be performed on ice for 15–30 min with occasional gentle shaking.

3 Transfer the cell lysate to a 1.5 ml microcentrifuge tube and pellet cellular debris with centrifugation for 10 min at 13 000 g at 4°C. Transfer the supernatant to a clean 1.5 ml microcentrifuge tube and determine the protein concentration on an aliquot of the cleared lysate.

4 Transfer 0.1–1.0 mg total sample protein to a new tube and add 1.0 µg of antibody. Mix the solution gently by inversion and incubate on ice for 30 min to 2 h.

5 Prepare a 10% slurry of protein A- or protein G–Sepharose beads in lysis buffer. Add 100 µl of this slurry to the cell extract and incubate at 4°C with gentle mixing for 1 h.

6 Collect the Sepharose beads with centrifugation at 10 000 g for 15–30 sec at 4°C. Remove as much of the lysis buffer as possible without disturbing the pellet.

7 Wash the pellet three times in 100 µl lysis buffer and remove as much of the buffer as possible following the final wash.

8 Estimate the pellet volume and add an equal amount of 2 × SDS–PAGE loading buffer. Heat the sample to 95°C for 5 min and vortex for 15 sec.

9 Briefly centrifuge the sample to clarify the solution and carefully remove the supernatant to load it in a well of a SDS–PAGE gel.

10 Resolve the proteins and perform Western blot analysis as described in Protocols 3 and 4.

5.2 Pull-down assays

Fusion proteins are also helpful in identifying suspected nuclear receptor–protein interactions (29). A nuclear receptor is inserted into a bacterial expression vector such that it is in-frame with sequence for another protein, such as glutathione-S-transferase (GST). Bacterial cell lysate containing the fusion protein is mixed with agarose or Sepharose beads conjugated to glutathione. The solution is pelleted to collect the beads that are then washed thoroughly. Because of its affinity for glutathione, the GST-fusion protein is retained in the pellet. A cell extract is then mixed with the beads. Proteins that bind the GST-fusion protein are retained while other proteins are eliminated by subsequent washes. Following the final wash, the pellet containing the beads, GST-fusion protein, and retained proteins are resuspended in 2 × loading buffer. The proteins are extracted as in *Protocol 6*, step 8 and analysed using SDS–PAGE.

Proteins can also be tagged with shorter antigenic peptides, such as the T7 Tag (Novagen), herpes simplex virus peptide (HSV tag; Novagen), or a His tag. The His tag is composed of a string of histidines (typically six), which has a strong affinity for metal ions such as nickel or cobalt. His-tagged proteins can be purified using immobilized metal affinity chromatography (IMAC). Cell lysate is passed over an IMAC column, and His-tagged protein is retained due to the interaction between the metal ions and the His tag. After thorough washing to remove other proteins, target cell lysate is passed over the column. Proteins that interact with the His-tagged nuclear receptor should be retained on the column. The column is again washed thoroughly and the protein complexes eluted using imidazole, a compound that acts as a competitor for binding sites on the IMAC column. Eluted proteins are analysed using SDS–PAGE.

5.3 Far Western analysis

Far Western analysis (30) is very similar to Western blot analysis in that total protein is resolved using SDS–PAGE and transferred to a membrane. However, immobilized protein is then renatured by successive incubations in buffers containing decreasing concentrations of urea or guanidine hydrochloride. In many cases, especially with small peptides, renaturation is not necessary. The membrane is blocked as in Western blot analysis and then incubated in buffer, containing a purified nuclear receptor, which can be directly labelled using a radioisotope or biotin. The blot is then washed to remove non-specific interactions. Nuclear receptor–protein interactions are detected by radiography if using radiolabelled probe, or an avidin-conjugated enzyme, if using a biotinylated probe. However, an antibody to the nuclear receptor probe is the most convenient means of detection. Far Western analysis can be modified and used to screen expression libraries for proteins that may interact with the GR or other nuclear receptors.

5.4 Yeast two-hybrid assay

Many transcription factors have defined and separable domains for DNA binding and transcriptional activation, with both domains necessary for transcriptional

(a)

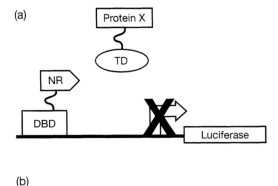

(b)

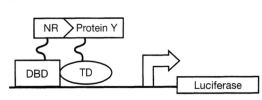

Figure 4 The two-hybrid system of identifying nuclear receptor–protein interaction. A region of a nuclear receptor (NR), thought to be involved in protein interactions, is fused to the DNA binding domain (DBD) of a transcription factor, while another protein (protein X or protein Y) is fused to a transactivating domain (TD) of a transcription factor. (a) In the absence of protein interaction, the DBD and TD remain separate and cannot induce transcription of the reporter gene. (b) Interactions between the fusion proteins bring the DBD and TD together, allowing transcription to occur.

activation. The yeast two-hybrid system (31) involves the creation of two hybrid proteins (*Figure 4*). The first hybrid contains the DNA binding domain of a transcription factor and the protein interaction domain of a nuclear receptor. The second hybrid contains the transcriptional activation domain of a transcription factor and protein X. Expression constructs containing the coding sequence for these two hybrids are co-transfected into yeast along with a reporter construct containing the specific binding sequence for the DNA binding domain of the first hybrid. If the two hybrid proteins interact, the necessary DNA binding and transcriptional activation domains are brought together and transcription of the reporter gene is increased. The strength of the interaction between the hybrid proteins is directly related to the increase in transcription. Two-hybrid systems are also useful in screening chemicals that can potentially enhance or disrupt DNA–protein or protein–protein interactions.

6 Conclusions

Because the actions of nuclear receptors are regulated on so many levels, it has been necessary to develop a wide variety of methods to study their function. While tried and tested methods such as RT-PCR and Western blot analysis still add much to our understanding of nuclear receptor function, new technologies such as the two-hybrid system and DNA microarray analysis have opened great new horizons of exploration. As these techniques lead to new discoveries, the volume of data will become prodigious. This is especially true when one considers the interactions of nuclear receptors with other signal transduction pathways that result in a complex integrative network. The processing of this information into a coherent, comprehensive representation of nuclear receptor action will rely heavily on computer modelling, and the field of informatics will become increasingly important. At this point, we will be able to understand better how nuclear receptor action contributes to complex biological, physiological, and pathological processes.

References

1. Chen, F., Watson, C. S., and Gametchu, B. (1999). *J. Cell Biochem.*, **74**, 418.
2. McCormick, J. A., Lyons, V., Jacobson, M. D., Noble, J., Diorio, J., Nyirenda, M., *et al.* (2000). *Mol. Endocrinol.*, **14**, 506.
3. Breslin, M. B. and Vedeckis, W. V. (2001). *Mol. Endocrinol.*, in press.
4. Weigel, N. L. (1996). *Biochem. J.*, **319**, 657.
5. Defranco, D. B. (2000). *Kidney Int.*, **57**, 1241.
6. Collingwood, T. N., Urnov, F. D., and Wolffe, A. P. (1999). *J. Mol. Endocrinol.*, **23**, 255.
7. Bamberger, C. M., Schulte, H. M., and Chrousos, G. P. (1996). *Endocr. Rev.*, **17**, 245.
8. Sambrook, I., Fritsch, E. F., and Maniatis, T. (ed.) (1989). *Molecular cloning: a laboratory manual* (2nd edn). Cold Spring Harbor Laboratory Press, NY.
9. Gossen, M. and Bujard, H. (1992). *Proc. Natl. Acad. Sci. USA*, **89**, 5547.
10. Hillen, W. and Berens, C. (1994). *Annu. Rev. Microbiol.*, **48**, 345.
11. Triezenberg, S. J., Kingsbury, R. C., and McKnight, S. L. (1988). *Genes Dev.*, **2**, 718.
12. Kain, S. R. and Ganguly, S. (1999). In *Current protocols in molecular biology* (ed. F. M. Ausubel, R. Brent, R. E. Kingston, D. D. Moore, J. G. Seidman, J. A. Smith, and K. Struhl), p. 9.6.3. John Wiley & Sons, Inc., Boston.
13. Vanderbilt, J. N., Miesfeld, R., Maler, B. A., and Yamamoto, K. R. (1987). *Mol. Endocrinol.*, **1**, 68.
14. Sapolsky, R. M., Romero, L. M., and Munck, A. U. (2000). *Endocr. Rev.*, **21**, 55.
15. Digman, J. D., Lebovitz, R. M., and Roeder, R. G. (1983). *Nucleic Acids Res.*, **11**, 1475.
16. Picard, D., Kumar, V., Chambon, P., and Yamamoto, K. R. (1990). *Cell Regul.*, **1**, 291.
17. Brenowitz, M., Senear, D. F., and Kingston, R. E. (1996). In *Current protocols in molecular biology* (ed. F. M. Ausubel, R. Brent, R. E. Kingston, D. D. Moore, J. G. Seidman, J. A. Smith, and K. Struhl), p. 12.4.1. John Wiley & Sons, Inc., Boston.
18. Maxam, A. and Gilbert, W. (1980). In *Methods in enzymology*, (ed. L. Grossman and K. Moldave. Vol. 65, p. 499. Academic Press, New York.
19. Prestridge, D. S. (1991). *Cabios*, **7**, 203.
20. Siebenlist, U. and Gilbert, W. (1980). *Proc. Natl. Acad. Sci. USA*, **77**, 122.
21. Gilman, M. (1993). In *Current protocols in molecular biology* (ed. F. M. Ausubel, R. Brent, R. E. Kingston, D. D. Moore, J. G. Seidman, J. A. Smith, and K. Struhl), p. 4.7.1. John Wiley & Sons, Inc., Boston.
22. Verma, I. M. (1977). *Biochim. Biophys. Acta*, **473**, 1.
23. Higuchi, R., Fockler, C., Dollinger, G., and Watson, R. (1993). *Biotechnology*, **11**, 1026.
24. Diatchenko, L., Chenchik, A., and Siebert, P. (1998). In *RT-PCR methods for gene cloning and analysis* (ed. P. Siebert and J. Larrick), p. 213. BioTechniques Books, Natick.
25. Liang, P. and Pardee, A. B. (1995). *Science*, **267**, 1186.

26. Schena, M., Shalon, D., Davis, R. W., and Brown, P. O. (1995). *Science*, **270**, 467.
27. Rosewell, D. F. and White, E. H. (1968). In *Methods in enzymology* (ed. M. A. DeLuca), Vol. 57, p. 409. Academic Press, New York.
28. Whitehead, T. P., Kricka, L. J., Carter, T. J., and Thorpe, G. H. (1979). *Clin. Chem.*, **25**, 1531.
29. Melcher, K. and Johnston, S. A. (1995). *Mol. Cell. Biol.*, **15**, 2839.
30. Edmondson, D. G. and Roth, S. Y. (1998). *Methods*, **15**, 355.
31. Gyuris, J., Golemis, E. A., Chertkov, H., and Brent, R. (1993). *Cell*, **75**, 791.

Chapter 12

In vivo gene transfer of antisense to receptors: methods for antisense delivery with adeno-associated virus

M. Ian Phillips, Keping Qian, and Dagmara Mohuczy
Department of Physiology, College of Medicine, Box 100274, University of Florida,
Gainesville, FL 32610-0274, USA.

1 Introduction

Ligand receptor binding is a critically important step in the process of many diseases. Over-production of a ligand can be as pathogenic as under-production or lack of a ligand and up-regulation of receptors can produce the same defect as over-production of a ligand. Likewise, down-regulation of receptors can result in the same effects as under-production of the ligand. For example, drugs used to treat hypertension, a chronic disease affecting millions, have included β-adrenergic receptor blockers, α-adrenergic receptor blockers and, more recently, angiotensin type 1 receptor (AT_1-R) blockers and Ca^{2+} channel blockers. All are aimed at reducing the ligand receptor binding interaction by drug receptor antagonism.

There are, however, limitations to this approach (1). Drugs that are based on small molecules are rarely, if ever, specific. At best, they are selective, meaning that while they are designed to interact with one receptor subtype, they can also interact with other subtypes of the same receptor. For example, β-blockers antagonize $β_1$- and $β_2$-adrenoceptors. $β_1$-adrenoceptors stimulate cardiac output and activate the renin angiotensin system (RAS) which leads to hypertension. However, stimulating $β_2$-adrenoceptors expands the airways of the lung. Therefore, $β_1$-adrenoceptor blockers, while effective as antihypertensive agents, cannot be used in patients prone to asthma because the non-specific inhibition of $β_2$-adrenoceptors in the lung will prevent airways from dilating. There are other problems too: drugs are short-lasting and require repeated dosage at least daily. They also have side-effects which might not be tolerated by the patient. Ideally, what is needed is a highly specific, long-lasting agent with few side-effects. We are approaching these objectives through the development of antisense inhibition of receptors or ligands (1, 2). By using a gene-based strategy, a high degree of specificity can be achieved. In practice, this strategy produces long-term inhibition of receptor function and, since the oligonucleotides do not cross the blood-brain barrier, side-effects that arise in the brain, such as impotence or insomnia, can be avoided (3).

The antisense approach to inhibiting receptors is quite different from the drug-antagonist strategy. Drugs compete with endogenous ligands for binding to receptors; this binding is saturable and is generally described by reversible dissociation kinetics. In contrast, antisense competes for the specific mRNA that translates the protein receptor, thereby reducing the total number of receptors in the membrane (1). That means that antisense inhibition does not produce total knockout of receptors, but reduces the total number which are physiologically active. It also means that antisense does not act as quickly as conventional drugs because synthesis inhibition is slower than receptor antagonism. On the other hand, antisense action outlasts drug action by many days and provides a stable and long-lasting reduction of ligand–receptor interaction; this is beneficial therapeutically.

Antisense inhibition has been developed, particularly in cell culture applications, to the point where it is being tested in clinical trials for treatment of HIV and cancer (4, 5). The pros and cons of its use have been reviewed elsewhere (6). Before 1992, however, antisense had not been applied *in vivo* with any success and there was much concern about the efficiency of cellular uptake of oligonucleotides. In 1992–3, three or four labs simultaneously and independently made and tested antisense oligodeoxynucleotides (AS-ODNs) in the brain (7–11). Cellular uptake was not a limiting factor in the central nervous system. The first demonstration of the potential use of antisense in the treatment of hypertension was antisense designed to inhibit angiotensinogen mRNA and angiotensin type-1 (AT$_1$) receptor mRNA (11).

AS-ODNs have many potentially attractive features as a new class of therapeutic agents. They are specific, effective, and have long duration (weeks) when given systemically with a lipid-protamine carrier in the appropriate ratio. To prolong the effect of antisense for months, DNA (partial or full-length) can be inserted in the antisense direction in viral vectors. This chapter discusses both approaches: first, the use of AS-ODNs and, secondly, the use of viral vectors with details of plasmid and recombinant adeno-associated virus vector (AAV) production.

2 Antisense oligodeoxynucleotides

2.1 Designing antisense molecules

Table 1 lists the characteristics of AS-ODNs that need to be incorporated into their design and use. The concept of antisense inhibition assumes that a short DNA sequence in the antisense direction binds to the specific mRNA of the target protein in the cytoplasm (12). These AS-ODNs are therefore targeted to the gene initiation codon (AUG), or part of the coding region downstream from it. Other

Table 1 Conditions for antisense oligonucleotide inhibition

- DNA sequence is specific and unique
- Uptake into cells is efficient
- Effect in cells is stable
- There is no non-specific binding to proteins
- Hybridization of the ODN is specific to the target mRNA
- The targeted protein and/or mRNA level is reduced
- The ODN is not toxic
- There is no inflammatory or immune response induced
- The ODN is effective compared to appropriate sense and mismatch ODN controls

approaches involve the formation of a triple helix with antisense constructed to the promoter region of a specified DNA.

AS-ODNs are short (15–20 bases) single strands of DNA. In contrast, longer or full-length DNA in the antisense direction is used in viral vectors. When designing antisense molecules, one has to consider two competing factors: the affinity of the ODN for its target sequence, which is dependent on the number and composition of complementary bases, and the availability of the target sequence, which is dependent on the folding of the mRNA molecule (13).

AS-ODNs targeted to different regions of the RNA have unequal efficiencies (14, 15). These differences may be related to the predicted secondary structure of the target mRNA (16–18) because folding of the mRNA affects target sequence availability. The RNA double helices, which are responsible for the secondary structure of the mRNA, incorporate a weaker G-U base pairing when they are next to A-U and G-C, and are generally short and rarely perfect. Therefore the design should avoid G-repeats. Burgess *et al.* (19) showed that when repeated G-sequences appear in the oligonucleotide, the effects produced can be due to non-antisense mechanisms. Proper testing of antisense requires a sense ODN and a mismatch ODN to serve as a control. An ODN that has strong (Watson–Crick) base pairing with 100% complementariness will form the more thermodynamically favourable structure with its target RNA (20).

AS-ODNs have several potential sites of action (*Figure 1*). AS-ODNs inhibit translation by hybridizing to the specific mRNA for which they are designed; this hybridization prevents either ribosomal assembly or ribosomal sliding along the mRNA. This kind of action assumes that AS-ODNs act in the cytosol and do not affect measurable mRNA levels. Indeed, there are several reports of antisense effects without any detectable change in target mRNA levels (9).

Alternatively, the mechanism can be by a reduction of mRNA. Decreased mRNA levels can occur by RNase H digestion of the RNA portion of the mRNA–antisense DNA hybrid. RNase H is found in the cytoplasm, as well as in the nucleus, and is normally involved in DNA duplication. The role of RNase H is to cleave RNA that has bound to DNA. Activation of RNase H is advantageous, because the enzyme leaves the AS-ODN intact and so it is free to hybridize with another mRNA, making the reaction catalytic rather than stoichiometric. Other

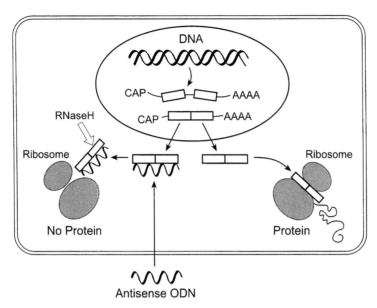

Figure 1 Antisense oligodeoxynucleotide (AS-ODN) inhibition. An AS-ODN is designed to hybridize to specific mRNA which will either prevent translation of the specific protein (by preventing read-through of mRNA in the ribosome), or stimulate RNase H which will destroy the mRNA hybridized to the AS-ODN. Either, or both, mechanisms will reduce levels of the specific protein.

possible antisense mechanisms of action have been proposed apart from inhibition of translation. Based on studies of cellular uptake of labelled ODNs, most AS-ODNs quickly migrate to the cell nucleus, suggesting an intranuclear site of action (21). Antisense DNA can hybridize to its target mRNA or pre-mRNA in the nucleus, forming a partially double-stranded structure which inhibits its transport out of the nucleus into the cytoplasm, thus preventing translation. AS-ODNs targeted to intron–exon junction sites prevent the splicing process and consequently the maturation of the transcript. As a result of this, antisense molecules might inhibit pre-RNA splicing or the transport of mRNA from the nucleus to the cytoplasm. An alternative anti-gene strategy is to target the DNA with triplex-forming ODNs to block DNA transcription. Effective AS-ODNs have been designed which target exon–intron splicing sites (22) or the major groove of the DNA (23) but triplex formation is corrected by DNA repair mechanisms.

Currently, the three regions that are considered to be the best targets for designing effective AS-ODNs are the 5′ cap region, the AUG translation initiation codon, and the 3′ untranslated region of the mRNA (24–26). Since most mRNAs have an AUG initiator codon site, targeting 12–15 of the neighbouring bases should produce inhibitory ODNs. All designed ODNs must be checked routinely with the GeneBank for existing sequences to avoid homology with other mRNAs. Essentially, antisense design initially involves trial and error testing in a model. A general rule suggested by our experience is that for a 15-mer oligo, three different sites need to be tested with the expectation that at least one will work.

Obviously, it is desirable to have a rapid screening test *in vitro* or *in vivo* for the specific protein that the AS-ODN has been designed to inhibit. Appropriate controls are sense ODN and mismatch (one or more nucleotides different from AS) or scrambled where the entire sequence is random.

2.2 Stability of oligonucleotides

Oligonucleotides in their natural form as phosphodiesters are subject to rapid degradation in the blood, intracellular fluid, and cerebrospinal fluid by exo- and endonucleases. The half-life of phosphodiester oligonucleotides in blood and tissue culture media is in the range of minutes. The half-life of ODNs is somewhat longer in cerebrospinal fluid, and intact ODNs can be detected 24 h after injection into the cerebral ventricles (26). Several chemical modifications have been proposed to prolong the half-life of ODNs in biological fluids and enhance uptake while retaining their activity and specificity.

The most widely used modified ODNs are phosphorothioates, where one of the oxygen atoms in the phosphodiester bond between nucleotides is replaced with a sulfur atom. These phosphorothioate ODNs have greater stability in biological fluids than normal oligodeoxynucleotides. The half-life of a 15-mer phosphorothioate ODN is 9 hours in human serum, 4 days in tissue culture media (27), and 19 h in cerebrospinal fluid (28). Phosphorothioate ODNs can be synthesized with automated DNA synthesizers but the product may require purification on an affinity gel.

One or more of the oxygen atoms in the phosphodiester bond can be replaced with a variety of other compounds, such as methyl groups (methylophosphoriate), alkyl phosphotriester, phosphoramidate, or boranophosphate all of which prolong the half-life of ODNs in *in vivo* experiments. Some new designs, in which the ODN has a dumbbell shape produced by a hairpin extension at the 3' end, are being attempted (29). It is hoped that these 3rd or 4th generation ODNs will provide longer-lasting stability, enhanced uptake kinetics, and affinity for the target (30).

2.3 Cellular uptake of oligonucleotides

In order to hybridize with the target mRNA, AS-ODNs have to cross cell membranes. Saturable uptake of ODNs occurs rapidly and reaches a plateau within 50 hours and, depending on the cells, the uptake can be efficient (31). Uptake is faster for shorter ODNs than for longer ones (31). Decreasing the temperature prevents oligonucleotide uptake, suggesting that it is an active uptake process. An 80 kDa ODN binding protein has been proposed to be the receptor molecule for ODN uptake (31). An efflux mechanism has also been described indicating temperature-dependent secretion of the ODNs from the cells to the extracellular space (27).

2.4 Pharmacology of AS-ODNs

Antisense inhibition resembles a drug–receptor interaction where the oligonucleotide is the drug and the target sequence is the receptor. For binding to occur between the two, a minimum affinity is required which is provided by hydrogen bonding between the Watson–Crick base pairs and base stacking with the double helix which is formed. In order to achieve pharmacological activity, a minimum number of 12–15 bases is required (32); longer sequences are more specific but, if they exceed 20 bases, problems of cell uptake begin to reduce their effectiveness.

One of the main advantages of antisense inhibition is the specificity of the AS-ODN target sequence interaction provided by the Watson–Crick base pairing. An oligonucleotide of 12–15 nucleotides is specific enough to be complementary to a single sequence (5). Increasing the length of the antisense should result in higher specificity, but it decreases uptake into cells (31). With viral vectors, however, the uptake problem is overcome because the virus freely enters cells by binding to viral receptors on cell membranes. Therefore, in a viral vector a full-length DNA antisense sequence can be used. The mechanism of action of antisense DNA is different from that of the AS-ODN. The antisense DNA produces an antisense mRNA which competes negatively with mRNA in the cytoplasm.

2.5 Toxicity

AS-ODNs can inhibit protein synthesis in cultured cells in the nanomolar concentration range. The therapeutic window for AS-ODNs is rather narrow (32): when testing for the optimal dose, small increments in the high nanomolar range should be tested (26). Higher concentrations may produce non-specific binding to cytosolic proteins and give misleading results.

Phosphodiester oligonucleotides are degraded relatively quickly to their naturally occurring nucleotides and so no toxic reaction is expected, even at high doses. Studies on phosphorothioated oligonucleotides in rats show that, following intravenous injection, phosphorothioated oligonucleotides are taken up from the plasma mainly by the liver, fat, and muscle tissues and appear in the urine within three days, mainly in their original form. There is an apparent mild increase in plasma LDH, and to a lesser extent indicators of a possible transient liver toxicity with very high doses of phosphorothioated oligonucleotides (33). Whole new classes of oligo backbone modifications are being developed to avoid the potential for liver toxicity of phosphorothioates in humans (30, 32).

2.6 Delivery of antisense

2.6.1 Naked DNA

Direct injection of the antisense DNA has been used in the experiments described below (33–36). For injections into the brain, naked DNA appears to be successful (34–36). In a number of studies, using different AS-ODNs, there has been an efficient uptake and effective reduction in protein synthesis and inhibition of the physiological parameters studied. In fact, uptake in the brain is so efficient that one difficulty with intracerebroventricular injections is that the DNA tends to be taken up close to the site of injection and does not spread to other parts of the brain. While this has little impact on hypertension therapy, it is an important consideration in antisense strategies for the treatment of brain diseases, such as Parkinson's disease, thalamic pain, Alzheimer's disease, and gliomas. In contrast to brain delivery, systemic injection of naked DNA in the periphery has not been shown to be effective: delivery of DNA in liposomes is necessary.

2.6.2 Liposomes

Liposomes, which are self-assembling particles of bilipid layers, have been used for encapsulating antisense ODN for delivery via the bloodstream. Antisense directed at angiotensinogen mRNA in cationic liposomes has had successful results in our laboratory (37). Tomita *et al.* (38) used liposome encapsulation of angiotensinogen antisense and a Sendai virus injected in the portal vein. Blood pressure decreased for several days, but the results were not compared with those for naked DNA. Wielbo *et al.* (2) compared liposome encapsulated antisense with naked DNA, given intra-arterially and found that only liposome encapsulation was effective. 24 h after injection of 50 mg liposome encapsulated antisense ODN, blood pressure decreased by 25 mm Hg. Neither empty liposomes nor liposomes that encapsulated scrambled ODN had any significant action. Unencapsulated antisense ODN also did not affect blood pressure. Confocal microscopy of rat liver tissue 1 h after intra-arterial injection of 50 mg unencapsulated FITC-labelled antisense or liposome-encapsulated FITC-labelled conjugated antisense, showed intense fluorescence in liver tissue sinusoids with the liposome-encapsulated ODN. Levels of protein (angiotensinogen and angiotensin II in the plasma) were significantly reduced in the liposome encapsulated ODN group. Antisense alone, lipids alone, and scrambled ODN in liposomes were all without effect on protein levels.

Liposome DNA complex, with cationic lipids, enables high transfection efficiency of plasmid DNA. Here, the short, single-stranded AS-ODNs are not actually encapsulated but instead are complexed with milamellar vesicles by electrostatic interactions. This simplifies the production of the antisense delivery system and enables a variety of routes of delivery including aerosol nasal sprays and parenteral injections.

3 Viral vectors for antisense DNA delivery

There are several viruses which have been tested for gene delivery. Each has its advantages, but none is an ideal viral vector. To be the perfect vector, a virus must fulfil all of the following criteria:

(a) The vector should be safe: either it cannot be a virus known to cause disease, or it has to be re-engineered to be harmless.

(b) The viral vector must not elicit an immune or inflammatory response.

(c) It must not integrate randomly into the genome, and thereby carry the risk of disrupting other cellular genes and mutagenesis.

(d) The virus also has to be replication-deficient to prevent it spreading to other tissues or infecting other individuals.

(e) An ideal vector would deliver a defined number of gene copies into each infected cell.

(f) The vector must be taken up efficiently by target tissue. The virus has to infect the target cells with high frequency in order to achieve the desired biological effect.

(g) The vector must be easy to manipulate and produce in pure form. The virus must be able to accommodate the gene of interest, along with its regulatory sequences, and the recombinant DNA has to be packaged with high efficiency into the viral capsid proteins.

3.1 Retroviruses

These have been used primarily because of their high efficiency in delivering genes to dividing cells (39). Retroviruses permit insertion and stable integration of single copy genes. Although effective in cell culture systems, they randomly integrate into the genome, which raises concerns about their safety for practical use *in vivo*. Retroviruses can act only in dividing cells which makes them ideal for tumour therapy but less desirable where other cells, that need to be protected, are dividing. In hypertension research they are being investigated in developing spontaneously hypertensive rats (SHR). Antisense DNA to the AT_{1b} receptor mRNA was injected in 5 day-old SHR (40). These rats did not become hypertensive in adult life. However, this has no possibility as a therapy. First, the injection was in the heart of neonatal animals; in humans there is no genetic marker to predict that a neonate will grow up to be hypertensive. Secondly, the vector delivered antisense to the brain because, in the neonate, the blood-brain barrier is still open. The effects of a retrovirus in the neonatal brain which is still developing is completely unknown and potentially damaging. Thirdly, the retrovirus passed into the germline and therefore passed on to future generations. So, for theoretical, practical, and ethical reasons, the use of retrovirus in these experiments on hypertension is unlikely to lead to any therapy.

3.2 Adenoviruses

Adenovirus vectors have been tested successfully in their natural host cells, the respiratory endothelia, as well as other tissues such as vascular smooth muscle, striated muscle, and brain (41–43). Adenovirus is a double-stranded DNA with 2700 distinct adenoviral gene products. The virus infects most mammalian cell types because most cells have membrane receptors. They enter the cell by a receptor-induced endocytosis and translocate to the nucleus. Most adenovirus vectors in their current form are episomal: i.e. they do not integrate into the host DNA. They provide high levels of expression but the episomal DNA will variably become inactive. In some species, e.g. mice, this may take a long time compared to their life span, but in humans it is a limitation of the virus as a vector. Repeated infections result in an inflammatory response with consequent tissue damage. This is because the adenovirus expresses genes which lead to immune cell attacks. This further limitation makes current recombinant adenovirus unsuitable for long-term treatment and several gene therapy trials using adeno-virus vectors have failed to produce acceptable results. Preliminary studies with adenovirus vectors for delivery of AT_1-R in mRNA antisense have been tested in rats and reduce the development of hypertension in SHR (44). The adenovirus is easy to produce and therefore useful for animal studies. As a vector it has too many limitations at present to be successful in human gene therapy. Indeed, clinical trials using the adenovirus had to be stopped in 1999 because of a fatality in a young man and the fatality may preclude adenovirus ever being used in humans. Further engineering of the adenovirus may eventually avoid these limitations.

3.3 Adeno-associated virus (AAV)

The AAV has gained attention because of its safety and efficiency (45). It has been used successfully to deliver antisense RNA against α-globin (46) and HIV LTR (long terminal repeat) (47), and it is our vector of choice for delivering antisense targeted to the AT_1-receptor in hypertensive rat models.

AAV is a parvovirus, discovered as a contaminant of adenoviral stocks. It is a ubiquitous virus (antibodies are present in 85% of the US population), which has not been linked to any disease. It is also classified as a dependovirus, because its replication is dependent on the presence of a helper virus, such as adenovirus or herpes virus. Five serotypes have been isolated, of which AAV-2 is the best characterized. AAV has a single-stranded linear DNA which is encapsidated into capsid proteins VP1, VP2, and VP3 to form an icosahedral virion of 20–24 nm in diameter (45).

The AAV DNA is approximately 4.7 kilobases long. It contains two open reading frames and is flanked by two inverted terminal repeats (ITR). There are two major genes in the AAV genome: *rep* and *cap*. *Rep* encodes proteins respon-sible for viral replication, whereas *cap* encodes capsid proteins VP1–3; each ITR forms a T-shaped hairpin structure. These terminal repeats are the only essential *cis*-components of the AAV for chromosomal integration, and so the AAV can be used as a vector with all viral coding sequences removed and replaced by the cassette of genes for delivery. Three viral promoters have been identified and named p5, p19, and p40, according to their map position. Transcription from p5 and p19 results in production of rep proteins, and transcription from p40 pro-duces the capsid proteins (45). For more powerful expression we have inserted a cytomegalovirus promoter. Other promoters are being tested which are specific to certain cells, including arginine vasopressin (AVP) promoter for cells synthesiz-ing AVP, neuron-specific enolase (NSE), and glial fibrillary acid protein (GFAP).

Upon infection of a human cell, the wild-type AAV (wtAAV) integrates to the q arm of chromosome 19 (48, 49). Although chromosomal integration requires the terminal repeats, the viral components responsible for site-specific integra-tion have been targeted recently to the rep proteins (50). With no helper virus present, AAV infection remains latent indefinitely. Upon superinfection of the cell with helper virus, the AAV genome is excised, replicated, packaged into virions, and released into the extracellular fluid. This process is the basis of recombinant AAV (rAAV) production for research.

Several factors have prompted researchers to study the possibility of using recombinant AAV as an expression vector. One is that the requirement for delivering a gene is surprisingly modest: all that is necessary is the 145 bp inverted terminal repeats, which comprise only 6% of the AAV genome. This leaves room in the vector to assemble an insertion of up to 4.4 kb DNA. While this carrying capacity may limit the AAV to delivering large genes, it is amply suited to

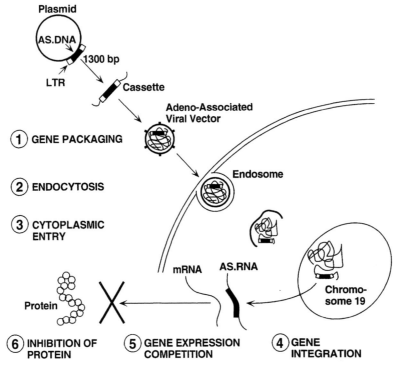

Figure 2 Gene delivery of antisense with adeno-associated viral vector (AAV). 1. **Gene packaging**: The plasmid containing the terminal repeats (TR) characteristic of AAV with cDNA subcloned in the antisense direction. The final cassette also contains promoters such as CMV and TK promoter. The packaging cell line is transfected with AAV-based plasmid, helper plasmid with *rep* and *cap* genes, and transduced with adenovirus as a helper virus. 2. **Endocytosis**: The viral vector fuses with the cell membrane by binding to adhesion molecules and becomes an endosome within the bilipid layer. 3. **Cytoplasmic entry**: The vesicle opens in the cytoplasm releasing the vector which is transported to, and enters, the nucleus. 4. **Gene integration**: AAV integrates with chromosome 19. It is not known if the addition of foreign DNA interferes with this integration. 5. **Gene expression competition**: Genomic message in chromosome 19 produces an antisense RNA. This competes with the natural mRNA and prevents it from producing its product. 6. **Inhibition of protein**: A binding of antisense RNA to the mRNA prevents translation through the ribosomal assembly to produce protein. The result of this gene delivery system should be a reduction of the protein specifically targeted by the antisense DNA.

delivering small genes and antisense cDNA. It is also sufficient to ensure a specific response, a potent promoter, and a selective marker such as a neomycin resistance gene (*neo*^r).

The second characteristic that makes AAV a good vector candidate is its safety. There is a relatively complicated rescue mechanism. Not only adenovirus (wild-type) but also AAV genes are required to mobilize the rAAV and the spread of recombinant AAV vectors to non-target areas can be limited to certain tissues. AAV is not pathogenic and not associated with disease. The removal of viral coding sequences in producing an +AAV minimizes immune reactions to viral gene expression, and therefore rAAV does not evoke an inflammatory response (in contrast to the recombinant adenovirus). Finally, AAV is a good candidate for gene therapy because it has a broad host range. AAV appears to infect all mammalian tissues except vascular endothelial cells (51) and it remains intact for long periods of time.

The limitation of AAV lies in its production. Although it can be purified and concentrated, which are advantages, it also has to be rendered free of adenovirus and therefore production is more complicated than for other vectors.

Overall, the advantages, particularly its safety, make AAV currently one of the most promising candidates for delivery of genes for long-term therapy. Recently, Flotte and colleagues (52) have established gene therapy Phase I trials for cystic fibrosis using AAV gene delivery in patients.

4 Methods for antisense delivery by AAV

The general concept for antisense gene delivery in the AAV vector and the steps involved are shown in *Figure 2*. To illustrate these steps, a brief description is given that is applicable to hypertension. Further details are presented in ref. 60.

4.1 Construction of plasmids

After subcloning the target gene into AAV-based vector, highly purified plasmid is needed for virus packaging. To reach this goal, we recommend the following protocol (*Protocol 1*).

Protocol 1

Large scale plasmid preparation

Equipment and reagents

- Optiseal centrifuge tube (4.9 ml) (Beckman Instruments, Inc.)
- Ultracentrifuge rotor NVT 90 (Beckman Instruments, Inc.)
- LB (Luria-Bertani) media: 10 g Bacto tryptone, 5 g Bacto yeast extract, 10 g NaCl, in 1000 ml of deionized H_2O pH 7.0
- LB media with 100 µg/ml ampicillin
- Lysozyme buffer: 25 mM Tris–HCl [Tris (hydroxymethyl) aminomethane] pH 7.5, 10 mM EDTA, 15% sucrose or glucose
- Lysozyme: 12 mg/ml in lysozyme buffer
- 0.2 M sodium hydroxide (NaOH), 1% SDS (sodium dodecyl sulfate)

Protocol 1 continued

- 3 M sodium acetate, pH 4.8–5.2
- 40% PEG 8000 (polyethylene glycol 8000)
- 5.5 M lithium chloride
- Phenol/chloroform/isoamyl alcohol (25:24:1, by vol.)
- 1 × Tris-EDTA (TE) buffer: 10 mM Tris–HCl pH 7.4, 1 mM EDTA pH 8.0
- Cesium chloride (CsCl)
- 10 mg/ml ethidium bromide

Method

1 Grow bacteria containing your plasmid in 1 litre of LB media with 100 µg/ml ampicillin.

2 Pellet bacteria at 3000 g at 4°C for 15 min.

3 Resuspend bacteria in 20 ml of lysozyme buffer.

4 Add 4 ml of lysozyme (12 mg/ml in lysozyme buffer), mix.

5 Incubate on ice for 5 min until the mixture becomes viscous.

6 Add 48 ml of 0.2 M NaOH, 1% SDS. Mix with a glass pipette, using it as a rod.

7 Incubate on ice for 5–10 min.

8 Add 36 ml of 3 M NaAc, pH 4.8–5.2, mix with the same pipette.

9 Add 0.2 ml of chloroform, mix.

10 Incubate on ice for 20 min.

11 Spin at 3000 g at 4°C for 20 min.

12 Transfer supernatant into a fresh bottle.

13 Add 33 ml of 40% PEG 8000, mix, and incubate on ice for 10 min.

14 Spin at 14 000 g at 4°C for 10 min.

15 Discard supernatant, dissolve pellet in 10 ml of sterile water, then add 10 ml of 5.5 M LiCl.

16 Incubate on ice for 10 min.

17 Spin at 14 000 g at 4°C for 10 min.

18 Save supernatant, transfer equal amounts (each 10 ml) into two 30 ml Corex (or plastic) tubes, add 6 ml of isopropanol to each tube, mix.

19 Incubate at room temperature (RT) for 10 min.

20 Spin at 10 000 g at RT for 10 min.

21 Dissolve each pellet in 3.7 ml of 1 × TE.

22 Add 4.2 g of CsCl to each tube, mix and dissolve, add 0.24 ml of ethidium bromide (10 mg/ml), mix.

23 Transfer the solution in each tube into a 4.9 ml Optiseal centrifuge tube (Beckman, Cat. No. 362185).

24 Spin in NVT90 rotor (Beckman) at 485 000 g, 15°C for 4 h. Three bands should be apparent: the upper one contains protein, the middle one is nicked and linear DNA, and the lowest one is closed circular plasmid DNA.

Protocol 1 continued

25 Carefully remove the plasmid band using a 3 ml syringe with an 18 gauge needle and transfer it to an Eppendorf tube (two tubes, each about 0.5 ml).

26 Extract three to four times with equal volume of isoamyl alcohol, discarding the organic phase (top layer) every time until all pink colour is removed.

27 Transfer into one 30 ml Corex (plastic) tube, add 2.5 volume (2.5 ml) of H_2O, mix, and add 2 of the combined volumes of ethanol (7 ml).

28 Incubate on ice for 30 min.

29 Pellet plasmid DNA at 12 000 g for 15 min at 4°C.

30 Discard supernatant, dissolve pellet in 500 μl of 1 × TE.

31 Transfer to an Eppendorf tube, add 500 μl of phenol/chloroform/isoamyl alcohol, vortex, spin for 2 min at top speed in a microcentrifuge.

32 Save the top aqueous layer, transfer into a fresh Eppendorf tube.

33 Add 50 μl of 3 M sodium acetate buffer, pH 4.8–5.2, mix, add 1 ml ethanol, mix.

34 Pellet the plasmid DNA at top speed in microcentrifuge for 5 min at 4°C.

35 Discard the supernatant, wash the pellet with 75% ethanol, vacuum dry.

36 Dissolve the pellet in 1 ml of 1 × TE.

We constructed plasmids for both AT_1 receptor antisense (paAT$_1$) and angiotensinogen antisense (paAo) in the AAV-derived expression vector. Initially we used a plasmid containing AAV genome and 750 bp cDNA inserted into the AAV in the antisense direction downstream from the AAV promoter. The NG108-15 cells or hepatoma H4 cells were transfected with AAV-based plasmid (PaAT$_1$) or PaAo, respectively, using lipofectamine (53, 54). In both cases, there was a significant reduction in the appropriate protein, namely AT_1 receptor and angiotensinogen. To test that the cells expressed AAV, we used the rep gene product as a marker. Immunocytochemical staining with a rep protein antibody showed that the majority of cells in culture fully expressed the vector. A further development of the AAV was the insertion of promoters, more powerful and specific than the p40 promoter. AAV with CMV promoter and *neo*[r] gene as a selectable marker is now being used in our current experiments. The AAV cassette contains either 750 bp of rat AT_1 cDNA in the antisense direction (*Figure 3*), or markers, green fluorescent protein (*gfp*) gene (63) or *lacZ* gene (56). The NG108-15 cells transfected with AAV plasmid containing the *gfp* and *neo*[r] genes were selected by antibiotic, G418 (600 μg/ml), and the selected clones viewed for *gfp* expression. Two weeks after transfection all cells were expressing *gfp*: transfection efficiency of this pAAV-*gfp* in different cell lines including ATt20 (mouse pituitary cells), L929 (mouse fibroblasts), 239 (human embryonic kidney cells), and NG108-15 (neuroblastoma-glioma cells) is over 50% (57).

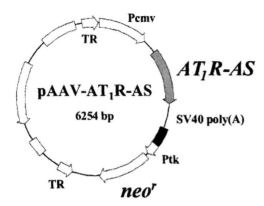

Figure 3 Schematic diagrams of recombinant AAV vectors (pAAV). In pAAV-*gfp* vector almost all of the parental wAAV genome has been deleted, except the terminal repeats, and replaced with *gfp*, driven by a CMV promoter, and *neo*[r] genes. In pAAV-AS vector the *gfp* gene is replaced by a 750 bp fragment of the AT$_1$-R gene with cDNA in antisense orientation. TR indicates AAV terminal repeat; Pcmv, human CMV early promoter; *gfp*-h, *A. victoria* green fluorescent protein gene (humanized); Ptk, thymidine kinase promoter; and *neo*[r], neomycin phosphotransferase gene from Tn5.

4.2 Preparation of pAAV-AT$_1$R-AS

The 749 bp fragment of the angiotensin receptor cDNA (–183 to 566) was amplified using polymerase chain reaction (PCR) and ligated to an AAV-derived vector in the antisense orientation, in place of *gfp*. The resulting plasmid vector (pAAV-AT$_1$R-AS) contained adeno-associated virus terminal repeats (TR), a cytomegalovirus promoter (Pcmv), the DNA encoding AT$_1$ receptor mRNA in the antisense direction, and a neomycin resistance gene (*neo*[r]) (*Figure 3*). The plasmid DNA was purified on a CsCl gradient.

pAAV-AT$_1$R-AS was tested for AT$_1$ receptor inhibition *in vitro*, using NG108-15 (58) and vascular smooth muscle cells (59). The cells had a significant ($p < 0.01$) decrease in angiotensin II AT$_1$ receptor number compared with the control cells: there were no effects on AT$_2$ receptors.

AAV-based plasmids are used to prepare recombinant virus. The method has been developed by Zolotukin *et al.* (60) and is used in the University of Florida Gene Therapy Center.

4.3 Method to prepare recombinant adeno-associated virus (rAAV)

To prepare recombinant AAV (rAAV), human embryonic kidney (HEK) 293 cells are transfected with plasmid vector containing the gene of interest in antisense orientation and AAV terminal repeats (pAAV-AS), together with helper plasmid delivering *rep* and *cap* genes (necessary for AAV replication) in *trans* using the calcium phosphate method. 8 hours after transfection, adenovirus is added at multiplicity of infection (MOI) of 5 (*Protocol 2a*). Alternatively, rAAV vector can be co-transfected with the helper plasmid, pDG, carrying both the AAV genes (*rep* and *cap*), and only the adenovirus genes required for rAAV replication and packaging (*Protocol 2b*) (ref. 60).

Protocol 2a

Production of rAAV by calcium phosphate transfection

Equipment and reagents

- Tissue culture plates
- 293 tissue culture cell line (ATCC)
- Medium: DMEM (Dulbecco's modified Eagle's medium)
- Heat inactivated FBS (fetal bovine serum)
- Penicillin (5000 U/ml), streptomycin (5000 μg/ml)
- AAV helper plasmid (such as ATCC No. 68066)

- Adenovirus (Ad5) stock with known titre
- 2 M calcium chloride (CaCl$_2$)
- 0.1 × TE buffer (1 × TE shown in *Protocol 1*)
- 2 × HeBS (Hepes-buffered saline) pH 7.05: 16.4 g NaCl, 11.9 g Hepes acid, 0.21 g Na$_2$HPO$_4$, 800 ml H$_2$O, pH to 7.05 with 5 M NaOH, add H$_2$O to 1 litre. Filter sterilize through a 0.45 μm nitrocellulose filter, test for transfection efficiency, and store at −20°C in 12.5 ml aliquots.

Method

1. Split low passage number HEK 293 cells grown in DMEM with 10% heat inactivated FBS and penicillin/streptomycin into 15 cm cell culture plates, so they are about 70% confluent on the next day.

2. Mix (per 1 plate): 20 μg of plasmid (containing gene of interest), 20 μg of helper plasmid (containing *rep* and *cap* genes), in a final volume 876 μl of 0.1 × TE.

3. Add 1 ml of 2 × HeBS, mix.

4. Add 124 μl of 2 M CaCl$_2$ by drops, mixing.

5. Incubate 15–20 min at room temperature to form precipitate, mix once by pipetting.

6. Add 2 ml mixture to the plate with cells, swirl gently, return cells to the incubator.

7. 8 h after transfection, change medium to a fresh one and add adenovirus at molecules of infection (MOI) of 5.

Protocol 2b

Production of rAAV by modified AAV helper plasmid

Equipment and reagents

- Tissue culture plates
- 293 tissue culture cell line (ATCC)
- DMEM medium
- Heat inactivated FBS (fetal bovine serum)

- Modified helper plasmid, pDG (ref. 68)
- 2.5 M CaCl$_2$
- 2 × HeBS: see *Protocol 2a*

Protocol 2b continued

Method

1 Split low passage number HEK 293 cells into 1:2 the day before the experiment, check cell batch for confluency, 75% is ideal.

2 Thaw 2 × HeBS and keep at 37 °C until ready to use.

3 Pre-warm media (DMEM/10% FBS).

4 Prepare transfection mixture:

   (a) Calculate how much input DNA (ml) to add. Input DNA (pDG + rAAV vector plasmid) needed: 60 μg/plate (45 μg pDG: 15 μg rAAV vector plasmid).

   (b) Calculate volume (X) of total input DNA for one cell batch (ten 15 cm plates) in ml.

   (c) To a 50 ml conical tube add (in order): (11.5 − X) ml H_2O, X ml input DNA, 1.25 ml of 2.5 M $CaCl_2$, and 12.5 ml 2 × HeBS. Mix the transfection mixture well.

5 Let the mixture incubate for 1 min to precipitate.

6 Transfer the mixture into pre-warmed 200 ml of DMEM/10% FBS.

7 Dispense 22 ml of medium from step 6 immediately into each plate (discard old medium from each plate before adding 2 × HeBS in step 4c).

8 Equalize and incubate at 37 °C (two to three days).

The transfected cells are harvested after three days. The rAAV is usually purified by conventional CsCl gradient centrifugation. However, it requires up to two weeks to complete and often results in poor recovery and poor quality virus. There are novel rAAV purification strategies that involve the use of non-ionic iodixanol gradients followed by heparin affinity chromatography (*Protocol 3*). The method results in more than 50% recovery of rAAV from a crude lysate and routinely produces virus that is more than 99% pure. More importantly, the new purification procedures consistently produce rAAV stocks with particle-infectivity ratios of less than 100, which is significantly better than conventional methods, such as CsCl purification protocol. The new protocols increase the yield of infectious rAAV by at least tenfold (see also *Protocol 4*) and allows for the complete purification of rAAV in one working day.

rAAV-AT$_1$R-AS was tested for AT$_1$ receptor inhibition *in vitro* using vascular smooth muscle cells (61). Transduced cells, without G418 selection, expressed the transgene for at least 8 weeks, had a lower number of AT$_1$ receptors, and reduced Ca^{2+} response to angiotensin II stimulation.

Expression *in vivo* was tested by direct injection into the brain. An AAV with an arginine vasopressin promoter (AVP) to drive a *lacZ* gene was constructed. The vector expressed galactosidase in neurons of the paraventricular nucleus and supraoptic nucleus and was found in magnocellular cells which normally express AVP (64). The expression was observed at one day, one week, and one month with no diminution of signal. This is an example of how AAV can be developed for specific tissue and/or cell gene expression and shows that AAV vectors can deliver foreign genes into adult brain for long periods of time.

Protocol 3

Iodixanol density gradient followed by heparin affinity chromatography

Equipment and reagents

- Quick-Seal Ultra-Clear 25 × 89 mm centrifuge tube (Beckman)
- Syringe with a 1.27 × 89 mm spinal needle
- A variable speed peristaltic pump, Model EP-1 (Bio-Rad)
- Type 70 Ti rotor (Beckman)
- Syringe with 18 gauge needle
- 2.5 ml heparin–agarose type I column (Sigma)
- 5% phenol red: 2.5 µl of a 0.5% stock solution/per ml of the iodixanol solution

- Iodixanol (5,5'[(2-hydroxy-1-3-propanediyl)-bis(acetylamino)] bis [N,N'-bis(2,3-dihydroxypropyl-2,4,6-triiodo-1.3-benzenecarboxamidel], prepared using a 60% (w/v) sterile solution of OptiPrep (Nycomed)
- 1 × PBS (phosphate-buffered saline): 137 mM NaCl, 0.7 mM KCl, 4.3 mM Na$_2$HPO$_4$, 1.4 mM KH$_2$PO$_4$ pH ~ 7.3
- PBS-MK buffer (phosphate-buffered saline-Mg^{2+}-K$^+$): 1 × PBS, 1 mM MgCl$_2$, 2.5 mM KCl

Method

1 Transfer 15 ml of the clarified lysate into Quick-Seal Ultra-Clear 25 × 89 mm centrifuge tubes (Beckman) using a syringe with a 1.27 × 89 mm spinal needle: avoid bubbles.

2 Using a variable speed peristaltic pump, Model EP-1 (Bio-Rad) to underlay in order:
   (a) 9 ml of 15% iodixanol and 1 M NaCl in PBS-MK.
   (b) 6 ml of 25% iodixanol in PBS-MK buffer containing phenol red.
   (c) 5 ml of 40% iodixanol in PBK-MK buffer.
   (d) 5 ml of 60% iodixanol in PBK-MK buffer containing phenol red.

3 Seal the tubes.

4 Centrifuge in a Type 70 Ti rotor (Beckman) at 350 000 g for 1 h at 18 °C.

5 Puncture the tubes on the side with a syringe equipped with an 18 gauge needle with the bevel uppermost and collect a total of about 4 ml of the clear 40% fraction.

6 Equilibrate a column (pre-packed 2.5 ml heparin–agarose type I) with 20 ml of PBS-MK buffer under gravity.

7 Apply the rAAV iodixanol fraction (from step 5) to the pre-equilibrated column under gravity.

8 Wash the column with 10 ml of PBS-MK buffer under gravity.

9 Elute rAAV with PBS-MK buffer containing 1 M NaCl under gravity, discard the first 2 ml of the eluant.

Protocol 3 continued

10 Collect the virus in the subsequent 3.5 ml of elution buffer.

11 Concentrate and desalt virus centrifugation through the BIOMAX 100 K filter (Millipore).

Protocol 4

Virus titre assay

Equipment and reagents

- Vacuum filter apparatus
- 0.45 μm nylon filter
- X-ray film and exposure cassette
- 293 tissue culture cell line (ATCC)
- Adenovirus (Ad5) stock with known titre
- wtAAV (wild-type AAV) stock with known titre
- Trypsin/EDTA

- 1 × PBS: see *Protocol 3*
- 0.5 M NaOH, 1.5 M NaCl
- 20 × SSC (sodium chloride/sodium citrate): 3 M NaCl, 0.3 M Na$_3$citrate.2H$_2$O pH 7.0
- 1 M Tris–HCl pH 7.0/2 × SSC
- Random prime labelling system (Amersham Pharmacia Biotech)
- ^{32}P-labelled probe specific for the gene of interest

Method

Titre of the virus was calculated using HEK 293 cells or PC12w, wtAAV, and adenovirus.

1 Seed 5×10^4 cells in each well of a 24-well plate and incubate 24 h.

2 Infect all but one well with adenovirus at MOI of 20.

3 Infect all but one well (different one) with wtAAV at MOI of 4.

4 Make serial 5 × dilutions of the rAAV and infect 8–10 wells, which already contain adenovirus and wtAAV. Infect cells (with no adenovirus or no wtAAV) with undiluted rAAV. Incubate cells 24 h.

5 Spin down the medium, discard supernatant, combine pellet with pre-washed and trypsinized cells (100 μl trypsin/EDTA per well), add 10 ml of 1 × PBS, and disperse into single cell suspension.

6 Transfer cell suspension onto a nylon filter pre-soaked in PBS and apply low vacuum.

7 Place filter on 3MM Whatman paper soaked in 0.5 M NaOH with 1.5 M NaCl for 5 min at room temperature.

8 Transfer filter to the top of Whatman paper soaked in 1 M Tris–HCl pH 7.0 with 2 × SSC for 5 min at room temperature.

9 Air dry the filter.

10 Hybridize the filter with ^{32}P-labelled probe specific for your gene of interest (random primed according to the kit instruction).

11 Expose the filter to X-ray at $-80\,^\circ$C. Count the spots and multiply by dilution factor for each well.

To test for effectiveness *in vivo*, rAAV-AT$_1$R-AS was microinfused into the lateral ventricles of adult male SHR. Control rats received AAV with *gfp* reporter gene but without the AS gene (mock vector) in a vehicle of artificial cerebro-spinal fluid. Blood pressure was measured by tailcuff method. There was a significant decrease in systolic blood pressure (SBP) in rats which received the rAAV-AS vector, but no effect was observed in the controls. SBP decreased by 23 $\pm$ 2 mmHg in the first week after administration. This fall was prolonged in four rats for nine weeks; this was considerably longer than the longest effect observed with AS-ODN. No reduction in blood pressure was seen in the controls. rAAV-*gfp* expression in hypothalamus of controls was detectable by RT-nested PCR 11 months after injection. Further, intracardiac injection of rAAV-AS in SHR, significantly reduced blood pressure and slowed the development of hyper-tension for several weeks (61). This result demonstrates that rAAV-AS in a single application, is effective in reducing hypertension chronically. It encourages further research on gene regulation in hypertension, and exploration of the most effective routes of delivery applicable to humans.

Acknowledgements

This work has been supported by NIH grant HL 39154. Thanks to Serge Zolotukhin, Nick Muzyczka, and their colleagues who have stimulated and guided us in the development of these methods and who have established the AAV as a safe, stable, efficient vector.

References

1. Phillips, M. I. (1997). *Hypertension*, **29**, 177.
2. Wielbo, D., Simon, A., Phillips, M. I., and Toffolo, S. (1996). *Hypertension*, **28**, 147.
3. Zhang, Y. C., Kimura, B., Shen, L., and Phillips, M. I. (2000). *Hypertension*, **35**, 219. Hoffman, W. E., Dietz, R., Schelling, P., and Ganten, D. (1977). *Nature*, **270**, 445.
4. Wagner, R. W. (1994). *Nature*, **372**, 333.
5. Crooke, S. T. (1992). *Annu. Rev. Pharmacol. Toxicol.*, **32**, 329.
6. Stein, C. A. and Cheng, Y.-C. (1993). *Science*, **261**, 1004.
7. Chiasson, B. J., Hooper, M. L., Murphy, P. R., and Robertson, H. A. (1992). *Eur. J. Pharmacol.*, **277**, 451.
8. Wahlestedt, C., Pich, E. M., Koob, G. F., Yee, F., and Heilig, M. (1993). *Science*, **259**, 528.
9. Wahlestedt, C., Golanov, E., Yamamoto, S., Yee, F., Ericson, H., Yoo, H., *et al.* (1993). *Nature*, **363**, 260.

10. McCarthy, M. M., Masters, D. B., Rimvall, K., Schwartz-Giblin, S., and Pfaff, D. W. (1994). *Brain Res.*, **636**, 209.

11. Gyurko, R., Wielbo, D., and Phillips, M. I. (1993). *Reg. Pep.*, **49**, 167.

12. Simons, R. W. (1988). *Gene*, **72**, 35.

13. Stull, R. A., Taylor, L. A., and Szoka, F. C. (1992). *Nucleic Acids Res.*, **20**, 3501.

14. Cowsert, L. M., Fox, M. C., Zon, G., and Mirabelli, C. K. (1993). *Antimicrob. Agents Chemother.*, **37**, 171.

15. Wakita, T. and Wands, J. R. (1994). *J. Biol. Chem.*, **269**, 14205.

16. Lima, W. F., Monia, B. P., Ecker, D. J., and Freier, S. M. (1992). *Biochemistry*, **31**, 12055.

17. Rittner, K. and Sczakiel, G. (1991). *Nucleic Acids Res.*, **19**, 1421.

18. Jaroszevski, J. W., Syi, J. L., Ghosh, M., Ghosh, K., and Cohen, J. S. (1993). *Antisense Res. Dev.*, **3**, 339.

19. Burgess, T. and Farrell, C. (1995). *Proc. Natl. Acad. Sci. USA*, **92**, 4051.

20. Singer, M. and Berg, P. (1992). In *Genes and genomes*, p. 54. University Science Books, Mill Valley, CA.

21. Iversen, P. L., Zhu, S., Meyer, A., and Zon, G. (1992). *Antisense Res. Dev.*, **2**, 211.

22. Colige, A., Sokolov, B. P., Nugent, P., Baserge, R., and Prockop, D. J. (1993). *Biochemistry*, **32**, 7.

23. Helene, C. (1991). *Anticancer Drug Res.*, **6**, 569.

24. Bennett, C. F., Condon, T. P., Grimm, S., Chan, H., and Chiang, M. Y. (1994). *J. Immunol.*, **152**, 3530.

25. Bacon, T. A. and Wickstrom, E. (1991). *Oncogene Res.*, **6**, 13.

26. Wahlestedt, C. (1994). *TIPS*, **15**, 42.

27. Li, B., Hughes, J. A., and Phillips, M. I. (1996). *Neurochem. Intl.*, **31**, 393.

28. Campbell, J. M., Bacon, T. A., and Wickstrom, E. (1990). *J. Biochem. Biophys. Methods*, **20**, 259.

29. Takaku, H. (1996). *Neurotide*, **15**, 519.

30. Stein, C. A. (1996). *Chem. Biol.*, **3**, 319.

31. Loke, S. L., Stein, C. A., Zhang, X. H., Mori, K., Nakanishi, M., Subasinghe, C., *et al.* (1989). *Proc. Natl. Acad. Sci. USA*, **86**, 3474.

32. Wagner, R. W., Matteucci, M. D., Lewis, J. G., Gutierrez, A. J., Moulds, C., and Froehler, B. C. (1993). *Science*, **260**, 1510.

33. Iversen, P. L., Mata, J., Tracewell, W. G., and Zon, G. (1994). *Antisense Res. Dev.*, **4**, 43.

34. Phillips, M. I., Wielbo, D., and Gyurko, R. (1994). *Kidney Intl.*, **46**, 1554.

35. Ambuhl, P., Gyurko, R., and Phillips, M. I. (1995). *Reg. Pep.*, **59**, 171.

36. Wielbo, D., Sernia, C., Gyurko, R., and Phillips, M. I. (1995). *Hypertension*, **25**, 314.

37. Zhang, Y. C., Bui, J. D., Shen, L., and Phillips, M. I. (2000). *Circulation*, **10**, 682.

38. Tomita, N., Morishita, R., Higaki, J., Kaneda, Y., Mikami, H., and Ogihara, T. (1994). *Hypertension*, **24**, 397.

39. Mulligan, R. C. (1993). *Science*, **260**, 926.

40. Katovich, M. J., Lu, D., Lyer, S., and Raizada, M. K. (1996). *Exp. Biol.*, Abstract 1588.

41. Brody, S. L., Jaffe, H. A., Eissa, N. T., and Daniel, C. (1994). *Nature Genet.*, **1**, 42.

42. Quantin, B., Perricaudet, L. D., Tajbakhsh, S., and Mandel, J.-L. (1992). *Proc. Natl. Acad. Sci USA*, **89**, 2581.

43. Le Gal La Salle, G., Robert, J. J., Berrard, S., Ridoux, V., Stratford-Perricaudet, L. D., Perricaudet, M., *et al.* (1993). *Science*, **259**, 988.

44. Lu, D., Yu, K., and Raizada, M. K. (1995). *Proc. Natl. Acad. Sci. USA*, **92**, 1162.

45. Muzyczka, N. and McLaughin, S. (1988). In *Current communications in molecular biology: viral vectors* (ed. Y. Gluzman and S. H. Hughes), p. 39. Cold Spring Harbor Laboratory Press, Cold Spring Harbor, NY.

46. Ponnazhagan, S., Nallari, M. L., and Srivastava, A. (1994). *J. Exp. Med.*, **179**, 733.

47. Chatterjee, S., Johnson, P. R., and Wong, K. K. (1992). *Science*, **258**, 1485.

48. Muzyczka, N. (1992). In *Current topics in microbiology and immunology*, Vol. 158, p. 97. Springer–Verlag, Berlin.

49. Samulski, R. J., Zhu, X., Xiao, X., Brook, J. D., Housman, D. E., Epstein, N., *et al.* (1991). *EMBO J.*, **10**, 3941.

50. Linden, R. M., Winocour, E., and Berns, K. I. (1996). *Proc. Natl. Acad. Sci. USA*, **93**, 7966.

51. Lebkowski, J. S., McNally, M. M., Okarma, T. B., and Lerch, B. (1988). *Mol. Cell. Biol.*, **8**, 3988.

52. Flotte, T. R., Carter, B., Conrad, C., Guggino, W., Reynolds, T., Rosenstein, B., *et al.* (1996). *Hum. Gene Therapy*, **7**, 1145.

53. Gyurko, R. and Phillips, M. I. (1995). *Exp. Biol.*, Abstract 1915.

54. Gyurko, R., Wu, P., Sernia, C., Meyer, E., and Phillips, M. I. (1994). American Heart Association 48th Annual Council for High Blood Pressure (Abstract).

55. Zolothukhin, S., Potter, M., Hauswirth, W. W., Guy, J., and Muzyczka, N. (1996). *J. Virol.*, **70**, 4646.

56. Wu, P., Du, B., Phillips, M. I., Bui, J., and Terwilliger, E. F. (1998). *J. Virol.*, **72**, 5919.

57. Mohuczy, D. and Phillips, M. I. (1996). *FASEB J.*, **10**, A447.

58. Zelles, T., Mohuczy, D., and Phillips, M. I. (1996). *Soc. Neurosci.*, 41.18, 83.

59. Mohuczy, D., Gelband, C., and Phillips, M. I. (1998). American Heart Association 52nd Council for High Blood Pressure (Abstract).

60. Zolotukhin, S., Byrne, B. J., Mason, E., Zolotukhin, I., Potter, M., Chestnut, K., *et al.* (1999). *Gene Therapy*, **6**, 973.

61. Phillips, M. I., Mohuczy-Dominiak, D., Coffey, M., Wu, P., Galli, S. M., and Zelles, T. (1997). *Hypertension*, **29**, 374.

List of suppliers

Affinity BioReagents, Inc.,14818 West 6th Avenue, Suite 10A, Golden, CO 80401, USA.
Tel: 001 800 527 4535 or 303 278 4535
Fax: 001 303 278 2424
URL: http://www.bioreagents.com

Agilent Technologies, PO Box 10395, Palo Alto, CA 94303, USA.
URL: http://www.agilent.com

Aldrich, The Old Brickyard, New Road, Gillingham, Dorset SP8 4XT, UK.
Tel: 01747 822 221
Fax: 01747 823 779
URL: http://www.sigma-aldrich.com

Allegience Healthcare Corp., SP Laboratory Products and Service, 1450 Waukegan Road, McGaw Park, IL 60085, USA.

Ambion, Inc., A2130 Woodward Street, Austin, Texas 78744-1832, USA.
Tel: 001 800 888 8804
Fax: 001 512 651 0201
URL: http://www.ambion.com

Amersham, Amersham Pharmacia Biotech UK Ltd., Amersham Place, Little Chalfont, Buckinghamshire HP7 9PA, UK.
Tel: 01870 606 1921
Fax: 01494 544 350
URL: http://www.apbiotech.com

Amersham Pharmacia Biotech Inc., 800 Centennial Avenue, PO Box 1327, Piscataway, NJ 08855, USA.
Tel: 001 800 526 3593
Fax: 001 877 295 8102
URL: http://www.apbiotech.com

American Type Culture Collection (ATCC), 10801 University Boulevard, Manassas, VA 20110-209, USA.
Tel: 001 800 638 6597
Fax: 001 703 365 2750

Amicon – see Millipore

Anderman and Co. Ltd., 145 London Road, Kingston-upon-Thames, Surrey KT2 6NH, UK.
Tel: 0181 541 0035
Fax: 0181 541 0623

Beckman Coulter (UK) Ltd., Oakley Court, Kingsmead Business Park, London Road, High Wycombe, Buckinghamshire HP11 1JU, UK.
Tel: 01494 441 181
Fax: 01494 447 558
URL: http://www.beckman.com
Beckman Coulter Inc., 4300 N Harbor Boulevard, PO Box 3100, Fullerton, CA 92834-3100, USA.
Tel: 001 714 871 4848
Fax: 001 714 773 8283
URL: http://www.beckman-coulter.com

Beckman Instrument, Inc., Bioanalytical Systems Group, 2550 Harbor Boulevard, E-27-C, Fullerton, CA 92634-3100, USA.
Tel: 001 800 742 2345
Fax: 001 800 643 4366
URL: http://www.beckman.com

Becton Dickinson UK Ltd., 21 Between Towns Road, Cowley, Oxford OX4 3LY, UK.
Tel: 01865 748 844   Fax: 01865 781 627
URL: http://www.bd.com/support/contact
Becton Dickinson USA Ltd., 1 Becton Drive, Franklin Lakes, NJ 07417-1883, USA.
Tel: 001 201 847 6800
URL: http://www.bd.com/support/contact

Bio 101 Inc., c/o Anachem Ltd., Anachem House, 20 Charles Street, Luton, Bedfordshire LU2 0EB, UK.
Tel: 01582 456 666   Fax: 01582 391 768
URL: http://www.anachem.co.uk
Bio 101 Inc., PO Box 2284, La Jolla, CA 92038-2284, USA.
Tel: 001 760 598 7299
Fax: 001 760 598 0116
URL: http://www.bio101.com

Bio-Rad Laboratories Ltd., Bio-Rad House, Maylands Avenue, Hemel Hempstead, Hertfordshire HP2 7TD, UK.
Tel: 020 8328 2000
Fax: 020 8328 2550
URL: http://www.bio-rad.com
Bio-Rad Laboratories Ltd., Division Headquarters, 2000 Alfred Noble Drive, Hercules, CA 94547, USA.
Tel: 001 510 741 1000
Fax: 001 510 741 5800
URL: http://www.discover.bio-rad.com

BIO 101 – see Qbiogene

Biowhitaker, Inc., 8830 Biggs Ford Road, Walkersville, MD 21793, USA.
Tel: 001 800 638 8174
Fax: 001 301 845 8388

Boehringer Mannheim, 9115 Hague Road, PO Box 50414, Indianapolis, IN 46250-0414, USA.
Tel: 001 800 262 1640
Fax: 001 800 428 2883

Brandel, 8561 Atlas Drive, Gaithersburg, MD 20877, USA.
Tel: 001 800 948 6506
Fax: 001 301 869 5570

Calbiochem – Novabiochem Corp., PO Box 12087, La Jolla, CA 92039-2087, USA.
Tel: 001 800 854 3417
Fax: 001 800 450 9600
URL: http://www.calbiochem.com

Christison Scientific, Albany Road, Gateshead NE8 3AT, UK.
Tel: 0191 478 8120   Fax: 0191 490 0549
URL: http://www.christison.com

Clontech Laboratories, Inc., Unit 2, Intec 2, Wade Road, Basingstoke, Hampshire RG24 8NE, UK.
Tel: 01256 476500   Fax: 01256 476499
URL: http://www.clontech.co.uk
Clontech Laboratories, Inc., 1020 East Meadow Circle, Palo Alto, CA 94303-4230, USA.
Tel: 001 800 662 2566
Fax: 001 650 424 1352
URL: http://www.clontech.com

Corning Life Sciences, Dealer Corporate Locations
URL: http://www.corning.com

Covance Inc., 210 Carnegie Center, Princeton, NJ 08540-6233, USA.
Tel: 001 609 452 4440
Fax: 001 609 452 9375
URL: http://www.covance.com

CP Instrument Co. Ltd., PO Box 22, Bishop Stortford, Hertfordshire CM23 3DX, UK.
Tel: 01279 757 711   Fax: 01279 755 785
URL: http://www.cpinstrument.co.uk

DNSTAR Inc., 1228 S. Park Street, Madison, WI 53715, USA.
Tel: 001 608 258 7420
Fax: 001 608 258 7439
URL: http://www.dnstar.com

Dupont (UK) Ltd., Industrial Products Division, Wedgwood Way, Stevenage, Hertfordshire SG1 4QN, UK.
Tel: 01438 734 000   Fax: 01438 734 382
URL: http://www.dupont.com
Dupont Co. (Biotechnology Systems Division), PO Box 80024, Wilmington, DE 19880-002, USA.
Tel: 001 302 774 1000
Fax: 001 302 774 7321
URL: http://www.dupont.com

Eastman Chemical Co., 100 North Eastman Road, PO Box 511, Kingsport, TN 37662-5075, USA.
Tel: 001 423 229 2000
URL: http://www.eastman.com

Erithacus Software Inc., PO Box 274, Horley, Surrey RH6 9YJ, UK.
Tel: 01342 841 938
Fax: 01342 841 939
URL: http://www.erithacus.com

Falcon
(Falcon is a registered trademark of Becton Dickinson and Co.)

Fisher Scientific UK Ltd., Bishop Meadow Road, Loughborough, Leicestershire LE11 5RG, UK.
Tel: 01509 231 166   Fax: 01509 231 893
URL: http://www.fisher.co.uk
Fisher Scientific, Fisher Research, 2761 Walnut Avenue, Tustin, CA 92780, USA.
Tel: 001 714 669 4600
Fax: 001 714 669 1613
URL: http://www.fishersci.com
Fisher Scientific Co. USA, 711 Forbest Avenue, Pittsburgh, PA 15219-4785, USA.

Fluka, PO Box 2060, Milwaukee, WI 53201, USA.
Tel: 001 414 273 5013
Fax: 001 414 273 4979
URL: http://www.sigma-aldrich.com
Fluka Chemical Co. Ltd., PO Box 260, CH-9471, Buchs, Switzerland.
Tel: 0041 81 745 2828
Fax: 0041 81 756 5449
URL: http://www.sigma-aldrich.com

FMC BioProducts, 191 Thomaston Street, Rockland, Maine 04841, USA.
Tel: 001 800 341 1574
Fax: 001 800 362 5552
URL: http://www.bioproducts.com

Gibbes BRL Products, Grand Island, NY, USA.
Tel: 001 301 840 4027
Fax: 001 301 258 8238
URL: http://www.lifetech.com

Gibco BRL - see Life Technologies/Gibco BRL

GraphPad Software Inc., 5755 Oberlin Drive, Suite 110, San Diego, CA 92121, USA.
Tel: 001 858 457 3909
Fax: 001 858 457 3141
URL: http://www.graphpad.com

Hybaid Ltd., Action Court, Ashford Road, Ashford, Middlesex TW15 1XB, UK.
Tel: 01784 425 000
Fax: 01784 248 085
URL: http://www.hybaid.com
Hybaid US, 8 East Forge Parkway, Franklin, MA 02038, USA.
Tel: 001 508 541 6918
Fax: 001 508 541 3041
URL: http://www.hybaid.com

HyClone Laboratories, 1725 South HyClone Road, Logan, UT 84321, USA.
Tel: 001 800 492 5663
Fax: 001 800 533 9450
URL: http://www.hyclone.com

Invitrogen Corp., 1600 Faraday Avenue, Carlsbad, CA 92008, USA.
Tel: 001 760 603 7200
Fax: 001 760 603 6500
URL: http://www.invitrogen.com
Invitrogen BV, PO Box 2312, 9704 CH Groningen, The Netherlands.
Tel: 00800 5345 5345
Fax: 00800 7890 7890
URL: http://www.invitrogen.com
Invitrogen Corporation, c/o British Biotechnology Products Ltd., 4-10 The Quadrant, Barton Lane, Abingdon, Oxon OX14 3YS, UK.

Jasco UK, 18 Oak Industrial Park, Chelmsford Road, Great Dunmow, Essex CM6 1XN, UK.
Tel: 01371 876 988
Fax: 01371 875 597
URL: http://www.Jasco.co.uk
Jasco Inc., 8649 Commerce Drive, Easton, MD 21601, USA.
Tel: 001 410 822 1220
Fax: 001 410 822 7526
URL: http://www.Jascoinc.com

Labsystems, Thermo BioAnalysis Companies, Action Court, Ashford Road, Middlesex TW15 1XB, UK.
Labsystems Affinity Sensors, Saxon Way, Bar Hill, Cambridge CB3 8SL, UK.
Tel: 01954 789 976
Fax: 01954 789 417
URL: http://www.affinity-sensors.com

Life Sciences International, Unit 5, The Ringway Centre, Edison Road, Basingstoke, Hampshire RG21 6YH, UK.

Life Technologies Ltd. / Gibco BRL (UK), 3 Fountain Drive, Inchinnan Business Park, Paisley PA4 9RF, UK.
Tel: 0141 814 6100
Fax: 0141 814 6287
URL: http://www.lifetech.com

Life Technologies Inc. / Gibco BRL (USA), 9800 Medical Center Drive, PO Box 6482, Rockville, MD, USA.
Tel: 001 800 338 5772
Fax: 001 800 331 2286
URL: http://www.lifetech.com

Macherey-Nagel, Hirsackerstrasse 7, Postfach, CH-4702 Oensingen, Switzerland.
Tel: 0041 62 388 55 00
Fax: 0041 62 388 55 05
URL: http://www.macherey-nagel.ch

MatTek Corporation, 200 Homer Avenue, Ashland, MA 01721, USA.
Tel: 001 508 881 6771
Fax: 001 508 879 1532

Merck, Frankfurter Str. 250, 64293 Darmstadt, Germany.
Tel: 0049 6151 72 0
Fax: 0049 6151 72 2000

Merck Sharp & Dohme, Research Laboratories, Neuroscience Research Centre, Terlings Park, Harlow, Essex CM20 2QR, UK.
URL: http://www.msd-nrc.co.uk
MSD Sharp and Dohme GmbH, Lindenplatz 1, D-85540, Haar, Germany.
URL: http://www.msd-deutschland.com

Microbix Biosystems Inc., 341 Bering Avenue, Toronto, ON M8Z 3A8, Canada.
Tel: 416 234 1624   Fax: 416 234 1626
URL: http://www.devicelink.com

Millipore (UK) Ltd., The Boulevard, Blackmoor Lane, Watford, Hertfordshire WD1 8YW, UK.
Tel: 01923 816 375   Fax: 01923 818 297
URL: http://www.millipore.com/local/UK.htm
Millipore Corp., 80 Ashby Road, Bedford, MA 01730, USA.
Tel: 001 800 645 5476
Fax: 001 800 645 5439
URL: http://www.millipore.com

Molecular Devices Corporation (UK), 135 Wharfedale Road, Winnersh Triangle, Wokingham, Berkshire RG1 5RB, UK.
Tel: 0118 944 8000   Fax: 0118 944 8001
URL: http://www.moldev.com
Molecular Devices Corporation (USA), 1311 Orleans Avenue, Sunnyvale, CA 94089-1136, USA.
Tel: 001 800 635 5577
Fax: 001 408 747 3602
URL: http://www.moldev.com

Molecular Probes Inc., PO Box 22010, Eugene, OR 97402-0469, USA.
Tel: 001 541 465 8300
Fax: 001 541 344 6504
URL: http://www.probes.com
Molecular Probes Europe BV, Poortgewouw, Rijnsburgerweg 10, 2333 AA Leiden, The Netherlands.
Tel: 0031 71 523 33 78
Fax: 0031 71 523 34 19
URL: http://www.probes.com

National Diagnostics, 305 Patten Drive, Atlanta, GA 30336, USA.
Tel: 001 404 699 2121
Fax: 001 404 699 2077

New England Biolabs, 32 Tozer Road, Beverley, MA 01915-5510, USA.
Tel: 001 978 927 5054
Fax: 001 978 921 1350
URL: http://www.neb.com

Newport Instruments, 4320 First Avenue, Newbury Business Park, London Road, Newbury, Berkshire RG14 2PZ, UK.
Tel: 01635 521 757   Fax: 01635 521 348
URL: http://www.newport.com

Nikon Inc., 1300 Walt Whitman Road, Melville, NY 11747-3064, USA.
Tel: 001 516 547 4200
Fax: 001 516 547 0299
URL: http://www.nikonusa.com

Nikon Corp., Fuji Building, 2-3 Marunouchi, 3-chome, Chiyoda-ku, Tokyo 100, Japan.
Tel: 00813 3214 5311
Fax: 00813 3201 5856
URL: http://www.nikon.co.jp/main/index_e.htm

Novagen, 601 Science Drive, Madison, Wisconsin 53711, USA.
Tel: 001 800 526 7319
Fax: 001 608 238 1388
URL: http//www.novagen.com

Nycomed Amersham plc, Amersham Place, Little Chalfont, Buckinghamshire HP7 9NA, UK.
Tel: 01494 544 000
Fax: 01494 542 266
URL: http://www.amersham.co.uk
Nycomed Amersham, 101 Carnegie Center, Princeton, NJ 08540, USA.
Tel: 001 609 514 6000
URL: http://www.amersham.co.uk

Nycomed Pharma Holding AS
Langebjerg 1, PO Box 88, DK-4000 Roskilde, Denmark.
Tel: 0045 4677 1111   Fax: 0045 4675 6640
URL: http://www.nycomed.com

OLYMPUS America Inc., Corporate Headquarters, Two Corporate Center Drive, Melville, NY 11747-3157, USA.

Omega, PO Box 573, 3 Grove Street, Brattelboro, VT 05302, USA.
Tel: 001 802 254 2690
Fax: 001 802 254 3937
URL: http://www.omegafilters.com

Packard BioScience Ltd., Brook House, 14 Station Road, Pangbourne, Berkshire RG8 7AN, UK.
Tel: 01189 844 981
Fax: 01189 844 059

Pel-Freez Biologicals, PO Box 68, Rogers, AR 72757, USA.
Tel: 001 800 643 3426
Fax: 001 501 636 3562
URL: http://www.pelfreez-bio.com

Perkin Elmer Ltd., Post Office Lane, Beaconsfield, Buckinghamshire HP9 1QA, UK.
Tel: 01494 676161
URL: http://www.perkin-elmer.com

Pharmacia Biotech (Biochrom) Ltd., Unit 22, Cambridge Science Park, Milton Road, Cambridge CB4 0FJ, UK.
Tel: 01223 423 723
Fax: 01223 420 164
URL: http://www.biochrom.co.uk
Pharmacia and Upjohn Ltd., Davy Avenue, Knowlhill, Milton Keynes, Buckinghamshire MK5 8PH, UK.
Tel: 01908 661 101   Fax: 01908 690 091

Pierce, PO Box 117, Rockford, IL 61105, USA.
Tel: 001 815 968 0747
Fax: 001 815 968 7316
URL: http://www.piercenet.com

Princeton Instruments, Roper Scientific, PO Box 1192, 43 High Street, Marlow, Buckinghamshire SL7 1GB, UK.
Tel: 01628 890 858
Fax: 01628 898 381
URL: http://www.prinst.com
Princeton Instruments, (USA East Coast), 3660 Quakerbridge Road, Trenton, NJ 08619, USA.
Tel: 001 609 587 9797
Fax: 001 609 587 1970
URL: http://www.prinst.com

Promega UK Ltd., Delta House, Chilworth Research Centre, Southampton SO16 7NS, UK.
Tel: 0123 8076 0225
Fax: 0123 8076 7014

URL: http://www.promega.com
Promega Corp., 2800 Woods Hollow Road, Madison, WI 53711-5399, USA.
Tel: 001 608 274 4330
Fax: 001 608 277 2516
URL: http://www.promega.com

Qbiogene (UK), Salamander Quay West, Park Lane, Harefield, Middlesex UB9 6NZ, UK.
Tel: 01895 453 700
Fax: 01895 453 705
Qbiogene (USA), 2251 Rutherford Road, Carlsbad, CA, USA.
Tel: 001 800 424 6101
Fax: 001 760 918 9313
URL: http://www.qbiogene.com

Qiagen UK Ltd., Boundary Court, Gatwick Road, Crawley, West Sussex RH10 2AX, UK.
Tel: 01293 422 911
Fax: 01293 422 922
URL: http://www.qiagen.com
Qiagen Inc., 28159 Avenue Stanford, Valencia, CA 91355, USA.
Tel: 001 800 426 8157
Fax: 001 800 718 2056
URL: http://www.qiagen.com

Roche Diagnostics Ltd., Bell Lane, Lewes, East Sussex BN7 1LG, UK.
Tel: 01273 480 444
Fax: 01273 480 266
URL: http://www.roche.com
Roche Molecular Biochemicals, 9115 Hague Road, PO Box 50414, Indianapolis, IN 46250-0141, USA.
Tel: 001 800 428 5433
Fax: 001 800 428 2883
URL: http://www.biochem.roche.com
Roche Diagnostics GmbH, Sandhoferstrasse 116, 68305 Mannheim, Germany.
Tel: +49 621 759 4747
Fax: +49 621 759 4002
URL: http://www.roche.com

Schleter and Schuell Inc., PO Box 2012, Keene, NH 03431A, USA.
Tel: 001 800 245 4024

SEMAT Technical (UK) Ltd., One Executive Park, Hatfield Road, Hertfordshire AL1 4TA, UK.
Tel: 01727 841 414
Fax: 01727 843 965

Shandon Scientific Ltd., 93-96 Chadwick Road, Astmoor, Runcorn, Cheshire WA7 1PR, UK.
Tel: 01928 566611
URL: http://www.shandon.com

Sigma-Aldrich Co. Ltd., The Old Brickyard, New Road, Gillingham, Dorset XP8 4XT, UK.
Tel: 01747 822 211
Fax: 01747 823 779
URL: http://www.sigma-aldrich.com
Sigma-Aldrich Co. Ltd., Fancy Road, Poole, Dorset BH12 4QH, UK.
Tel: 01202 733 114
Fax: 01202 715 460
URL: http://www.sigma-aldrich.com
Sigma Chemical Co., PO Box 14508, St Louis, MO 63178, USA.
Tel: 001 800 325 3010
Fax: 001 800 325 5052
URL: http://www.sigma-aldrich.com

Skatron - see Molecular Devices

Sorvall, 31 Peeks Lane, Newton, CT 06470-2337, USA.
Tel: 001 800 522 7746
Sorvall, Stevenage, UK.
Tel: 01438 342 911

Stratagene Inc., 11011 North Torrey Pines Road, La Jolla, CA 92037, USA.
Tel: 001 800 424 5444
Fax: 001 512 321 3128
URL: http://www.stratagene.com

Stratagene Europe, Gebouw California, Hogehilweg 15, 1101 CB Amsterdam Zuidoost, The Netherlands.
Tel: 00800 9100 9100
URL: http://www.stratagene.com

Teflabs, 9503 Capital View Drive, Austin, Texas, TX 78747, USA.
Tel: 001 512 280 5223
Fax: 001 512 280 4997
URL: http://www.teflabs.com

Textco, 27 Gilson Road, West Lebanon, NH 03784, USA.
Tel: 001 603 643 1471
Fax: 001 603 643 1771
URL: http://www.textco.com

TomTec., 1000 Sherman Avenue, Hamden, CT 06451, USA.
Tel: 001 203 281 6790
Fax: 001 203 248 5724
URL: http://www.tomtec.com

United States Biochemical, PO Box 22400, Cleveland, OH 44122, USA.
Tel: 001 216 765 5000
URL: http://guide.labanimal.com/company/382.html

Vector Laboratories Inc., 30 Ingold Road, Burlingame, CA 94010, USA.
Tel: 001 650 697 3600
Fax: 001 650 697 0339
URL: http://www.vectorlabs.com

Vivascience, 2 Park Drive, No. 5, Westford, MA 01886, USA.
Tel: 001 978 392 0222
Fax: 001 877 289 8482

VQT, Pierre Krug, Rue de Neuchâtel 2A, St.Blaise, 2072 Switzerland.
Tel: 0041 38 33 51 55
Fax: 0041 38 68 46

Warner Instrument Corporation, 1141 Dixwell Avenue, Hamden, CT 06514, USA.
Tel: 001 203 776 0664
Fax: 001 203 776 1278
URL: http://www.mrainternational.com

Waters Technology Ltd., Corporate Headquarters (for worldwide distribution), 34 Maple Street Milford, MA 01757, USA.
Tel: 001 508 478 2000
Fax: 001 508 872 1990
URL: http://www.waters.com

Whatman Inc., 6 Just Road, Fairfield, NJ 07004, USA.
Tel: 001 201 882 9277
Fax: 001 201 882 5134
URL: http://www.whatman.com

Carl Zeiss, PO Box 78, Woodfield Road, Welwyn Garden City, Hertfordshire AL7 1LU, UK.
Tel: 01707 871 200
URL: http://www.zeiss.co.uk

Index

absorption spectra 11
acceptor-sensitized fluorescence 125
acidification rate 176
adeno-associated virus (AAV) 257, 265
adeno-associated virus (AAV) helper plasmid 271
adeno-associated virus (rAAV) recombinant 270
adenoviruses 264
adenylyl cyclase 51, 158, 170,
α_1-adrenoceptors 65
β_2-adrenoceptor 29, 158
affinity probes 41
agonist intrinsic activity 163
Alexa dyes 1
alkylating reagents 118
amiloride 215
aminonaphthalene 119
angiotensin II fluorescent probes 215, 225
anisotropy 24, 25, 27
annealling 68
antibodies 48, 197
antibody affinity chromatography 99
anti-peptide antibodies 91, 92
antisense delivery 257
antisense oligodeoxynucleotides 258, 259, 261
arrestin 167
atrial natriuretic factor 141
autofluorescence 151

BCECF 189, 216
binding sites, 2
bioluminescence 57

bis-axonal, 218
BLAST algorithm 44
blue fluorescent protein (BFP) 122
BODIPY FL C_5-ceramide 215
BODIPY-conjugated endothelin-1 225

calcein 228
calcium green 188, 212
calcium measurement 56
calcium mobilization 8, 13, 51
calcium orange 212
calcium phosphate transfection 147, 217
calcium sensitive fluorescent dyes 184
calcium, cytosolic free 55
calmodulin 197
calmodulin inhibitor peptide 197
calyculin A 168
cAMP assays 52
cAMP response element-binding protein (CREB) 57
candidate protein kinase identification 207
carboxy SNARF-1 216, 217
cassette replacement 67
cell culture 77, 78
CGRP/adrenomedullin receptor, 48
channel dyes 225
chemokine receptors 2, 9, 17, 28
chimeric G proteins 43, 47, 51
chimeric receptors 65, 185
chloride currents 31

CHO cells 9, 14, 30, 32, 47, 53, 54, 56, 88, 178, 186
CL-NERF 216
cloning plasmids; 145
co-culture 210, 211
competitive binding 85, 86,
computer-based homology screening 44
concanavalin A (Con A) 166, 169
confocal imaging 50, 142, 152, 209
confocal microscopy 209
conserved transmembrane (TM) regions 43
coral-derived red fluorescent protein (DsRed) 142
corticotrophin releasing factor (CRF) receptor 177
COS cells 80, 82
CP 96345 28
cultured neuronal cells 195
cyan fluorescent protein 137
cyanine dyes and derivatives 1, 3, 119

DEAE-dextran transfection 80
Di-4-ANEPPS 218
Di-8-ANEPPS 215, 218, 219, 222
diacylglycerol 197
diamino acids 6
Dibac(3,4) 189
Dil fluorescent labelling 227
DiOCn 215, 218, 222, 223
directed or point mutagenesis 65
displacement binding curves 13
DM-NERF, 216

DNA array technology 246
DNase I footprinting 238
Dns 3, 8
donor quenching 123, 124
double-stranded DNA
    sequencing 76

Edman degradation 9
electronic domain mapping
    139
electrophoretic mobility shift
    assay 240, 242
electrospray mass spectrometry
    9
endogenous gene expression
    243
endogenous ligands 41
endogenous ligands for orphan
    receptors 52
endoplasmic reticulum staining
    223
endosomes 156, 161
enhanced chemiluminescence
    250
enhanced cyan fluorescent
    protein (ECFP) 142
enhanced green fluorescent
    protein (EGFP) 142, 145,
    150,
enhanced yellow fluorescent
    protein (EGYP) 142
epitope tagging 92, 101, 106,
    197
expression 146
expression of mutant receptors
    77
expression vector 73

Far Western analysis 253
fingerprint mapping 201
FITC 7, 8, 9
FLAG 48
flow cytometry 14
Fluo-3 55, 184, 188, 190, 210,
    212, 214, 216, 219, 220,
    222, 223
Fluo-4 55, 188, 189, 214
fluorescein 3, 6, 9, 119
fluorescence energy transfer
    measurements (FRET) 2,
    29, 30, 33, 34, 36, 37,
    101, 113, 142
fluorescence proteins 1
fluorescence quenching 19

fluorescence sensitivity 2
fluorescence spectroscopy 1
fluorescence-activated cell
    sorting (FACS) 14,17
fluorescent antibodies 120, 121
fluorescent fusion proteins 137,
    138, 144, 145
fluorescent probes 116, 209
fluorescent report groups 2
fluorescent semiconductor
    nanocrystals 119
fluorimetric imaging plate
    reader (FLIPR) 56, 175,
    184
forskolin 197
Forster radius 115, 116, 119,
    132
freeze-shock transformation of
    bacteria 235
FRET see fluorescence energy
    transfer.
fura red 212
fura-2 9, 13, 188

G protein 163, 165, 168, 170
G protein-coupled receptors
    (GPCRs) 1, 13, 41, 49, 56,
    155, 193
GABAa receptor 91, 101, 126,
    127
galanin 3
gene expression systems 236
Gene Splicing through Over-lap
    Extension (SOEing) 144,
    146
gene transfer 257
genistein 197
GFP see green fluorescent
    protein
glucocorticoid receptor (GR)
    233
glucocorticoid responsive
    element (GRE) 233
glutamate-gated ion channel
    GLUR 1 139
GR 94800 6, 8, 14, 19
green fluorescent protein (GFP)
    30, 121, 137, 138, 142,
    145, 151
green fluorescent protein uv
    (GFPuv) 138

heat shock protein 90 233
HEK 293 cells 49, 53, 148, 150

heparin affinity
    chromatography 273
homogeneous time-resolved
    FRET assay 132
HPTS 216
hydroxysuccinimide 2
hyper-osmolar sucrose 166

IANBD 29
IBMX 197
immunoaffinity purification
    92, 96, 98, 99
immunofluorescence
    microscopy 49
immunoprecipitation 92, 195,
    251, 252
immunoprecipitation of NMDA
    receptor 97
influx pinocytic cell loading
    reagent kit 226
inhibitors of endocytosis 165
inhibitors of recycling 165, 167
inositol triphosphate (InsP3) 55,
    87, 163, 165
insertional mutagenesis 104
insurmountable receptor
    antagonists 160
intermolecular distances 2
intracellular ions 209, 219
iodoacetamide 2
ion channel 148
ion exchange chromatography
    62
ionotropic receptors 194
isothiocyanates 2, 117
isothiocyanates of fluorescein
    (FITC) 117
isothiocyanates of
    tetramethylrhodamine
    (TRITC) 117

JC-1 215, 218, 219, 223

labelled peptide ligands 6
lavendustin 197
ligand binding 82
ligand kinetics 17
ligand screening 60
ligand-gated ion channels 193
ligand-receptor interactions 2
ligation 144
ligation of PCR fragments 74
lipid-mediated transfection 147

Lipofectamine 2000 150
liposomes 263
lodixanol density gradient 273
low stringency hybridization 43
luciferase 57, 59
luciferase reporter gene 57
lysosomes 155, 156

maleimide esters 2
mammalian bicistronic
    expression vectors 45
mammalian expression vectors
    106
melanin-concentrating
    hormone 45
metabolic rate 176
metabotropic receptor 148
microphysiometry 175
misacylated tRNA 30, 31
mito fluor green 215
mitochondria staining 224
mitogen-activated protein (MAP)
    kinases 171, 197
m-maleimidobenzoic acid
    N-hydroxysuccinimide
    ester (MBS) 94
monensin 167, 168
muscarinic acetylcholine
    receptor 158, 159, 160,
    177, 179
myc tag 48

native receptors 162
NBD 3, 6, 7, 8, 32
NBDC6-ceramide 222
negative glucocorticoid
    responsive element
    (NGRE) 233
neurokinin A (NKA) 3, 8
neurokinin receptor ligands 2,
    3, 4, 5, 6, 7, 8, 11, 13, 15,
    19, 27, 28, 31
neuronal transfection 150
nigericin 167, 169
NMDA receptor 91
NMDA receptor subunit 91
normalized score. 140
Northern blot analysis 243
nuclear hormone receptors 177
nuclear receptor function 233

oligonucleotide
    phosphorylation 68

oligonucleotides 261
oocyte 30, 31, 32, 33, 91, 101
orphan' GPCRs 41
orphan receptor strategy 41, 42
orphanin receptor 51, 62
osmotic lysis 225

patch-clamp technique 148
PCR mutagenesis 68
peptide affinity
    chromatography 95
peptide agonists 3
peptide conjugates 10
peptide ligands 6, 8
peptide map analysis 201
perfusion chambers 152
Perrin plots 24, 25, 26
pH fluorescent probes 216
phenylarsine oxide (PAO) 166
pHlourin f 138
phorbol esters 197
phosphatases 193
phosphoamino acid
    identification 202
phosphoinositide hydrolysis
    158
phospholamban 129
phospholipase Cβ 51, 55
phosphopeptide mapping 206
phosphoproteins 199
phosphorylation consensus
    sequences 206
phosphorylation site
    identification 205
phosphorylation site-specific
    antibodies. 206
phosphospecific antibodies 207
photobleaching 229
photobleaching FRET
    techniques 131
PKA 57
PKA inhibitor peptide 197
PKC inhibitor peptide 197
plasmid preparation 76, 267
plasmid expression vectors 45
pluronic F-127C, 212, 216
point mutations 65, 66
polymerase chain reaction (PCR)
    43, 65, 138, 143, 145
pore loops 141
potassium channel 141
potassium channel, shaker 143
primer sites 142
prolactin-releasing peptide
    receptor 45

propyl-benzilylcholine mustard
    (PrBCM) 158
protein kinase activators 197
protein kinases 193, 198, 199
protein phosphorylation 193
protein sequence 139
protein sorting signals search
    engine (PSORT) 141
protein synthesis 156
protein-DNA interaction 237
protein-protein interaction 251,
    252
proteolytic digestion 199
pull-down assays 253
PYY fluorescent probe 215, 225

quantum yield 2, 11
quencher molecules 2, 21, 22

radioligand binding 83, 155
RAMPS 48
random labelling 116
random mutagenesis 65
Rantes proteins 6
rate constants 162
RBL cells 184
receptor architecture 29
receptor conformation 156, 161
receptor degradation 225
receptor desensitization 167
receptor dyes 225
receptor endocytosis 155, 158,
    159, 172
receptor family, class A or
    rhodopsin-like 45
receptor family, class B or
    secretin-like 45
receptor family, class C 45
receptor function 175
receptor internalization 158
receptor phosphorylation 194
receptor recycling 155, 159,
    161
receptor resensitization 167
receptor reserve 171
receptor stoichiometry 113,
    126, 132
receptor subunit antibodies. 92
receptor tagging 48
receptor trafficking 137, 155
receptor transfected cell 13
receptor turnover 156, 165
receptor tyrosine kinases 193

receptor-coupled ion channels 193
red fluorescent protein 137
reporter gene assay 57
reporter protein 237
restriction 144
restriction digestion of DNA 72
restriction sites 68, 142, 143, 144, 145
retroviruses 264
reverse transcription polymerase chain reaction (RT-PCR) 244
reversed-phase chromatography 62
RFP tagged protein kinase C gamma 148
RH2g2 218
rhodamine 119
rhodopsin receptor 43
rhodopsin-like receptor family 45
RNase protection assay 244

sarcolemma perforation 222
Sculptor 104
SDS-PAGE 197, 199, 249
second messenger 42
second messenger activation 65
second messenger assays 87
secretin-like receptor family 45
sequence homologies 43

signal transduction 177
signalling pathway 47
site-directed mutagenesis 104, 143, 205
SNARF-1 215, 216, 217
SNARL 216
sodium green 215, 216
sodium ion fluorescent probes 216
SOEing 144, 146
somatostatin/opioid receptor family 45
staurosporine 197
stop codon suppression 30, 31
substance P 3, 5, 28
succinimidyl esters 2, 117
sulfonyl chlorides 117
SWISSPROT 139, 140, 141
synthetic CRE sequences 59
Syto-11 215, 222

tachykinin receptors see neurokinin receptors
tagging 49
tag-specific antibodies 48
Tet repressor 236
tetramethylrhodamine 218
texas red 129
thiazole orange 218
time-lapse microscopy 138, 152, 153
TMR 3, 33

transfected cells 14, 47
transfected gene expression constructs 234
transfected receptors 162
transfecting neurons 151, 153
transfection 65, 148, 150
transfection, transient 80
transformation into bacteria 75
transgenic animals 147
trypsin cleavage 200
turnover kinetics 155
tyrphostins 197

vascular endothelial cells 211
vascular smooth muscle cells 211
vasopressin receptor antagonists 8
vasopressin receptors 2
viral promoters 45
viral vectors 263
virus titre assay 274
voltage clamp technique 21

Western blot 194, 207, 248, 250

yeast two-hybrid assay 253
yellow fluorescent protein (YFP) 122